Handbook of Laser Pulses

Handbook of Laser Pulses

Edited by **Juan Landers**

$\mathcal{CL}$LANRYE
INTERNATIONAL

New Jersey

Published by Clanrye International,
55 Van Reypen Street,
Jersey City, NJ 07306, USA
www.clanryeinternational.com

Handbook of Laser Pulses
Edited by Juan Landers

© 2015 Clanrye International

International Standard Book Number: 978-1-63240-275-2 (Hardback)

Contents

Preface

It is often said that books are a boon to mankind. They document every progress and pass on the knowledge from one generation to the other. They play a crucial role in our lives. Thus I was both excited and nervous while editing this book. I was pleased by the thought of being able to make a mark but I was also nervous to do it right because the future of students depends upon it. Hence, I took a few months to research further into the discipline, revise my knowledge and also explore some more aspects. Post this process, I begun with the editing of this book.

This book aims to present the latest developments in the field of laser pulses. It explains the characteristics of laser pulse creation, classification and applications. It even illustrates some accomplishments made in designs, experiments and theories. The book deals with its physical principles and illustrates examples regarding laser operations carried out under several extreme circumstances. The topic of mode-locking, based on optical noise perception, is analyzed within the book. Furthermore, it also discusses the paradoxes of ultra-short laser pulse arrangement. This book will help students and engineers to understand laser technology better.

I thank my publisher with all my heart for considering me worthy of this unparalleled opportunity and for showing unwavering faith in my skills. I would also like to thank the editorial team who worked closely with me at every step and contributed immensely towards the successful completion of this book. Last but not the least, I wish to thank my friends and colleagues for their support.

Editor

Introduction

Time and Light

Igor Peshko

Additional information is available at the end of the chapter

1. Introduction

1.1. From seconds to attoseconds

This book is devoted to Laser Pulses. In the modern laser world, the word "Pulse" covers pulse durations from microseconds (free-running laser) to tens of femtoseconds (1fs is 10^{-15} s) (mode-locked laser). It is possible to generate attosecond pulses (1as is 10^{-18} s) by using non-linear processes. Recently, a new time range was discussed in publications: zeptosecond (1zs is 10^{-21} s). To generate pulses from milliseconds to femtoseconds, hundreds of different laser systems have been developed. They can typically generate pulses of specific durations, which are due to laser principles of operation, specific construction, parameters of gain medium, type of modulator, and so on.

In the solid-state free-running laser the parameters of the electromagnetic field interaction process with an inversed population of the gain medium play a dominant role in shaping the laser spikes. These characteristic parameters limit the pulse duration from the long side of the range. A chaotic sequence of such spikes can be as long as the pumping source could effectively excite the gain medium. In some technological applications this can be hundreds of milliseconds envelop. To achieve laser pulse duration of a few seconds, it is necessary to use for modulation the processes with characteristic times of the same order of magnitude. On the short side of achievable durations, another limitation exists. At certain conditions, waves, interacting with solids, can shape a single peak of energy, propagating "alone". In optics, such a single wave is called a "soliton". A tsunami is an example of a mechanical soliton. To propagate in a crystal, an "optical tsunami" can excite the medium and get back the energy at some conditions. In a vacuum, there is no medium that can accumulate energy and support existence of a relatively short, lossless wave. Hence, in a vacuum, a single pulse could be shaped as a wave-package, which is a result of interference of many independent electromagnetic waves, propagating co-axially in the same direction. Since light is an electromagnetic wave repeatable in space and time, a single wave period with minimal length (among all present in the wave set) is the minimal possible duration of an energy pack. Out of the pulse, the negative interference suppresses the energy presence.

Pulse measurement techniques were being developed together with each type of laser and are based on principles depending on the specific range of pulse duration. The shorter the achieved pulse durations, the more difficult the problem of how to measure such pulses. The answer is inside the laser pulses. Light itself contains information about Time.

In summary, we can say that lasers, generating "light in time", combine these two categories as light is a periodical, cyclic process and can be a measure of time and of length. Since this is an introductory chapter, let us consider some historical milestones on the way of Time/Light understanding first.

1.2. In the past

On the way of analysing and understanding Nature, ancient scientists interpreted Light and Time as two absolutely different categories. To this day, it is not clear what Time is - is it a characteristic of processes, or an independently existing parameter? Theoretical physics operates with 4D space: 3 space coordinates and the fourth as time. To measure any distance, a researcher can compare some etalon of length with the object of interest. However, historically, the etalon of length was voluntarily chosen: 1 m is not a "natural" unit of length. At the same time, the light specific wavelength is a natural, repeatable etalon of length. The same is true for time – any process running with macro-objects being involved is just approximately repeatable. Since the speed of light (in vacuum) is the same everywhere, it can be used for measuring time and length: the same number of light waves with specific length corresponds to the same distance in any corner of the Universe – at least, we think so.

If existence of Light is absolutely evident, what supposed to be Time is not very clear until now. A lot of serious discussion and speculation was done but the problem still has to be solved. Time is not a single example of a "questionable" phenomenon. Very often, scientists propose a phantom model that helps make a very helpful device or theoretical approximation. The most popular parameter in optics - "index of refraction" - is probably the most investigated phenomenon that does not exist in Nature at all. This is a very useful model that helps provide theoretical research and find technological solutions in the fields of optics, photonics, and other related areas of activity.

We see or measure light as a result of photon interaction with more "solid" matter: atoms in our eyes or other detectors. In the case of time, the situation is more complicated. Since the early ages of humanity, people observed some regularly repeatable events, like day-night, winter-summer, etc., that, de-facto, were connected with complete or partial appearance and disappearance of light. Probably the first instrument for estimation of day time was the sundial – a vertically installed bar that showed the time period relative to specific points when the position of the sun over the horizon is maximally high and bar shadow is maximally short (midday). However, in cloudy days or at night, the system was useless. The next step in time measuring devices was the sand or water clock (Clepsydra): a spring of water or sands running through a small hole filled or emptied some calibrated tank. This is the most ancient experimental 3D demonstration of the basic formula of time: $T = M/R$ where M is mass and R is rate of mass changes. A very well-known 1D variant of this formula reads: $T = L$ (length)$/V$ (velocity). In case of a "floating clock", M is the total amount of water in the tank and R is the amount of water that drops through the hole during a unit of time. In this case, to build two identical clocks, one needs to make holes with ideally identical cross-sections. So, accuracy of time measurement is linked to accuracy of length measurement.

The next step was a mechanical clock with a pendulum. This device periodically transforms kinetic energy of motion into potential energy of a pendulum in the gravitation field. The electronic clocks use a periodic process of transformation of the electrical energy stored in a capacitor into magnetic energy of a coil. Each repeatable process can be used for measuring time. This could be a planet's rotation around a star, or a planet's rotation around its own axis, or mechanical vibrations – sound, or electromagnetic waves – light.

As of today, the history of pulsed lasers spans about 50 years. Hundreds of books devoted to this subject have been published during this period. More and more techniques go to practical use every year. Shorter and shorter pulses are routinely applicable. New spectral ranges, like deep UV, middle and far IR, including TeraHertz bands became a reality. This book demonstrates new achievements in theory, experiments, and commercial applications of pulsed lasers.

1.3. What is this chapter about?

Until the mid-eighties of the previous century, most of the lasers generating ns- and ps-pulses operated with passive or active Q-modulator. However, it became more and more clear that such non-linear and multi-parametric system as a laser could support self-effecting, self-modulating mode of operation without any external devices or internal elements. Starting from the time of the laser's invention, the theoretical models of laser operation include mode spectrum genesis. This is logically understandable, as the open Fabry-Perot cavity is one of the main components of the laser system. However, the picosecond or, moreover, femtosecond pulse generator has thousands and thousands of modes. Theoretical description of few modes is possible and can be used for analysis of laser operation. However, solving, for example, fifty thousand equations that describe all operating modes is a problem and the results could be very far from the reality.

The author of this chapter considers another approximation. The logic of this is as follows. The laser generation develops from luminescence radiation, which is de-facto, an optical noise. Let us analyze how this noise is being modified propagating through the gain medium after the laser threshold is achieved. At each moment of time, the spectral width and averaged power level determine the statistical properties of radiation. So, depending on laser parameters, the exceeding of the maximal, stochastically appeared optical spike over the next smaller one irradiated during a cavity round-trip period, could be found. At some conditions, the multiple, small-amplitude spikes saturate the amplification of the gain medium and stop the in-cavity power accumulation. Only the highest spikes that are able to saturate the non-linear absorber continue growing.

Around the 1960s, the properties of noise in radio-band were very well investigated on the way of radio rangefinders development. This theory was applied to describe the mode-locking process. Some difference was just in mathematical description: in radio-band the measured value is amplitude (strength) of the electromagnetic wave, and in optics, the square of this value, namely, intensity (power) of light. This approximation can estimate statistical parameters of generation: probability of single-pulse formation on the round-trip cavity period, depending on gain media spectrum width, rate of gain increase, cavity length, output mirror reflectance, and other laser practical parameters.

In this chapter, one can find interesting concepts, theoretical models, and experimentally proven techniques that for different reasons were not revealed in public at the moment they were proposed and demonstrated. Two different laser systems will be discussed: 1) single-frequency laser that, at certain conditions, is capable of slow self-modulation initiated by thermal processes in the cavity; 2) multi-mirror laser that, at certain conditions, is capable of self-mode-locking and generation of ultrashort pulses without any modulator.

This chapter represents philosophy of creation of laser pulses of different gradations:

1. "Slow" pulses of second duration
2. Microsecond pulses in millisecond envelope
3. Nanoseconds: passive and active Q-modulation
4. Picoseconds: passive, active, and natural auto-mode-locking
5. Femtosecond lasers
6. Attosecond and Zeptosecond

For today, hundreds of books describing the laser operational modes, design, constructions, and technologies are available. The condensed description can be found in an open-access encyclopaedia by Dr. Rüdiger Paschotta [1]. In this chapter, we often refer to the encyclopaedia pages, which typically contain a wide list of publications, discussing specific area of research. This chapter focuses mainly on some specific laser regimes and theoretical concepts, which are not considered in traditional books and in on-line encyclopaedia, and can be useful for the design of simple, low power-consumption, and environmentally-stable systems.

2. "Slow" pulses of second's duration

First of all, the initial question should be answered: how to slowly modulate such a fast operating and sensitive system as a laser. In principle, this can be provided by slow "delicate" changes of the gain or losses (or both) in a stationary operated laser with narrow spectral band. Such modulation is very difficult to provide in case of a multi-mode laser because of mode competition process: if output power of one mode is decreased, the others immediately grow up. Hence, first of all, a single-frequency operation has to be provided.

Let us consider an example of specific single-frequency laser [2, 3] that is capable of providing cavity losses periodical self-modulating. This regime has been achieved in the diode-pumped single-frequency Nd:YVO$_4$, Nd:YAG, and Nd:YLF lasers with metallic thin-film (10nm) selector. The main idea of the absorbing thin film selector operation is the following. The metallic film, with thickness Δd significantly smaller than the standing wave period ($\Delta d<\lambda/100$), is placed in the linear cavity. If a thin film plane is adjusted to the node surface area of any mode, the losses for this mode become close to zero and single longitudinal mode starts to operate. A detailed description of the interferometer with an absorbing mirror can be found in [2, 3].

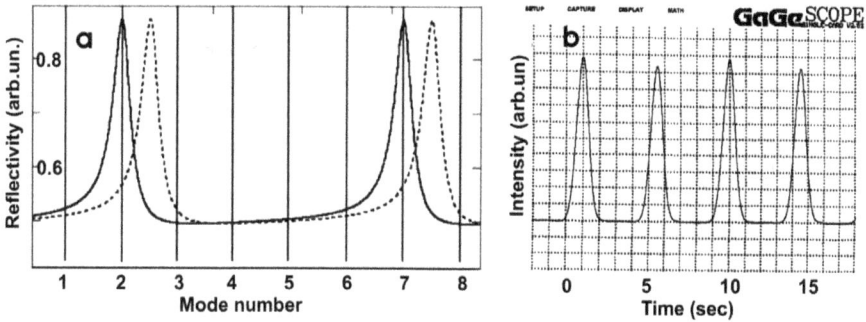

Figure 1. Mode switching mechanism: a) Vertical lines demonstrate laser cavity modes. The laser operates when the interferometer maxima (solid curve) coincide with the laser modes spectral positions, and does not – when the interferometer modes are located between the cavity modes (broken line); b) Oscillogram of the laser output power. One division of time scale is 1 sec. The peak power is approximately 0.5W. The radiation maxima correspond to the spectral positions of the interferometer maxima shown by the solid line in Figure 1a.

The pulses with duration of approximately 1 to 3 s and period of about 3 to 10 s (see Figure 1b), depending on pump power and thermo-optical properties of the cavity and gain crystal, have been observed. The effect has been explained by the thermal changes of the cavity length connected with the difference of the heat generation rate for operating and non-operating laser [4].

The self-pulsation regime of operation can be explained by periodical modulation of losses of a cavity with a thin-film selector caused by the thermally induced changes of optical length of an active medium. Let us assume that the thin-film selector is placed between the two nodes of neighbouring modes and the pump level is slightly below the threshold for both of the modes. At this moment, the laser does not operate; the maximum possible portion of pump power is transformed into heat and the temperature of the crystal increases. Because of the thermal elongation of active medium (and a total cavity), the positions of some nodes move to the absorbing thin-film location and the losses for this mode fall down. The mode switching mechanism is explained in Figure 1. Relative positions of the modes and the absorbing interferometer reflection peaks are shown. The interferometer is formed by a thin-film selector and an output coupler. The solid curve demonstrates resonance position that provides laser operation. The laser starts to operate in the single-frequency regime. At maximum output power, the heat dissipation rate becomes minimal and crystal temperature starts decreasing. Further, the thin-film location "comes out" from the node position and interrupts the laser emission process (dashed curve in Figure 1a). After the laser action break-off, the temperature rises up again and the process repeats.

3. Microsecond pulses in millisecond envelop

The operation of a free-running laser can be described and understood in terms of gain-loss dynamics [5,6]. When a pumping source intensively irradiates the laser medium, the gain grows until it becomes equal to all losses. Starting from this moment, the number of newly "born" (generated) photons is greater than those "dying" (absorbed or scattered). The light intensity grows very fast (avalanche conditions) and the laser starts generating. Very soon, the stimulate emission becomes so strong that the pumping radiation cannot refill the inversion population - in other words, the gain drops down to the level when generation is interrupted. The laser pulse is finished and during the next generation pause the inversion population grows up again until the gain achieves the loss level and the next cycle of generation begins.

When counter-propagating narrow-bandwidth light waves are superimposed, they form a so-called standing-wave interference pattern, the period of which is half the wavelength [7]. This results in so-called spatial and spectral "hole-burning": a) in maxima of standing wave the inversion population is falling down, shaping periodical spatial grating of the gain along the active medium; b) the operating modes are "eating" the gain at their spectral positions and other modes start generating. These have various consequences for the operation of lasers:

1. Difficulties in achieving a single-frequency operation or operation with stable and repeatable parameters when generating in linear cavities;
2. The optical bandwidth and spectral structure of a free-running laser radiation is different when the gain medium is located in different resonator areas;
3. Spatial hole-burning can reduce the laser efficiency, when the excitation in the nodes cannot be utilized.

The ring cavities support running waves and spatial hole-burning can be eliminated.

4. Nanosecond pulses: Passive and active Q-modulation

The Q-switching is a technique which provides generation of laser pulses with extremely high peak power: MegaWatt to GigaWatt [5,8]. Here Q means a quality-factor of the laser cavity. Typical durations of such pulses are between several nanoseconds to few tens of nanoseconds. This effect is achieved by modulating the intracavity losses. Initially, the resonator losses are kept at such a high level that the laser cannot operate at that time. Then, the losses are suddenly reduced by modulator to a small value. So, the laser becomes highly "over-excited" and starts generating just after several flights of light along the cavity. When the intracavity power has reached the value of the gain saturation, the laser radiation intensity drops down to the luminescence level. The resonator losses can be switched in different ways: through active and passive Q-switching.

For active Q switching, the losses are modulated with an active control element, typically either an acousto-optic or electro-optic modulator. Initially, there were also mechanical Q switches such as spinning mirrors or prisms. In any case, shorter cavities and more intensive pumping provide shorter lasing pulses.

For passive Q switching, the losses are automatically modulated with a saturable absorber. This can be a specific dye dissolved in liquid or solid matrix, some doped crystal or glass, and some bulk elements demonstrating the Kerr-type non-linearity. The pulse is formed as soon as the energy stored in the gain medium has reached a level sufficient to keep the absorbing particles in an excited state. The recovery time of a saturable absorber is ideally longer than the pulse duration. At this condition additional unnecessary energy losses would be avoided. However, the absorber should be fast enough to prevent the lasing when the gain recovers after the pulse termination.

5. Mode-locking

Mode-locking is a group of methods for laser generation of ultrashort pulses [9-13]. Typically, the pulse duration is roughly between 50 fs and 50 ps. To provide synchronous (phase-locked) operation of different modes, the laser cavity should contain either an active or a nonlinear passive element.

Figure 2 and Figure 3 demonstrate the difference between mode intensity and phase distributions for: (a) radiation that just is emitted at close to threshold conditions and (b) for already mode-locked radiation at the moment close to gain medium saturation. Simulation was done just for 10 modes to clearly illustrate the difference. In reality the number of operating modes may be tens or hundreds of thousands. The red line in Figure 2 shows the averaged mode intensity distribution. It is the same for any number of modes and any phase distributions. In any case, the initial gain medium luminescent radiation is a stochastic optical noise – a random sequence of occasionally irradiated spikes, which have different spectral content. The passive saturable absorber automatically selects (emphasizes) some spikes with relatively high energy and with relatively short duration. This results in some temporal uncertainty of the generation appearance and in generation of few ultrashort pulses on the cavity round-trip period because the gain medium continuously generates new noise patterns. The temporal distribution of radiation is "breathing" and is not ideally repeatable even on the neighbour round-trip periods of time. Moreover, the general view of time-intensity distribution of generation is unrepeatable from shot to shot – it could be one, or two, or even more high intensity pulses on the round-trip period.

In Figure 4 [14] two oscillograms of the laser output radiation are depicted: (a) an initial stage of generation development and (b) view of the mode-locked generation after non-linear absorber action. The arrows on the oscillogram (a) indicate the positions of strong spikes (peaks are out of the picture) that were transformed later to the single ultrashort pulses. The oscillograms have been acquired with coaxial vacuum photodiode and analogue wide-band oscilloscope.

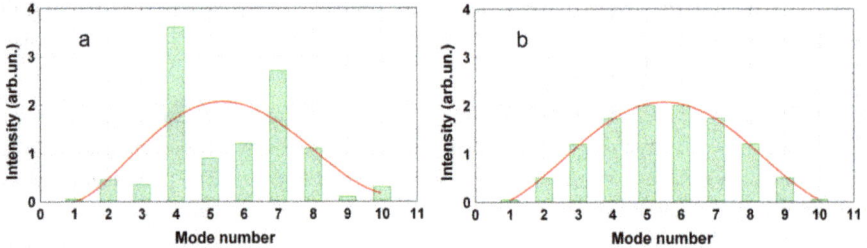

Figure 2. Spectral distribution of intensity of the modes: (a) at the beginning of generation development randomly modulated spectrum of a single noise patter; (b) at moment of pulse train emission. In both cases the red bell-shape line shows the averaged intensity distribution. Distribution (a) is chaotically being changed at each laser shot and during the generation development until achieves the distribution (b). The same is for the phases in Figure 3.

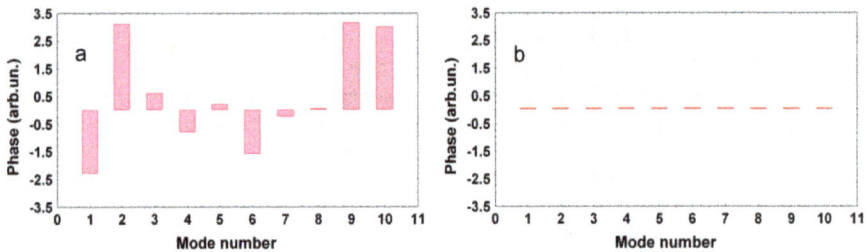

Figure 3. Spectral distribution of the phases of the modes: (a) at the beginning of generation development; (b) at moment of pulse train emission.

Each initial noise spike can be shaped by interference of the modes that belong to different parts of the spectrum and have different phase distribution. During the process of intracavity power increase even the neihbour (in time) spikes coud be modified differently. The ellipses in Figure 4 shows such case when only during two round trip periodes the first spike among three, initially being the smallest, became the highest.

In case of active modulation, the maximum of the modulator transparency very rarely coincides with occasional generated maximal spike of luminescence. However, since any spike is in-time repeatable on the round-trip period and, if coinciding with periodical maximal cavity Q-value, it will grow faster than even bigger spikes irradiated out of high Q-window. After hundreds of round-trips, this initially, maybe minimal pulse achieves the maximal amplitude and is generated as a train of laser pulses. Typically, at active modulation the pulse duration is longer but the jitter of the pulse appearance is much less as compare with passive modulation.

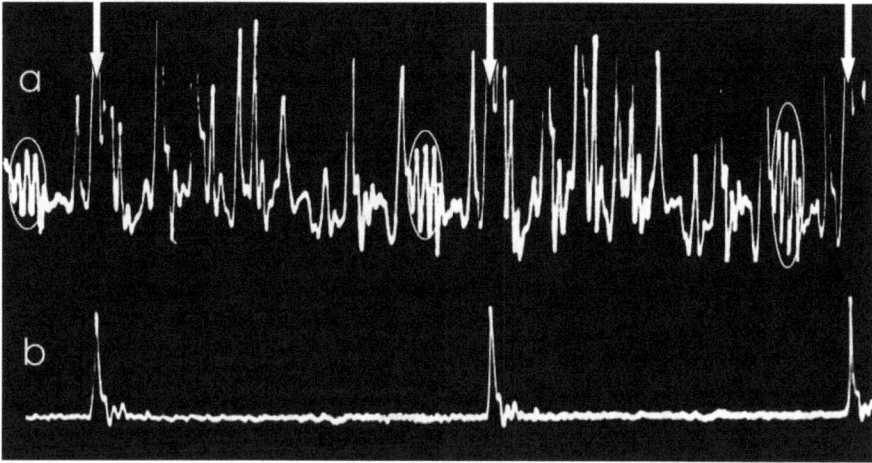

Figure 4. Oscillograms of the laser output radiation: (a) on the initial stage of generation development and (b) after non-linear absorber action. The arrows on the oscillogram (a) indicate the positions of strong spikes (their peaks are out of picture) that were transformed later to the single ultrashort pulses.

6. Propoperties of the optical noise

6.1. Mode formation and radiation dynamics

The phase and amplitude relations between the modes respond to the statistics of a normal (Gaussian) process. At the absence of phase modulation, a minimal possible duration of light random spike is inversely proportional to the total luminescence spectrum width. The amplification band typically has some specific intensity modulation. For relatively narrow spectral bands, the most energetic central modes introduce the main income into the pulse shape formation. At non-stationary pumping, the intensity and spectral width of the gain band is changeable in time that results in changes of random spikes statistics. Moreover, spectral content of the spikes additionally is being deformed because of dispersion, absorption and scattering processes in the cavity elements. As a result, the final time-intensity distribution may be absolutely different from the initial optical random noise distribution that was emitted at threshold conditions.

For the radiation propagating within the linear cavity of length L, the correlation function defers from zero, at least, on the time intervals $Tc = 2L/c$ (for the ring cavity $Tc = L/c$). The existence of signals correlation on Tc intervals means that the operational set of the modes is formed. At this moment, the averaged intensity of circulating in the cavity radiation should be significantly higher than the intensity of firstly irradiated luminescence pattern. After that moment, the temporal intensity distribution does not change significantly after each round-trip through the cavity. However, because of spectral dispersion of an amplification band and different spectral content of each spike, the intensity growth rate for each spike is different. During hundreds of cavity round-trips, the general view of a random noise-like pattern (realization) changes and may be significantly reshaped by the moment of the final ultra-short pulses emission. The shortest spikes shaped during the total period of radiation development may be "late" and not have enough time to be "accelerated" to the maximum intensity. It means that the gain bandwidth may exceed the width, actually used at certain values of pumping, cavity length, material dispersion of the elements, and other laser parameters.

With a Fourier analysis, it is possible to transfer a signal from time domain to frequency and inversely. The mode-locking technology provides zero phase difference between all operational modes. At Gaussian shape of the operational spectral envelope, the pulse intensity has the same shape in time. If one modulates statically the spectrum (by interferometer, for example), the respective modulation appears in temporal intensity distribution. In case of the pulsed pumping, it is possible to modify the gain spectrum by variations of pump rate, cavity length, dispersion of the cavity elements, losses of the cavity to achieve finally different ultrashort pulse durations.

6.2. Statistical properties of optical noise

The theory of electro-magnetic noise was initially developed for the radio band. The main parameter considered in this range is electro-magnetic wave amplitude (E). In optics, because of very high optical signals carrying frequency, the intensity (I) that is a square of field amplitude (E) is considered and measured. Mathematical analysis of the optical stochastic signals with formal exchange E^2 to I has been done in [14]. For detailed description of the mode-locked process, let us repeat here this analysis with taking into account the parameters of the laser cavity. At the beginning, let us rewrite the formulas of random process theory [15] but in terms of intensity, instead of the amplitudes.

The famous Wigner's formula demonstrates the connection between autocorrelation function $k(\tau)$ (time domain) and spectral density $S(w)$ (spectral domain) [16]:

$$k(\tau) = \frac{1}{2\pi} \int_{-\infty}^{\infty} S(w)e^{jw\tau}dw, \tag{1}$$

$$S(w) = \int_{-\infty}^{\infty} k(\tau)e^{-jw\tau}d\tau. \tag{2}$$

The commonly used optical parameter Δf_e is an energetic spectral width of the random process:

$$\Delta f_e = \frac{1}{S_0} \int_{-\infty}^{\infty} S(w)\frac{dw}{2\pi}, \tag{3}$$

here $S_0 = S(w_0)$ is spectral density at maximum of spectral density distribution.

The luminescence field of a solid-state laser has a view of an optical noise with relatively narrow spectral width ($\Delta v/v \sim 10^{-1} \div 10^{-3}$). During evolution of the laser radiation on initial stage (after the pumping applied), it becomes "more and more harmonical". The spectral width decreases for one to three orders of magnitude because of dispersion of the amplification coefficient [17].

The narrow band process is the stationary random process with a zero average value, the spectral density of which is concentrated near specific frequency f_0:

$$f_0 \gg \Delta f_e \qquad (4)$$

A narrow-band stationary random process that has symmetrical spectral density with relation to f_0 is named a quasi-harmonic one. Luminescence bands of most laser ions are shaped by a superposition of several spectral lines, even in such homogeneously broaden system as, for example, Nd:YAG.

This results in important conclusions that typically are missed in books and papers:

- Typically, a wide gain band is not symmetrical relatively to the maximum and may have several local maxima. In principle, luminescence radiation of gain crystal is not quasi-harmonic signal. However, it transforms to one because of significant narrowing during generation evolution.
- Zeros of the dispersion curves do not coincide with intensity maxima. Non-monotonic dispersion curve (within the operational spectrum band) stimulates compression or extraction of the equidistant mode spectrum. This means the variations of round-trip time (for different frequencies), and, automatically, the extraction of the single ultrashort pulse duration.

The correlation function of the narrow band random process may be represented as:

$$k(\tau) = \sigma^2 \rho(\tau) \cos\left[w_0 \tau + \gamma(\tau) \right], \qquad (5)$$

here σ^2 is a dispersion of the process, $\rho(\tau)$ and $\gamma(\tau)$ are slowly modified functions as compare with $\cos(w_0 \tau)$, $\rho(0) = 1$ and $\gamma(0) = 0$.

Correlation function of the wide band process typically is represented as:

$$k(\tau) = \sigma^2 \rho(\tau) , \qquad (6)$$

here $\rho(\tau)$ is normalized correlation function. As it follows from (1) - (6):

$$\sigma^2 = \frac{1}{2\pi} \int_{-\infty}^{\infty} S(w)dw = \overline{I} \qquad (7)$$

Evidently, the parameter σ^2 represents total power of radiation and may be used for the normalization of the amplitudes or intensities of the random spike process. From a general view of the formula, describing probability density for the set of random values ξ_n [15], it is easy to get an expression for electromagnetic wave density distribution with substitution $n = 1$.

$$w_1(\xi_1) = \frac{1}{\sigma_1\sqrt{2\pi}}\exp\left[-\frac{(\xi_1-m_1)^2}{2\sigma_1^2}\right].$$

(8)

For the alternative electromagnetic field, the average value $m = 0$, and the final expression for the function, describing a density of probability of the amplitude A for the output laser radiation with an average intensity value of $\bar{I}$, is:

$$W(A) = \frac{A}{\sqrt{\bar{I}}}\exp(-A^2/2\bar{I})$$

(9)

At that, the phase is distributed homogeneously on the interval $-\pi \leq \phi \leq \pi$.

6.3. Statistics of the laser optical noise pattern

From the statistics point of view, the different stages of the generation evolution may be classified as:

1. Luminescent field below the threshold: wideband random optical field;
2. First part of "linear" generation development (mode formation period): summarizing of the wideband and narrowband random signals;
3. Second part of "linear" generation development: quasi harmonic quasi periodical signal formation;
4. Nonlinear stage – gain or losses saturation.

Respectively, the statistical properties of radiation are different on each stage of generation genesis. By manipulating the initial internal radiation and laser external parameters, it is possible to achieve different results in the end of the path.

The sum of harmonic and quasi-harmonic signals [15] may be represented as follows:

$$S(t)+\xi(t) = A_m \cos(w_0 t+\phi_0) + A_c(t)\cos w_0 t - A_s(t)\sin w_0 t = V(t)\cos[w_0 t+\psi(t)],$$

(10)

here $A_c(t) = V(t)\cos\psi(t) - A_m\cos\phi_0$, $A_s(t) = V(t)\sin\psi(t) - A_m\sin\phi_0$.

A random function $V(t)>0$ is envelope, and a function $\psi(t)$ is a random phase of the sum of harmonic signal and quasi-harmonic noise. From (10) it is easy to achieve:

$$V(t) = \{[A_c(t) + A_m\cos\phi_0]^2 + [A_s(t) + A_m\sin\phi_0]^2\}^{1/2} \, V(t)>0,$$

(11)

$$\psi(t) = arctg\frac{A_s(t)+A_m\sin\phi_0}{A_c(t)+A_m\cos\phi_0}, -\pi \leq \psi(t) \leq \pi.$$

(12)

Envelope $V(t)$ and random phase $\psi(t)$ have one-dimension probability densities:

$$W(\upsilon) = \upsilon\exp\left(-\frac{a^2+\upsilon^2}{2}\right)I_0(a\upsilon), \ \upsilon = V/\sigma,$$

(13)

$$W(\psi) = \frac{1}{2\pi}\exp\left(-\frac{a^2}{2}\right)\left\{1 + \sqrt{2\pi}a\cos(\psi - \phi_0)\Phi\left[a\cos(\psi - \phi_0)\right]\exp\left[\frac{a^2\cos^2(\psi - \phi_0)}{2}\right]\right\}, \quad (14)$$

here $a = A_m/\sigma$ is signal to noise ratio, $I_0(z)$ is Bessel function of zero's order with imaginary argument.

$$I_0(av) = \frac{1}{2\pi}\int_{-\pi}^{\pi}\exp\left[\pm iav\cos(\psi)\right]d\psi \quad (15)$$

$$\Phi(z) = \frac{1}{2\pi}\int_{-\infty}^{z}\exp\left(-\frac{x^2}{2}\right)dx \text{ is a probability integral} \quad (16)$$

From a general consideration, it is clear that the higher the average radiation power, the higher the number of random spikes amplitude variants that can be realized. Increasing the width of the probability density curve means that the probability of generation of two spikes with the close amplitudes increases versus time. This is an undesirable effect, because the closer the amplitudes are, the more difficult separating them and suppressing lower spikes on the non-linear stage of generation development becomes. To achieve complete mode-locking, it is necessary to maximize the difference between the amplitudes of the random spikes that makes the single pulse separation on the round-trip period of the cavity easier.

In [15], it has been shown that to calculate the number of excesses over certain curve C, it is necessary to estimate the joint density of probability for the envelope and for the derivative of this function. The final formula for the average number of positive excesses of the envelope $V(t)$ of the sum of random (noise) and quasi harmonic processes in unit of time is:

$$N_1^+(C) = \sqrt{\frac{-\rho_0''}{2\pi}}W = \sqrt{\frac{-\rho_0''}{2\pi}}\frac{C}{\sigma}\exp\left[-\frac{a^2}{2} - \frac{C^2}{2\sigma^2}\right]I_0\left(\frac{aC}{\sigma}\right). \quad (17)$$

Respectively, for the quasi-harmonic signal ($a=0$) the view for (17) may be simplified:

$$n_1^+(C) = \sqrt{\frac{-\rho_0''}{2\pi}}\frac{C}{\sigma}\exp\left[-\frac{C^2}{2\sigma^2}\right]. \quad (18)$$

It is known that the square of the energetic width of the signal spectrum is proportional to the second derivative from the correlation coefficient. For the Gaussian spectral density function, taking into account (6), (7) it is possible to write:

$$\sqrt{\frac{-\rho_0''}{2\pi}} = \Delta f_e. \quad (19)$$

Thus, for quasi-harmonic optical signal that is correlated on time T_C (round-trip period), the average number of the intensity spikes that surely exceed the average radiation intensity level $\bar{I}$ may be found from the following:

$$n^+ = T_C \Delta f_e W\left(\frac{I}{\sqrt{\bar{I}}}\right) \qquad (20)$$

Versus the generation development, the value Δf_e decreases, but $\sqrt{\bar{I}}$ increases. Thus, temporal dependence of n^+ depends on the functions mentioned above and might have complicated non-monotonous behaviour. Formula (20) demonstrates that spectral width and total radiation power determine all properties (statistics) of the laser radiation noise pattern. Hence, calculating these values at any moment of generation development, one can evaluate the statistical parameters of radiation.

6.4. Statistical properties of radiation on the "linear" stage of the generation development

In [18-20], the process of mode-locking in solid-state lasers has been analysed. It was shown that at the end of the linear stage, the noise pattern is a superposition of a great number of patterns emitted after the threshold conditions have been achieved. For increase of the output parameters repeatability, it was proposed to decrease a total number of the operating modes but with keeping the same a total spectrum width. Such mode number thinning out results in increase of repetition rate of the pulses and improves the pulses parameters repeatability. To check these dependencies, a Nd:glass laser with round-trip period in range 20 ps - 20 ns has been built and studied experimentally.

Starting from (9) and (20) with n =1, 2, one can find two maximal amplitudes C_1 i C_2 (the first and the second) probably generated on the laser cavity round-trip period. Typically, for small relative difference $\Delta = (C_1 - C_2)/C_1 \ll 1$ and for significant number of spikes $q = f_e T_C \gg 1$, amplitude difference may be estimated as:

$$\Delta = \ln 2 / 2[1 + \sqrt{2\ln q} / (2\ln q - 1)]\ln q. \qquad (21)$$

Spontaneous luminescence of a gain laser medium starts practically at the same moment with pumping. If the gain medium is located in the optical cavity, its emission may be separated on the portions equal to the light round-trip time between the cavity mirrors. Each of these patterns is independent noise realization. The total field at the end of the linear stage is the superposition of a great number of such patterns. Because of the amplification coefficient spectral dispersion, the initial spectral-temporal distribution deforms even on, so called, "linear" stage. It is clear that this classical term cannot be applied to the generation evolution period, which is characterized by signal frequency transformations.

Let suppose that an averaged in time envelope of the spectral intensity distribution is Gaussian: $I(w) = I_0 exp(-\Delta w^2/\delta)$. Total power after N round-trips of light is an integral through the spectrum with taking into account the growth of amplification coefficient. After each round-trip, a "fresh" noise pattern is added to the previous amplified radiation pattern. However, for a new noise realization, amplification exceeds losses in wider and wider spectral range. Because of these small differences in starting conditions, the summarized radiation becomes non-Gaussian, even if each realization is Gaussian (but with different spectral width!). The function that describes average intensity change is the following:

$$\sigma_m^2 = \sigma_0^2 \exp\{\beta T_p (N-m)(N-m+1)/2\} / \sqrt{1+\alpha_0(N-m)}, \tag{22}$$

here σ_0^2 is average luminescence power, β is amplification coefficient growth rate; T_C is axial period of the cavity; N is number of round-trips from the moment of threshold completion for the central frequency; m is number of round-trips from the moment of the threshold up to m-th noise realization emission, α_0 is linear initial losses (dimensionless). Spectral width of m-th noise realization changes respectively:

$$\delta_m = \delta_0 \left[1 + \alpha_0 (N-m)\right]^{-1/2}, \tag{23}$$

where δ_0 is width of the luminescence of the gain medium. Formula (23) was acquired in assuption $\beta T_C N_{lin} << \alpha_0$, N_{lin} is a number of round-trips during total linear stage. Average intensity of final field accumulated in the cavity in the end of linear stage is as following:

$$\sigma^2 = \sigma_0^2 \exp\left[\beta T_C \frac{N_{lin}(N_{lin}+1)}{2}\right] \sum_{m=0}^{N_{lin}} \frac{\exp\left[-\beta T_C \frac{m(m+1)}{2}\right]}{\sqrt{1+\alpha_0(N_{lin}-m)}}. \tag{24}$$

It follows from (24) that the noise realizations with the numbers from $m = 0$ to

$$M=[2/\beta T_C]^{1/2}, \tag{25}$$

introduce the main income into the average intensity. They dictate the spectral parameters of the final field. For the typical laser parameters $\beta = 10^3 s^{-1}$, $T_C = 10^{-8}s$ one can estimate $M \approx 450$. Because of (23) the spectrum is not exactly Gaussian. However, difference in first actual M realizations is not significant, because usually $M << N_{lin}$.

Taking into account that on the "linear" stage, until there is no gain medium saturation, each realization develops independently, the final field may be considered as quasi harmonic random process with an average intensity:

$$\sigma^2 = \frac{\sigma_0^2 M}{\sqrt{1+\alpha_0 N_{lin}}} \exp\left[\beta T_C \frac{N_{lin}(N_{lin}+1)}{2}\right] \tag{26}$$

and the spectrum width

$$\delta = \delta_0 / \sqrt{1+\alpha_0 N_{lin}}. \tag{27}$$

Thus, the field formed up to the "linear" stage end is described by the statistics of quasi-harmonic random process. However, its evolution time is shorter as compared with classical estimations because of storage of energy by adding a significant number of optical noise realizations. It means that the spectral width is higher and the amplitude difference between maximal spike and the second one is less than that predicted in linear model. As a result, the final probability of complete mode-locking is much less than it was estimated before. For different gain media, activator doping concentrations, methods of pumping, and levels of losses into a cavity, the linear stage duration is different. Hence, in certain limits the final spectrum width is not proportional to the starting (luminescent) width when different systems are compared. Though, affecting some parameters of the laser system, it is possible to satisfy requirements, providing the guaranteed achievement of single pulse generation on the cavity round-trip period.

6.5. Spectral-temporal dynamics

The noise pattern spectrum envelope only in average is proportional to the gain profile. In each specific "shot" the number of quanta in each certain mode may significantly differ from the statistically averaged value. Initial mode phases are distributed homogeneously in the range $-\pi \leq \varphi \leq \pi$. Mode-locking process results in two actions: introducing certain constant phase relations between the modes and in transferring the energy between the randomly modulated modes, so the spectrum envelope becomes smooth without chaotic intensity modulation.

Let us write the modes oscillations in specific point of cavity as the series:

$$
\begin{aligned}
A*(w,t) = &A_0\sin(w_0 t + \varphi_0) + A_1\sin(w_1 t + \varphi_1) + \\
&A_{-1}\sin(w_{-1}t + \varphi_{-1}) + \ldots + A_n\sin(w_n t + \varphi_n) + A_{-n}\sin(w_{-n}t + \varphi_{-n})
\end{aligned}
\tag{28}
$$

here $w_{\pm n} = w_0 \pm \Delta w \cdot n$, w_0 is the central (maximal) frequency of the operational spectrum, $\Delta w = c/2L$ is an intermode frequency space, n is the mode number starting from the central one.

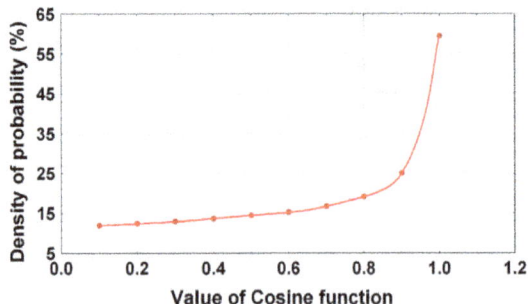

Figure 5. Probability to find a value of $\cos(\varphi)$ function in intervals $0 - 0.1, 0.1 - 0.2,\ldots 0.9 - 1$ at φ homogeneously distributed on the interval $-\pi/2 < \varphi < \pi/2$.

If the initial phase φ_n is not equal to zero, $\sin(w_n t + \varphi_n)$ may be represented as a series with the components: $A_c\sin(w_n t) + A_s\cos(w_n t)$, where $A_c = \cos\varphi_n$ and $A_s = \sin\varphi_n$ are the random values that changes slowly, as comparing with $\cos(wt)$. Let pay attention to the fact, that even in case when a random phase is distributed homogeneously on the range $-\pi \leq \varphi \leq \pi$, the distribution density of A_c or A_s relate to cosine or sinus functions; in other words, this is not a homogeneously distributed function any more on the range of existence $(-1 \leq A \leq 1)$. The probability to find a function on specific interval of meanings is inversely proportional to the rate of function changes, or to the derivative. For the function $A_c = \cos\varphi$ density of probability to find meaning into the interval $1 < A < -1$ is proportional to $\sin\varphi$. Thus, the most probable meaning for A_c is 1 in the interval $-\pi/2 < \varphi < \pi/2$ and -1 in the intervals $\pi/2 < \varphi < 3\pi/2$. For A_s the probable meaning in these intervals is close to zero. Figure 5 shows that in 58% attempts with phase distributed inside the range $-\pi/2 < \varphi < \pi/2$, the meaning of $\cos(\varphi)$ function is inside the range 0.9-1. Respectively, interval 0.8-1 involves about 85% of all noise realization. The joint probability that two neighbour modes are generated with close phase values within this range is 0.72. At small number of operating modes in most number of the laser shots the sum $\sum_n \sin(w_n t + \phi_n)$ may be exchanged to the $\sum_n \sin(w_n t)$.

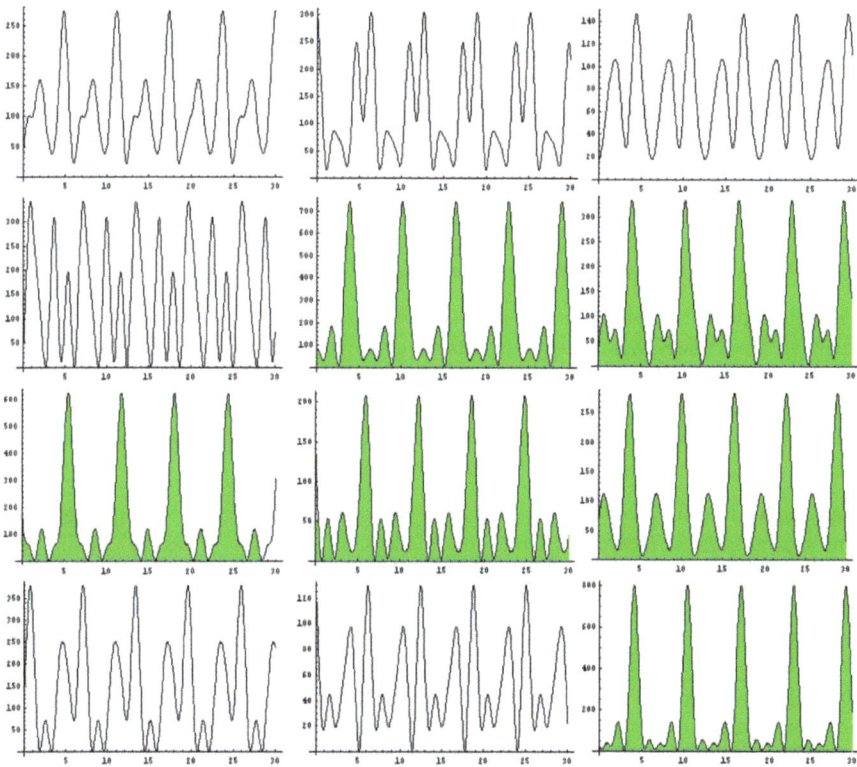

Figure 6. Computer simulation of the five mode interference (natural mode-locking process). Green oscillograms show the cases when the main peak contains above 80% energy of all pulses located on the round-trip period.

Figure 6 demonstrates a computer simulation of natural "mode-locking". The program generates random phases and amplitudes of five modes. Evidently, that half of the oscillograms contains 80% of round-trip period energy in a single pulse. This demonstrates "natural" mode-locking process, achieved without any non-linear elements placed into the cavity. Even with a weak non-linearity, such surely separated spike amplitudes may be easy additionally emphasized. However, there are more modes – less probability to generate "fully synchronized" spectrum. The solution is to keep a wide total spectral width but to decrease a total mode number. This can be done with low quality interferometer that does disturb the equidistant mode spectrum.

Moreover, in a real laser, the mode behaviour is not independent, especially if the amplification band is relatively narrow. Because of inversion hole-burning [7], two neighbour modes have maximal amplification because they are not overlapping in the centre of cavity where typically the gain medium is located. However, the next pair of the modes is overlapping with the first pair in the centre and near the mirrors. For the next 5th and 6th modes the amplification drops down because of spectral envelope's bell-like shape. Because of this phenomenon, practically anytime the laser output from the "pure" cavity (without selection) demonstrates modulation with axial period, which is a beating of two central high power modes. Taking into account the high probability that a "phase coefficients" A_c would be close to 1, one can understand why in big number of laser shots the emission demonstrates periodical structure, even without of any modulators or nonlinear elements.

6.6. Optimal spectral width of the ps-laser

It is well known that the wider the gain spectrum is, the shorter the pulses that can be generated are. However, some experiments have shown that this is not always true. This and the following sections analyse how the index of refraction and gain dispersions influence the duration of the spikes on the "linear" stage of generation evolution. The basic question is: what set of parameters provides the minimal duration of pulses before the non-linearity of the absorber or gain medium starts working.

In principle, the luminescence band of glass lasers has a sufficient width to generate femtosecond pulses. However, some effects result in an increase of the initial spike duration in the process of amplification and generation. First of all, because of the spectral dispersion of the amplification coefficient, the typical spike duration increases approximately 50 times [17]. To avoid this phenomenon, in [21], the interferometer was located in the cavity. It was aligned so that it flattened the maximum of the amplification band. The shortening of the pulses was about 1.5 to 2 times. However, it was much less than expected.

The next reason that makes it difficult to achieve short pulses in solid-state laser hosts is the dispersion of the material refractive index that results in broadening of the ultrafast pulses. At the conditions of low amplification near the threshold, the initial amplitude of the random short and intensive spike decreases during several round-trips through the cavity.

For the initially shorter pulses, the time of spike intensity development up to the absorber saturation energy is longer than for the relatively long pulses. Certainly, longer pulses survive in the process of generation evolution. Let us analyse the random spike spectrum genesis by taking into account amplitude-frequency transformations along the linear stage period.

To simplify the calculations, we assume that the spike spectrum shape is Gaussian, with an energy normalized to one. To study the process, we apply the Fourier transformation and follow all spectrum components developing.

$$I(t) = \frac{\delta}{\sqrt{\pi}} \exp(-\delta^2 t^2) = V(t)V^*(t), \tag{29}$$

Here δ is a luminescence spectrum width (dimension of δ is [s^{-1}]).

$$V(t) = \frac{1}{\sqrt{2\pi}} \int_{-\infty}^{\infty} \{\frac{\exp[-\frac{w-w_0}{\delta}]^2}{\delta\sqrt{\pi}} \exp[\beta T_C \frac{N(N+1)}{2} - N\alpha_0(\frac{w-w_0}{\delta})^2]\}^{1/2} \exp(-i\Delta\Phi)\exp(-iwt)dw \tag{30}$$

The first exponent in (30) describes the amplitude distribution of the spectral components of a pulse with maximum at w_0, the second one shows the amplification depending on the number N of the light field round-trips with the axial period T_C, initial (linear) coefficient of the losses α_0, and amplification coefficient increase rate β. An expression for the phase modification is as following:

$$\Delta\Phi = \frac{2\pi\Delta nL}{\lambda} = \frac{dn}{dw}|w-w_0|wT_qN, \tag{31}$$

here Δn is refractive index changes (at tuning out of w_0); L is a total length of the gain media passed by the light during the whole linear stage of generation; dn/dw is the refractive index dispersion; T_q is time of a light single pass through the gain medium.

Let us consider, at first, the pulse evolution that is represented by (29) – (31). The final expression for the single pulse intensity is as following:

$$I(t) = \exp[\beta T_C \frac{(N+1)N}{2} - \frac{\Delta(aw_0+t)^2}{2(\Delta^2+a^2)}][2\delta\sqrt{\pi(\Delta^2+a^2)}]^{-1}, \tag{32}$$

here the designations are: $\Delta = \frac{1+\alpha_0 N}{2\delta^2}$, $a = \frac{dn}{dw}T_qN$.

From (32), it follows that the final inverse duration is connected with the initial one as:

$$\delta' = \delta\left[1+\alpha_0 N\right]^{1/2}[(1+\alpha_0 N)^2 + (2\delta^2 \frac{dn}{dw}T_qN)^2]^{-1/2}. \tag{33}$$

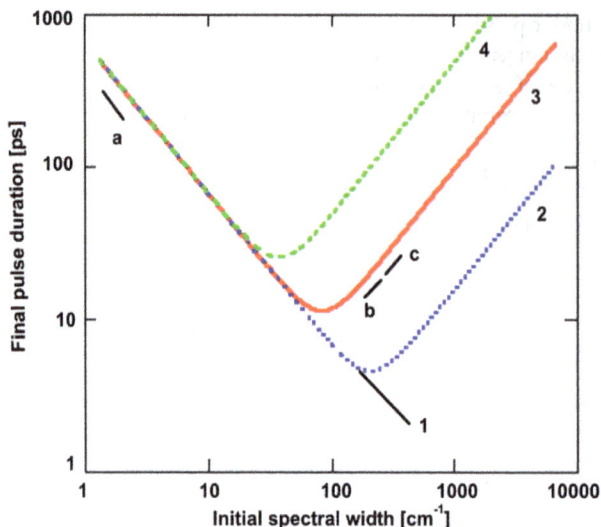

Figure 7. Dependence of the spike duration at the end of linear stage vs the initial noise pattern spectral width for different values of the refractive index dispersion $dn/dw = 0$ (1), $0.13 \cdot 10^{-17}$s (2), $0.8 \cdot 10^{-17}$s (3), $5 \cdot 10^{-17}$s (4).

Figure 7 demonstrates the dependency of the pulse duration at the end of the linear stage Δt on the start spectrum width δ of the random spike at different values of the product $T_q(dn/dw)$. For $(dn/dw) \neq 0$, the curves coincide in the range of small δ meanings. For wide δ, the final Δt depends on the specific value of the product $T_q(dn/dw)$. In Figure 7, the letters mark the spectrum width for (a) Nd:YAG, (b) Nd:phosphate glass, (c) Nd:silicate glass.

At the end of the linear stage of generation, $N\alpha_0 \gg 1$ and expression for the δ' may be simplified:

$$\delta' = \frac{\delta}{\sqrt{\alpha_0 N}} \text{ for the small } \delta; \tag{34}$$

$$\delta' = \frac{1}{\delta}\sqrt{\frac{\alpha_0}{N}} \left(2T_q \frac{dn}{dw}\right)^{-1} \text{ for the big } \delta. \tag{35}$$

From Figure 7, it is evident that the final spike duration Δt has a minimum. This area determines the certain start spectral width that, at some given set of the parameters (dn/dw, α_0, β, T_C, T_q), can provide the minimal duration of the spikes Δt_{min} at the end of the linear stage. From the (33) the conditions of the Δt_{min} achievement may be deduced:

$$2\delta^2 \frac{dn}{dw}T_q = \alpha_0 \tag{36}$$

It follows from (36) that, for example, for the Nd:silicate glass ($T_q = 10^{-9}$ s, $(dn/dw) = 8 \cdot 10^{-18}$ s) the optimal initial spectrum width is ≈ 70 cm^{-1} but the real one is about 250 cm^{-1}. Thus, to achieve minimal ultrashort pulses in the laser with Nd:silicate glass, it is necessary to narrow the initial luminescence spectrum width 3-4 times. The use of prisms is not appropriate in this case because with spectrum narrowing, it simultaneously results in the increase of the product $T_q(dn/dw)$. As an element with anomalous dispersion, an interferometer may be used. In this case, the etalon automatically plays the role of a gain spectral dispersion flattener.

By decreasing the losses (e.g., by improvement of cavity construction), the value Δt_{min} moves to the smaller spectral width values. If β stays the same, the relative amplification growth is higher in a system where α_0 is less. In this case, the spike spectrum width grows faster and natural spectrum selection process is restrained. If one decreases α_0 at relatively small initial spectrum $\Delta < 30$ cm^{-1}, the spike duration decreases because of the process of natural spectrum selection weakening. Inversely, for $\Delta > 60$ cm^{-1} the duration increases because of negative influence of the refractive index dispersion.

The starting losses α_0 include linear losses on cavity elements, on mirrors (output), and non-saturated losses in absorber. To achieve Δt_{min}, it is necessary to vary mirror reflection only, because decrease of the absorber density makes the pulse discrimination worse. For garnet type media, the highly reflecting mirrors are preferable; for glasses, it is better to use low reflecting mirrors.

Increase of the product βT_C (with the other parameters staying the same) results in Δt_{min} decrease. This phenomenon is connected with diminishing of the total number of the round-trips during the linear stage of the generation development. Having a low threshold and high amplification increase rate β, the garnets have lower located curves $\Delta t(\delta)$ as compared with glass hosts. This explains why these media, being with about two order of magnitude different luminescence bandwidth, demonstrate ps-pulses with durations just 3-5 times different. The phosphate glasses that have narrower luminescence bandwidth as compared with the silicate ones can provide shorter pulse generation. In all cases, possible variations of the laser parameters may minimize the random spike duration at the end of the linear stage of the generation development.

7. Experimental results

7.1. Spectrum modulation with dynamical interferometer

The mode-locking process supposes introducing of the certain amplitude-phase relations between all generating modes. In known techniques, phase locking is typically provided simultaneously along the whole actual spectral range. However, the condition of simultaneous phase regularization is not obligatory. In principle, the modes may be linked step by step, if a total mode number is relatively small or the time of generation development is relatively long [14]. In this case, the process of mode-locking spreads consequently from small group of modes to the whole spectrum. The simplest way to build such a system is to install into a laser cavity an additional mirror with a piezo-driver. This three-mirror cavity is a dual-interferometer system. Let us place an interferometer with base length l_{0i} into the cavity with an optical length $L_C = kl_{i0}$, where k is integer. A diagram in Figure 8 shows the case where $k = 10$. Because of different cavity and interferometer lengths, there are k cavity modes located between two peaks of the interferometer. Internal mirror should be with low reflectivity and do not disturb significantly the mode spectrum of the main cavity. For two interferometers with bases of different optical length, the velocity of the reflection peak motion in the frequency domain is higher for the interferometer, which has shorter optical base. In principle, it is possible to move the interferometer peak from mode to mode position exactly during the light round-trip time along the cavity. In this case, relatively fast modulation with intermode beating frequency is realized by the relatively slowly moving mirror. This system can be named as an "optical lever" technology.

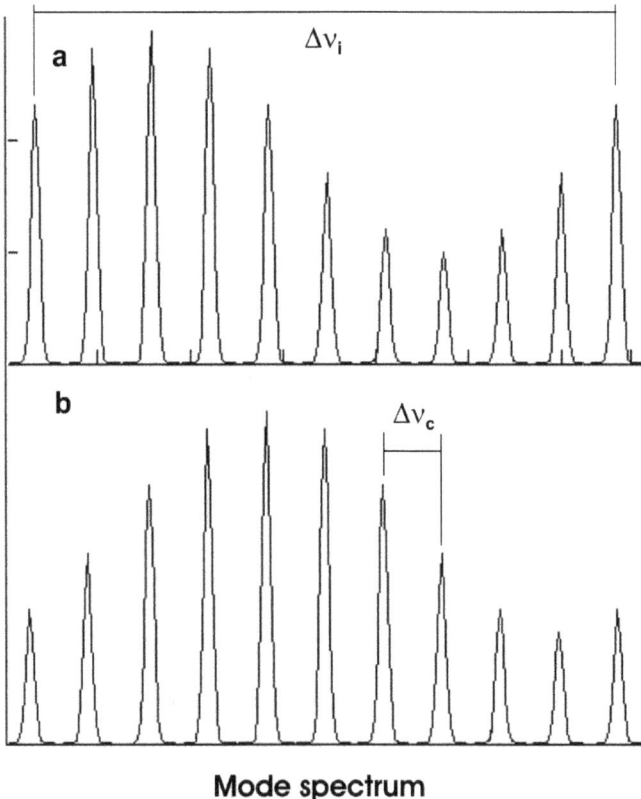

Mode spectrum

Figure 8. Examples of the mode spectrum fragments: (a) at time moment t and (b) at $t+2T_C$ – after two round-trip periods. Cavity length is equal to ten interferometer lengths: $L_C = 10l_i$. For illustration purpose the modulation depth is significantly increased.

To find the conditions of such mode-locking, let us suppose the optical length of the interferometer l_i changes with constant speed V. In this case, $l_i = l_{0i} + Vt$, here l_{0i} is the start interferometer length. The position of the n-th transparency maximum (of an unmovable interferometer) is described as:

$$v_n = 1/\lambda_n = n/2l_i, \text{ here } v = 2l_i/\lambda_{n0} \tag{37}$$

Differentiate (37) with respect to time and get n-th interferometer maximum frequency changes rate. With condition $Vt \ll l_{0i}$ (because $Vt \cong (1-10)\lambda_{n0}$, and $l_{0i} \cong 10^4\lambda_{n0}$) it is easy to achieve:

$$\frac{dv}{dt} = -\frac{nV}{2l_i^2} = -\frac{l_{i0}V}{\lambda_{n0}\left(l_{i0} + Vt\right)^2} \cong -\frac{V}{\lambda_{n0}l_{i0}}. \tag{38}$$

With the moving mirror the new cavity mode has the maximum quality consequently in each new T_C interval. Let us find such a mirror speed that provides the interferometer peak motion from the mode position to the next one for the time of the light round-trip in a linear cavity $T_C = 2L_C/c$:

$$\left(\frac{dv_n}{dt}\right)T_C = -\Delta v_C. \tag{39}$$

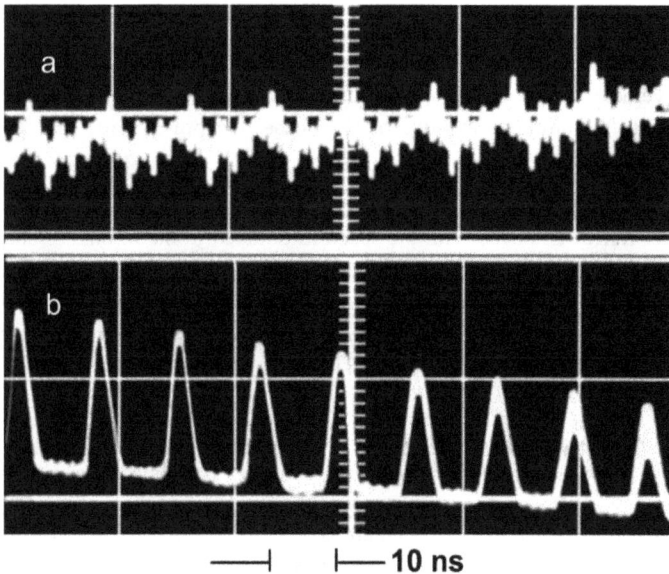

Figure 9. The oscillograms of laser output (a) with unmovable internal mirror and (b) with movable mirror.

After substitution into this equation dv/dt and T_c the value of V can be found. Intermode space Δv_i for the ring cavity is $\Delta v_i = 1/l_i$, and for the linear one – $\Delta v_i = 1/2l_i$. Introducing the coefficient $\gamma = 1/2$ for the linear cavity and $\gamma = 1$ for the ring one, finally the formula for the resonant mirror speed is:

$$V = \gamma \lambda_{n0} \Delta v_p^{\ 2} / \Delta v_i. \tag{40}$$

Figure 9 demonstrates the oscillograms of the laser output (a) without interferometer and (b) with scanning interferometer. The interferometer mirror was moved by the piezo-transducer. Cavity length L_c was 200cm, interferometer base l_i was 0.25 cm that provided $k = 800$. With laser operational wavelength $\lambda = 1.06$ microns, the mirror velocity was about $V = 5$ cm/s. For lower k values the necessary mirror velocity should be higher. Hence, the mechanical motion of the mirror could be hardly achievable. In this case the electro-optical system should be used to control the cavity modes properly. An inclined interferometer limits the total spectrum width and the ultrashort pulses are not achievable with this technique. To solve this problem, the normally installed mirror should be used. To avoid back reflection that results in parasitic selection, a thin film selector should be used as a mirror. The properties of such a system have been described in section 2.

7.2. Laser with anti-mode-locking: "Black" pulses generation

Any periodical signal may be represented as a sum of harmonics with specific amplitude and phase coefficients. The scanning interferometer mirror may be moved with changeable velocity, following the special function. In this case, because of interference of the modes with certain combinations of the phase, the pulse shape may be specially designed [14].

One of the interesting examples of such laser generation is "anti-mode-locking". If the interferometer mirror is moved with the speed as doubled to the resonance one, the zero-phase difference is dictated to each second mode. The neighbour modes are modulated in anti-phase conditions. Figure 10 demonstrates several examples of simulated (left column) and experimentally generated (right column) pulse trains. The cavity length was 525 cm. That corresponds to a 35 ns round-trip time. The interferometer base was about 1.5 cm, with a resonance speed of about 4 cm/s.

Figure 10a – resonance mirror motion; it is classical mode-locking. Figure 10b – average mirror speed is two times higher as compare to the resonance one; motion is with acceleration. Figure 10c – motion with acceleration two times higher as in case (b). Figure 10d – average mirror speed is two times higher compared to the resonance one; even modes are in anti-phase conditions to the odd modes. ("anti-mode-locking"). The picture is inversed to the classical mode-locking (Figure 10a). On the background of the quasi-cw generation there are narrow peaks of radiation absence, or "black pulses".

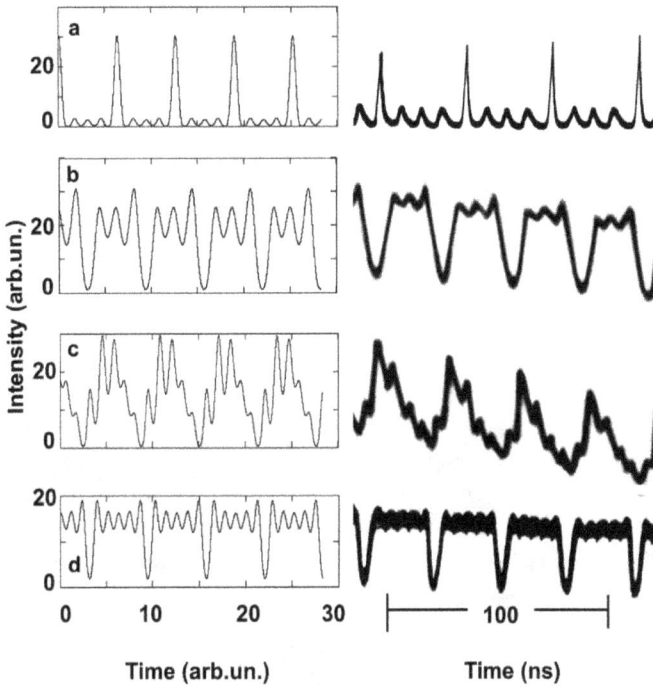

Figure 10. Simulated (left column) and experimental (right column) oscillograms in case of mode-locking (first row), anti-mode-locking (fourth row) and with some phase-shifted modes (second and third row).

7.3. High pulse repetition rates

In various applications such as telecommunications, pulse trains with multi-gigahertz pulse repetition rates are required. The resonators of the bulk lasers are usually too long to achieve such repetition rates with fundamental mode-locking. The high repetition rate pulse trains were obtained with harmonic mode locking [22, 23], when the active modulator worked at frequency multiply integer higher than the inverse cavity round-trip period. Similar research but at passive mode-locking has been provided in [24].

In this section, a 50-GHz repetition rate generation from the Nd:glass laser with 45-MHz cavity is described. The system was of modulator free [14]. This work was completed approximately nine years before the first self-starting Ti:sapphire laser was demonstrated. However, honestly saying, during Nd:glass laser's experiments, the role of the Kerr-type non-linearity was not understood.

The construction of the high repetition rate laser is very simple. The cavity contains just one additional low reflecting mirror that is installed so that the short base interferometer length (l) is exactly an integer multiple (n) of the full cavity length (L) (with heated gain medium): L

= *nl*. At this condition, the ultrashort pulse, running back-and-forth inside the interferometer, after *n* reflections, exactly coincides with the pulse that was running along the entire cavity. Since the total spectral width is the same, the pulse duration remains the same. The multiplication of ps-pulses results in periodical modulation of spectrum and, respectively, in a decrease of total mode number. At this condition, even without a non-linear absorber but with an aperture installed near the output mirror, the laser generated a train of ultrashort pulses. With a fourth mirror, which shapes a new interferometer with a base length integer to the first interferometer, the pulse repetition rate was increased again. Figure 10 demonstrates oscillograms of the mode-locked generation (a, c) and harmonic mode-locked (b, d). A photograph (e) shows the two-photon luminescence track achieved with a four-mirror cavity without a non-linear absorber. The pulse duration is 11±2 ps and the period of repetition is 20 ps. In all cases, the total cavity round-trip length was 6.6m (22ns).

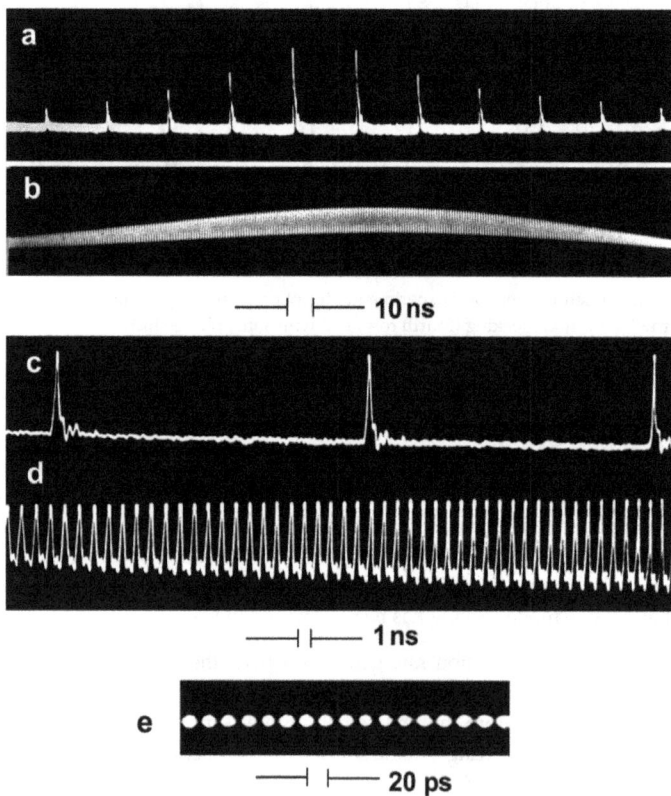

Figure 11. Oscillograms of the mode-locked generation: (a) total train; (c) three-pulse zoomed fragment. Three-mirror cavity: (b) total train; (d) zoomed fragment – period 0.98 ns. (e) Photo of two-photon luminescence track at four-mirror cavity – period 20 ps.

From my personal experiences, the idea to use an interferometer in the mode-locked laser cavity is typically rejected by project or paper reviewers because of supposed strong interferometer dispersion, which affects equidistant mode spectrum and makes short pulse generation impossible. In response, it should be specially emphasized that maximal exceeding of the gain over the losses in an free-running laser could be around 3-4%; hence, to shape the necessary spectral structure, the interferometer should provide about 5% modulation depth, which practically does not shift the modes. Moreover, it is possible to use the interferometer with absorbing mirror (see in section 2), which demonstrates a unique property – it does not reflect light at normal beam falling on the interferometer. In other words, there is no disturbance of the mode spectrum with such an interferometer.

8. Femtosecond pulses

From a theoretical point of view, the basic principles of picosecond and femtosecond pulses generation are the same [9-13]. The difference is in some parameter values. Femtosecond pulses have three orders of magnitude shorter duration than the picosecond ones. Hence, they require much wider operational spectral bands and faster mechanisms of modulation. Because of dispersion, the broadening of fs-pulses happens faster when propagating through the optical elements. Because of very high pulse peak power, the non-linear processes are routine problems for fs-lasers. Practically all types of ps-lasers were repeated in fs-lasers. They are bulk lasers with open cavity, fiber lasers, dye lasers, and semiconductor lasers. Disregarding the construction variety, principles of operation of all these systems are the same, or very similar. In the last five years, fiber lasers were the most dynamically developed segment of the fs-laser. The efforts of researchers are focused on suppression of non-linear effects, on increase of output average power, and on exploitation of new spectral ranges.

9. Attosecond and zeptosecond pulses on the way to yoctoseconds

The attosecond pulse generation and measurement are the hottest subjects of popular and solid scientific journals [25]. With achievement of this duration range, these super-ultra-short laser pulses became an instrument of intra-atomic process investigation. For the first time ever, researchers are capable of seeing how an electron "jumps" between atoms. The peak intensity of the as-pulses approaches and overcomes 10^{18} Wcm^{-2}. This is a powerful tool in studying the ultrafast phenomena, such as the chemical/biological transformations occurring on the femtosecond range of durations. An attosecond control of collective electron motion in plasmas has been provided in [26]. There are two ways to generate an attosecond pulse or pulse train [27]. One is through the nonlinear processes of the superposition of high order harmonics generated in the laser-gas atom interactions. The limitation is that the laser intensity should be low enough to avoid the ionization of atoms. Typically, the efficiency of harmonic emission from the atoms is low. The other way is to generate high order harmonics from the dense surface of plasma created by the high-intensity femtosecond laser. Till now, 12 attoseconds is the world record for shortest controllable time [28].

The femtosecond lasers are more and more used as the sources in high energy atomic and nuclear investigations – see, for example [29]. It was shown that in the scattering of 100-fs laser pulse with an intensity of around 10^{19} W/cm² by a counter propagating electron with an initial energy of 10 MeV, a crescent-shaped pulse with pulse duration of 469 as and the photon energy ranging from 230 eV to 2.5 keV is generated in the backward direction. Recently the Internet brought us a new word: Yoctosecond – 10^{-24} seconds. The explanation what this is and where it could be used is very simple and fundamental: this is time taken for a quark to emit a gluon [30].

10. Conclusions and epilogue

1. Theory of random processes, initially developed for the radio-band frequencies, has been applied to describe the properties of the narrow-band optical noise and dynamics of the linear stage of generation evolution.
2. Theoretical analysis of the optical noise properties demonstrated high probability of the "natural mode-locking", when the main part of light energy accumulated in the cavity may be concentrated in a single pulse.
3. It has been shown that for each set of the solid-state laser parameters, there is a specific width of the amplification band that provides minimal duration of random spikes at the end of linear stage of generation. It was shown that two media, having, at least, one order of magnitude difference in luminescence spectrum width, might generate pulses of the same duration.
4. The harmonic mode-locked generation, achieved in a three-mirror cavity without a non-linear absorber, has been demonstrated in 22-ns cavity of Nd:glass laser.
5. A new method for controllable mode-locking with the help of "optical lever" – movable mirror placed into the laser cavity – has been proposed. The anti-mode-locking process with "black pulses" generation has been realized. Generation of rectangular and triangular ns-pulses has been achieved.

During the 50 years of their history, the pulsed lasers passed from seconds to zeptoseconds or about 20 orders of magnitude into the short duration's side. This road was not smooth and easy. Every time, starting from huge, complex, ineffective, and very expensive machines, the lasers became elegant, economical, more powerful and smart instruments in science, technology, medicine, and everyday life. We hope this chapter and this book will be useful for a wide spectrum of specialists, for professors and students, and for those who are interested in history and in future of the laser technologies.

Author details

Igor Peshko
Department of Physics and Computer Science, Wilfrid Laurier University, Canada

Acknowledgement

I very much appreciate my former University of Toronto Master's program student Inderdeep Matharoo who did a great job as a technical editor and as a first reader. At last, this work would be impossible without excellent coffee that my wife Nataliya regularly supplied me. Moreover, only because of her punctuality and accuracy some old photographs with oscillograms used in this chapter survived the journey between several countries and continents where we were lucky to live and work.

11. References

[1] Encyclopedia of Laser Physics and Technology. http://www.rp-photonics.com/encyclopedia.html (accessed 6 May 2012).

[2] Jabczyński J, Firak J, Peshko I. Single-frequency, thin-film tuned, 0.6 W diode-pumped Nd:YVO4 laser. Applied Optics 1997; 36(12) 2484-2490.

[3] Peshko I, Jabczyński J, Firak J. Tunable Single- and Double-Frequency Diode-Pumped Nd:YAG Laser. IEEE J QE 1997; 33(8) 1417-1423.

[4] Peshko I, Jabczynski J. Thermally induced pulsation in the solid-state single-frequency diode pumped laser. Optica Applicata 1999; XXIX (3) 319-325.

[5] Tang C, Statz H, de Mars G. Spectral output and spiking behavior of solid-state lasers. J. Appl. Phys. 1963; 34 2289-2295.

[6] Zvelto O, Principles of Lasers. Heidelberg: Springer; 2010.

[7] Encyclopedia of Laser Physics and Technology. Spatial Hole Burning. http://www.rp-photonics.com/spatial_hole_burning.html (accessed 6 May 2012).

[8] Encyclopedia of Laser Physics and Technology. Nanosecond Lasers. http://www.rp-photonics.com/nanosecond_lasers.html (accessed 6 May 2012).

[9] Rullière C. Femtosecond Laser Pulses: Principles and Experiments. New York, USA: Springer; 2005.

[10] Diels J.-C, Rudolph W. Ultrashort Laser Pulse Phenomena: Fundamentals, Techniques, and Applications on a Femtosecond Time Scale. Technology & Engineering. Elsevier Science Publishing Co Inc Academic Press Inc; 2006.

[11] Kärtner F.X. Few-Cycle Laser Pulse Generation and its Applications. Berlin: Springer; 2004.

[12] Topics in Applied Physics: Ultrashort Light pulses Picosecond Techniques and Applications Ed.S.L.Shapiro. Berlin, Heidelberg, New York Springer-Verlag 1977.

[13] Encyclopedia of Laser Physics and Technology. Mode locking. http://www.rp-photonics.com/mode_locking.html (accessed 6 May 2012).

[14] Peshko I. Self-effecting processes in the solid-state lasers. Honorary Dr. of Sciences dissertation. Institute of Physics Kiev; 2003.

[15] Tikhonov V I. Excesses of random processes. Moscow: Nauka; 1970. (in Russian).

[16] Wigner distribution function. http://en.wikipedia.org/wiki/Wigner_distribution_function (accessed 6 May 2012).

[17] Sooy W. The natural selection of modes in a passive Q-switched laser. Appl.Phys.Lett. 1965; 2(6) 36-58.

[18] Peshko I, Khiznyak A, Soskin M. Mode-locked laser with controllable parameters. New York: Plenum-press; 1985. (Plenum Press bought the author's rights for initially published brochure: Peshko I, Khiznyak A, Soskin M. Laser of ultrashort pulses with tunable parameters. Kiev: Publishing of Institute of Physics, Academy of Sciences; 1984, #4).

[19] Peshko I, Khizhnyak A. Generation of the pulses of extreme short duration by solid-state lasers. Quantum Electronics 1987; 33, Kiev, Naukova Dumka p.14-20. (in Russian).

[20] Peshko I, Soskin M, Khiznyak A. Generation of the picosecond pulse train with controlable parameters. Quantum Electronics 1982; 9(12) 2391-2398. (in Russian).

[21] Graf F, Low C. Passively mode locked Nd-glass laser with partially suppressed natural mode selection. Opt. Communs. 1983; 47(5) 329-334.

[22] Becker M, Kuizenga D, Siegman A. Harmonic mode locking of the Nd:YAG laser. IEEE J. Quantum Electronics 1972; 8(8) 687-693.

[23] Encyclopedia of Laser Physics and Technology. Harmonic Mode Locking. http://www.rp-photonics.com/harmonic_mode_locking.html (accessed 6 May 2012).

[24] Zhan L. et al. Critical behavior of a passively mode-locked laser: rational harmonic mode locking. Opt. Lett. 2007; 32(16) 2276-xxx

[25] Attosecond pulse generation and detection www.mpq.mpg.de/lpg/research/attoseconds/attosecond.html (accessed 6 May 2012).

[26] Borot A, Malvache A, Chen X, Jullien A, Geindre J.-P, Audebert P, Mourou G, Quéré F, Lopez-Martens R. Attosecond control of collective electron motion in plasmas. Nature Physics 2012; 8 416-421.

[27] Zhu J, Xie X, Sun M, Bi Q, Kang J. A Novel Femtosecond laser System for Attosecond Pulse Generation. Advances in Optical Technologies 2012; Hindawi Publishing Corporation, article ID 908976, doi: 10.1155/2012/908976, 6 pages.

[28] Phys.Org.News. http://www.physorg.com/news192909576.html (accessed 6 May 2012).

[29] Lan P, Lu P, Cao W, Wang X. Attosecond and zeptosecond x-ray pulses via nonlinear Thomson backscattering. Phys. Rev. E 2005; 72, 066501[7 pages].

[30] Femtosecond, Attosecond, and Yoctosecond http://nextbigfuture.com/2009/10/attoseconds-zeptoseconds-and.html (accessed 10 May 2012).

Pulsed World

All Solid-State Passively Mode-Locked Ultrafast Lasers Based on Nd, Yb, and Cr Doped Media

Zhiyi Wei, Binbin Zhou, Yongdong Zhang, Yuwan Zou, Xin Zhong, Changwen Xu and Zhiguo Zhang

Additional information is available at the end of the chapter

1. Introduction

Mode-locking technique is a widely used method for generating ultrashort laser pulses. The mode-locked laser output is a sequence of equally spaced laser pulses. The pulse width is limited by the spectral range of the gain medium and inversely related to the bandwidth of the laser emission. Compared with the active mode-locking technique, passively mode-locked laser with saturable absorber is able to generate much shorter pulse with a simple configuration. In particularly, based on the passive mode-locking mechanism, the Kerr-lens mode locking (KLM) Ti:sapphire laser is recognized as the most important ultrafast laser source. Not only a series of commercial femtosecond lasers with Ti:sapphire crystal were released, but also lead to many innovations in science, such as frequency comb, laser wake field acceleration, attosecond science, laser micro-fabrication *etc*. However, the major drawback of Ti:sapphire laser is its green pump laser source. Currently available green lasers generated by frequency doubled Nd:YAG laser or by Argon laser are relatively bulky and expensive, which limits the practical application of ultrafast Ti:sapphire lasers.

With the remarkable progresses in the semiconductor saturable absorber mirror (SESAM) and laser diode technology, ultrafast laser sources with directly diode-pumped schemes and SESAM mode-locking have attracted more and more attentions because of the compactness and low cost compared to the well developed femtosecond Ti:sapphire laser. Nowadays, intra-cavity SESAM is widely used to start and maintain stable mode locking. Intense studies have been performed on this kind of lasers. Many efforts have been made to find new gain materials for ultrafast laser generations. Our research activity in this field is focused on the development of all solid-state passively mode-locked ultrashort lasers based on a variety of gain media with various wavelength ranges in the near-infrared. Several gain media doped with Nd, Cr or Yb, such as Nd:YVO4, Nd:GdVO4, Nd:LuVO4, Nd:GSAG,

Cr:forsterite, Cr:YAG, Yb:YAG, Yb:YGG, Yb:GYSO, Yb:LSO etc, have been tested. The results indicate a series of ultrafast laser sources with low cost, compact, simple and robust configuration in the picosecond and femtosecond regimes, which would find a wide range of practical applications.

2. Mode-locking mechanism passively mode-locked solid-state laser with a SESAM

By using an intra-cavity SESAM for passive mode locking, one has to choose the correct parameters of SESAM for a given laser system to get pure CW mode locking [2.1]:

$$\left|\frac{dR}{dI}\right|I < \frac{g_0}{l}\frac{T_R}{\tau_L}. \tag{1}$$

where R is the absorber reflectivity, I is the laser intensity, T_R is the cavity round trip time, g_0 is small signal gain of the laser, l is the total loss coefficient of the laser cavity, τ_L is the upper state lifetime of the laser. That is, the laser tends to operate in Q-switched mode-locking regime with a longer upper state lifetime, larger total loss, or shorter cavity length. In other words, the Q-switching can be suppressed for a large small-signal gain, or small loss, or large saturation intensity, or long cavity. For a fast saturable absorber, the condition is given by [2.1]:

$$\left|\frac{dR}{dE_p}\right|E_p < \frac{g_0}{l}\frac{T_R}{\tau_L}. \tag{2}$$

3. Overview for different solid-state laser materials

Solid-state mode-locked lasers are applied in various fields of physics, engineering, chemistry, biology and medicine *etc*, with application including ultrafast spectroscopy, metrology, superfine material processing and microscopy. To generate ultrafast laser pulses, the laser materials must meet a series of conditions. Firstly, a pump wavelength for which a good pump source is available. Secondly, small quantum defect and high gain of the materials can help get high laser efficiency. Furthermore, it is crucial for laser media to possess a board emission band, which is necessary to generate ultrafast pulses.

We divide materials for generation of ultrafast laser pulses into two types.

The first type is the ones that could only support picosecond laser pulses due to their limited gain bandwidth. Whereas, these laser materials commonly have excellent laser capabilities, such as good thermodynamic property and could be directly pumped by a diode laser with high power. The representative laser media are Nd^{3+} doped materials, which have been applied in diode-pumped energetic and efficient picosecond lasers and amplifiers. A SESAM mode-locked, Nd^{3+}:YAG laser with pulse width of 20 ps and output power of 27 W has been reported [3.1]. Also, an Nd^{3+}:YVO_4 laser with 20 ps of pulse width and 20W of laser power was achieved [3.2]. Malcolm *et al* reported an additive pulse mode-locking (APM) Nd^{3+}:YLF laser, and 1.5 ps pulses were generated [3.3], which is the shortest pulse result by Nd^{3+} doped crystals.

Also, the Nd^{3+} laser on the $4F_{2/3}$– $4I_{9/2}$ transition with emission wavelengths of around 900–950 nm has attracted more and more interest because the quasi three-level pulsed laser could be used for lidar detection of water vapor in atmosphere [3.4]. Up to now, only few works have been reported on passively mode-locked laser on quasi three-level operating of Nd^{3+} doped materials. By Nd:$YAlO_3$ crystal, laser pulses centered at 930 nm with 1.9 ps pulse duration and 410 mW average power has been demonstrated with a Ti:sapphire pump laser[3.5]. With Nd:YVO_4 crystal, 8.8 ps pulses with 87 mW of power [3.6] and 3 ps pulses with 140 mW of power [3.7] centered at 914 nm were achieved, pumped by diode laser and Ti:sapphire laser, respectively.

Other than Nd^{3+} doped crystals, a commonly used host material doped with Nd^{3+} ions is Nd:glass. Different with crystals, glasses are significantly cheaper and have a smoother fluorescence spectrum, which support femtosecond laser pulses generation. As a well known example, laser pulses as short as 60 fs with the output power of 80 mW at the center wavelength of 1053 nm have been achieved with Nd: glass laser [3.8]

The other type of gain media have relatively broad emission spectra, but suffer from one or more unpopular physical properties (Ti:sapphire crystal is the outstanding exception). Yb-doped materials belong to this group. Compared with Nd-doped laser media, Yb-doped materials have attracted even more and increasing interest over the past few decades because of their small quantum defect (resulting a low thermal loading), simple electronic structure (avoiding such unwanted processes as excited-state absorption, up-conversion, and concentration quenching), long fluorescence lifetime (particularly advantageous for Q-switched lasers and high-power ultrashort pulse amplification) and broad absorption and emission bands (compared with Nd^{3+}). These media have been successfully applied in high-power diode-pumped all-solid-state ultrafast laser sources. Until now, extensive mode-locking researches have been reported with various Yb-doped crystals, such as garnet (Yb:YAG [3.9-10]), vanadate (Yb:$GdVO_4$ [3.11]), oxyorthosilicates (Yb:LSO [3.12-13], Yb:YSO [3.13]), tungstates (Yb:KGW [3.14], Yb:KYW [3.15], Yb:KLW [3.16], Yb:NYW [3.17]), borates (Yb:YAB, Yb:LSB, Yb:LYB, Yb:BOYS, Yb:GdCOB, Yb:YCOB)[3.18-23], fluorite (Yb:YLF) [3.24], sesquioxide (Yb:Sc_2O_3, Yb:Lu_2O_3, Yb:$LuScO_3$) [3.25-27], silicate(Yb:SYS) [3.28], niobate(Yb:CN) [3.29]. The short pulse duration of 47 fs has been reported with a Yb^{3+}: $CaGdAlO_4$ crystal [3.30] and the shortest pulse width of 35 fs has been obtained by a Yb:YCOB crystal [3.20].

Transparent ceramics fabricated by the vacuum sintering technique and nanocrystalline technology have the advantages of high doping concentration, easy fabrication of large size samples and high thermal conductivity. This kind of materials has been intensively investigated for the use in ultrashort pulse lasers [3.31-36]. The first reported mode-locked ceramic laser is a Yb:Y_2O_3 laser, which generated 210 mW laser pulses with 450 fs pulse duration and a center wavelength of 1037 nm [3.32]. With the Yb:YAG ceramic, 286 fs pulses with 25 mW output power and 233 fs with 20 mW output power at center wavelength of 1030 nm and 1048.3 nm respectively were achieved [3.35],. Our recent work has boosted the output power from a femtosecond ceramic Yb:YAG laser to 1.9 W [5.19].

To further reduce the thermal effect, especially important for high power laser operation, the thin-disk gain media configuration has been developed. The geometry of the gain media is designed such that the thickness is considerably smaller than the laser beam diameter and the heat generated along with the lasing is extracted dramatically through the cooled end face [3.37-3.40]. Combined with the high absorption of the pump laser and long lifetime of the excited state of the Yb^{3+}-doped active materials, this thin disk laser operation has been a great success in a variety of Yb^{3+}-doped materials [3.41-3.47]. Thin-disk laser based on Yb-doped active materials is a good way to get efficient and high output power laser with excellent beam quality and will find wide applications.

Cr^{4+}:forsterite crystal is an important laser medium to generate ultrashort laser pulses around 1.3 μm. It could be pumped by Nd:YAG laser or Yb-doped fiber laser around 1064 nm. Cr^{4+}:forsterite crystal has wide luminescence spectrum from 1100 nm to 1500 nm [3.48]. Up to now, the shortest pulse duration with Cr^{4+}: forsterite laser is 14 fs [3.49]. Also, another Cr^{4+} doped material (Cr^{4+}:YAG) could be employed as the laser medium to obtain ultrafast laser pulses around 1.4 to 1.6 μm. The absorption spectrum of Cr^{4+}:YAG extends from 950 nm to 1100 nm, so it could share similar pump laser as the Cr^{4+}:forsterite laser. So far, the shortest pulse duration with Cr^{4+}:YAG laser is 20 fs and achieved by Kerr-lens mode locking technology [3.50].

4. Mode-locked Nd-doped lasers at quasi-three levels

4.1. Laser transition at quasi-three levels

Nd^{3+} ions was the first of the trivalent rare earth ions to be used in a laser, and Nd^{3+} doped materials have been remaining the most important laser medium until now. Laser operation has been obtained with this ion incorporated in at least 100 different host materials, and the commonly used host materials are YAG and glass. Take Nd: YAG for example, Figure 1 shows the energy level diagram of Nd^{3+} pumped by 808 nm. The ground state population was pumped to the energy level $^4F_{5/2}$, and then via radiationless transition transfers energy to $^4F_{3/2}$ level. The stimulated emission is commonly obtained at three different groups of transitions centered at 1.34, 1.06, and 0.946 μm, which result from $^4F_{3/2} \rightarrow ^4I_{13/2}$, $^4F_{3/2} \rightarrow ^4I_{11/2}$, and $^4F_{3/2} \rightarrow ^4I_{9/2}$ transitions, respectively.

When radiation at $^4F_{3/2} \rightarrow ^4I_{13/2}$ and $^4F_{3/2} \rightarrow ^4I_{11/2}$ transitions, the upper and lower energy levels of pump and laser are separate, this is called four-level operation. Because the lower laser level is well above the ground state $^4I_{9/2}$ and can be quickly depopulated by multi-phonon transitions, there is no appreciable population density ideally in the lower laser level at room temperature. In that way, re-absorption of the laser radiation is effectively avoided and a lower threshold pump power can be achieved.

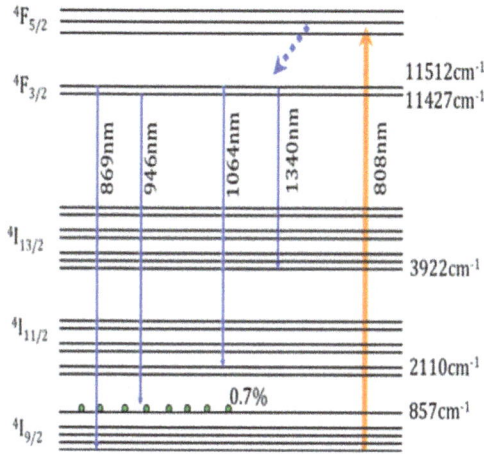

Figure 1. Energy level diagram of Nd:YAG

However, for $^4F_{3/2} \rightarrow \,^4I_{9/2}$ transition, the lower-laser level is the upper sub-level of the $^4I_{9/2}$ multiplet, so there is some population on the lower-laser level at room temperature. This is called quasi-three level operations. For example a thermal population of about 0.7 % in the lower laser level in Nd:YAG at room temperature. This sub-level is thermally coupled with the ground-state and should be efficiently populated with increasing temperature. This would induce a partial re-absorption loss of the laser radiation and cause an increase in the ground-state absorption loss, corresponding to an increasing in passive intra-cavity loss and laser threshold, result in the reducing of the laser output energy. In addition, the laser cross emission section of the quasi-three-level laser is dozens of times smaller than that of four-level system. Therefore, the laser operation with quasi-three levels is more difficult than that one with four levels.

As the typical Nd-doped laser medium, mode-locking researches on Nd:YAG laser for both quasi-three levels and four levels have been extensively reported, therefore, we only introduce our researches on mode-locking with some special Nd-doping media in this chapter.

4.2. Picosecond Nd:GSAG laser at 942nm for three levels operation

During the past few years, the Nd^{3+} laser on the 4F$_{2/3}$– 4I$_{9/2}$ transition has attracted wide interest because of emission wavelengths of around 900–950 nm. One important application for this quasi three-level pulsed laser is that it could be used for lidar detection of water vapor in atmosphere [3.4] because of characteristic absorption in the 935 nm, 942 nm, and 944 nm wavelength regions. Compared with optical parametric oscillators, generation of above wavelengths by Nd-doped crystal laser is much easier. Also, Frequency doubling of mode-locked quasi-three-level Nd^{3+} lasers generates picosecond blue pulses, which have potential application in many fields, such as life science, holography, and semiconductor inspection [4.1].

In the previous section, we have discussed the difficulty on quasi three-level operation compare with four-level operation. For Nd:GSAG here, a peak emission cross section of 2.7$\times$ 10^{-20} cm^2 at 942.7 nm [4.2] and a peak emission cross section of 3.2×10^{-19} cm^2 around 1.06 μm [4.3] were determined by Kallmeyer *et al.* and Brandle *et al.*, respectively. One can see the emission cross section at 942 nm is less than one-tenth the emission cross section at 1.06 μm. This leads to an increased threshold pump power and decreased optical conversion efficiency [4.3–5]. For the first time, we successfully a diode-pumped passively mode-locked Nd:GSAG laser with quasi-three-level operation at the central wavelength of 942.6nm [4.6]. The maximum output power is 510 mW at incident pump power of 16.7 W, and the pulses duration is short as 8.7 ps at a repetition rate of 95.6 MHz.

In the experiment, a standard z-fold resonator with end-pump configuration was employed as shown in Fig. 2. The pump source is a commercial fiber-coupled 808 nm LD with a core diameter of 200 μm, and an NA of 0.22. The crystal used in the experiment has dimensions of 4 mm$\times$4 mm and Nd^{3+} concentration of 1 at. %. The both facets of the crystal have been antireflection (AR) coated at 942 nm, 808 nm and 1061 nm. The crystal was wrapped with indium foil and mounted on a water cooled copper heat sink and the water temperature was maintained at 10°C. Passive mode locking was started by using a SESAM (BATOP GmbH, Germany) with saturable absorptance of 4% at 940 nm, a saturation fluence of 70 μJ/cm^2, and relaxation time of less than 10 ps. The laser spot size on the SESAM is estimated to be 27 μm $\times$20 μm (tangential direction sagittal direction) using the ABCD matrix calculation. The dependence of the total output power on the incident pump power is shown in Fig. 3. Stable CW mode-locked laser can be obtained at the pump power above 11.5 W. When the incident pump power reached 16.7 W, the output power rose to 510 mW, corresponding to a optical efficiency of 3.1%, and the pulse energy inside the cavity and outside the cavity are about 88.3 nJ and 2.65 nJ, respectively. The high threshold pump power and the low efficiency are mainly due to the relatively high total transmission of 6% brought by the folded output coupler.

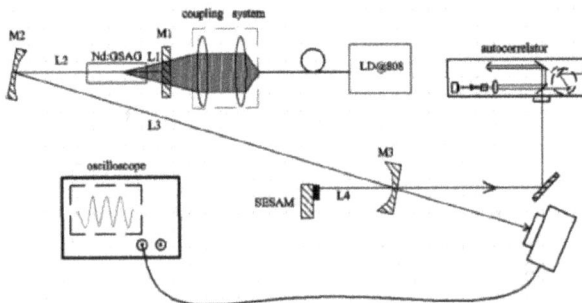

Figure 2. Schematic diagram of the experimental setup. Mirrors: M1: HT@808 nm and 1064 nm, HR@942 nm; M2: HR@942 nm, HT@1064 nm, radius of curvature (RoC)=500nm; M3: T=3%@942 nm, HT@1064 nm, RoC=100 nm.

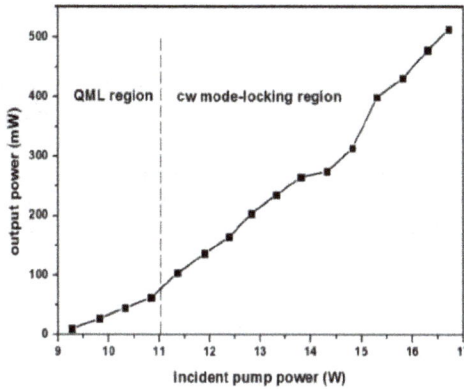

Figure 3. Dependence of the mode-locked laser output power on incident pump power.

The beam quality of the laser was measured with a CCD that could translate along a straight and slick track under incident pump power of 16.7 W. After fitting the measured data, as shown in Fig. 4 (a), we found that M^2 parameters were 1.83 and 1.55 for tangential direction and sagittal direction, respectively. The stable CW mode-locking regime held well for several hours and the repetition rate of the pulses was 95.6 MHz. The pulse duration was measured with an autocorrelator (FR-103MN, Femtochrome Research, Inc.) at the maximum output power, as shown in Fig. 4 (b). The pulse duration is about 8.7 ps assuming a Gaussian shape. Measured by an optical spectrum analyzer (AQ6315A, YOKOGAWA), the spectrum was centered at 942.6 nm which has an FWHM of about 0.65 nm, also shown in Fig. 4 (b). The time–bandwidth product of the pulses is 1.91, which is 4.3 times the transform limit for Gaussian pulses. The primary reason for this is that positive group-velocity dispersion introduced by crystal itself stretched the pulses.

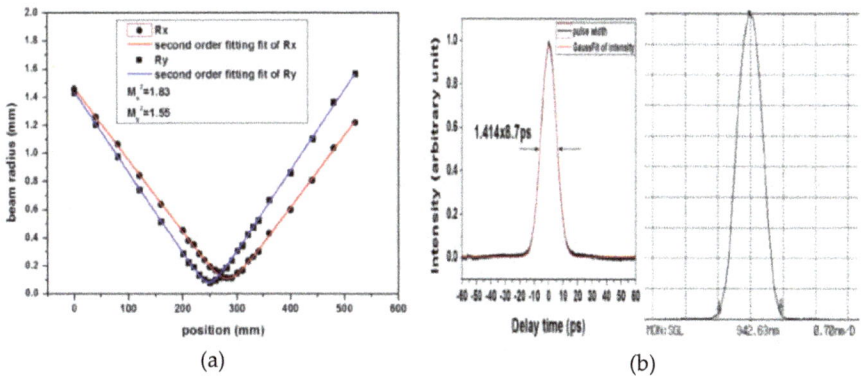

(a)

(b)

Figure 4. (a) Measured M^2 for tangential direction (Rx) and sagittal direction (Ry); (b) left: Intensity autocorrelation trace of the pulse; right: The laser spectrum of mode-locking operation.

4.3. Picosecond Nd:GGG lasers for both four levels and quasi-three levels

The Nd^{3+} doped gadolinium-gallium-garnet (GGG) crystal, a well known laser gain medium, was firstly reported in 1964 [4.7]. It has many desirable properties as a laser host material, such as excellent thermal conductivity, good mechanic properties, large thermal capacity, high doped concentration, and large size etc, as shown in Tab.1 [4.8-9]. With this laser medium, 30 kW average output power had been demonstrated from a solid-state heat capacity lasers (SSHCL) by Lawrence Livermore National Laboratory in 2004 [4.10]. Although Qin *et al* had obtained mode-locking operation at 1061 nm; the output power is only hundreds of milli-watt [4.11]. In this section, we described a diode pumped mode-locked Nd:GGG laser operation at both four levels and quasi-three levels transition. A laser pulse of 15 ps and 24 ps were generated with an average power of 3.2 W and 320 mW at 1062 nm and 937.5 nm, respectively [4.12-13].

Chemical formula	$Nd:Gd_3Ga_5O_{12}$
Crystal Structure	Cubic
Lattice Constant	a=1.2376 nm
Melting point	1725 $^\circ$C
Density	7.1 g/cm^3
dn/dT ($\times 10^{-6}$/K)	17
Thermal conductivity	7 W/mK
Heat capacity	0.38×10^3 Ws/(Kg$\cdot$K)
Thermal expansion coefficient	8×10^{-6}/$^\circ$C
Laser Transition	$^4F_{3/2} \rightarrow {}^4I_{11/2}$ 1062 nm $^4F_{3/2} \rightarrow {}^4I_{13/2}$ 1331 nm $^4F_{3/2} \rightarrow {}^4I_{9/2}$ 937.5 nm
Emission cross section at 1061 nm	$2.7\text{-}8.8 \times 10^{-19}$cm^{-2}
Fluorescence lifetime	240 μs
Absorption band	808 nm
Photon energy at 1.06 μm	hν = 1.86 $\times 10^{-19}$ J
Index of refraction	1.94

Table 1. Properties of Nd:GGG crystal

Laser operation was performed by using a 5-mm-long, antireflection-coated, 0.5 at. % doped Nd:GGG crystal. A high brightness fiber-coupled diode laser, with a core diameter of 200 μm and a numerical aperture of 0.22, was used as pump source. For ease of use, we designed a typical resonator as shown in Fig. 5 for four level running. Passive mode locking was started by a SESAM, which has a saturable absorption of 4% at 1060 nm and a saturation fluence of 70 μJ/cm^2. A plane-wedged mirror with transmission rate of 10 % was used as the output coupler (OC). With this configuration, the laser mode radii in the crystal and on the SESAM were calculated to be 98μm and 105 μm, respectively.

After the optimized alignment, stable mode-locking operation with single mode output was obtained when the incident pump power exceeded 15 W. We measured the intensity autocorrelation trace by using a commercial noncollinear autocorrelator (FR-103MN, Femtochrome Research, Inc.). As shown in Fig. 6 (a), the FWHM width of the autocorrelation trace is about 23 ps. If a sech^2-pulse shape is assumed, the mode-locked pulse duration is 15 ps. The central wavelength locates at 1062.5 nm with spectrum bandwidths of 0.25 nm. The time-bandwidth product was calculated to be 0.997, which was about 3 times the transform limit for Gaussian pulses.

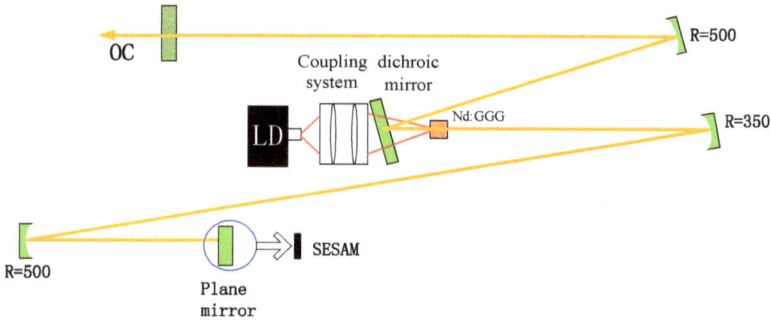

Figure 5. The Experimental layout of passively mode-locked for Nd:GGG laser operating at four level transition.

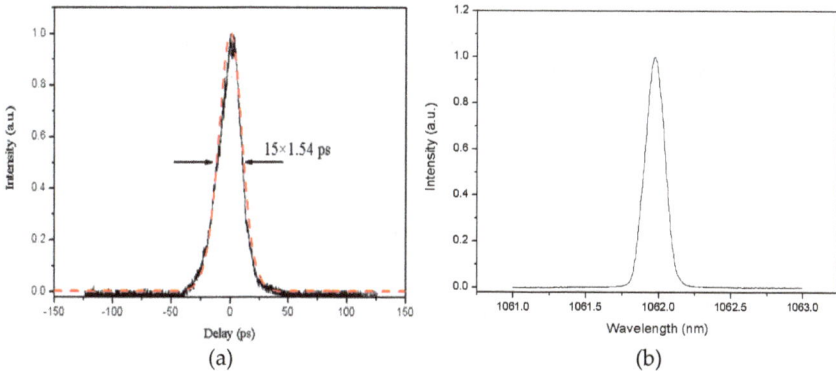

Figure 6. (a) Intensity autocorrelation trace of the pulse; (b) the laser spectrum of mode-locking operation

Mode-locking Nd:GGG laser at quasi-three level transition was realized with the same pumping source and the laser crystal. In order to reduce loss, a simple Z-folded cavity consisted of four mirrors was employed, as shown in Fig.7, the total cavity length was approximately 1.82 m corresponding to a repetition rate of about 82 MHz. To suppress parasitic oscillation at 1.06 μm, most of the cavity mirrors were antireflection coated at this wavelength. A similar SESAM from same company was used to start the mode-locking. For optimization of the cavity alignment, a plane mirror coated for high reflection at 940 nm was used as an end mirror. The waist radii of the laser mode in the crystal and on the SESAM were design to be 80 μm and 20 μm for mode match, respectively. The stable pulses train with the maximum output power of 320 mW has been obtained under the incident pump of 20 W.

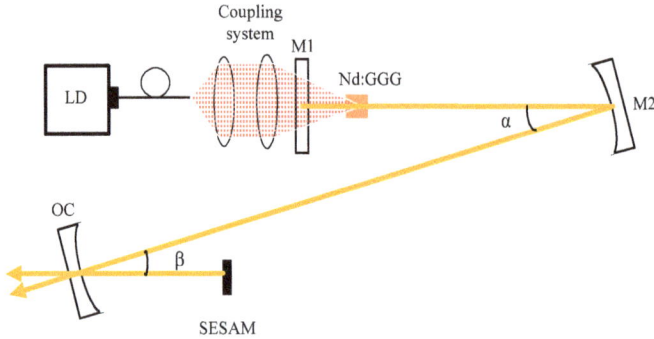

Figure 7. Scheme of Nd:GGG laser mode-locked at quasi-three-level transition.

Intensity autocorrelation trace shown that the FWHM width of the autocorrelation trace is about 39.3 ps (Fig. 8 (a)). If a Gaussian-pulse shape is assumed, the mode-locked pulse duration is 27.8 ps. Fig. 8 (b) depicts the output spectrum at the stable mode locking which was centered at 937.5 nm and had a bandwidth of 0.14 nm. This resulted in a time–bandwidth product of 1.33 which was 3 times the transform limit for Gaussian pulses. The high time-bandwidth product may be resulted from the high group delay dispersion of the laser crystal [4.14].

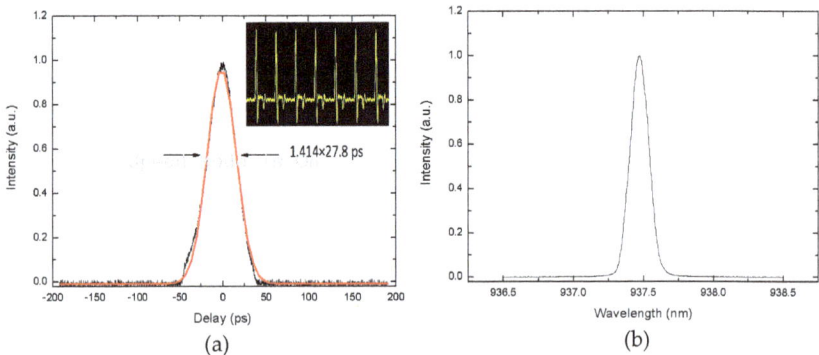

Figure 8. (a) Intensity autocorrelation together with Gaussian fit curve shows the pulse width of 24.7 ps. The inset is oscilloscope traces of the CW mode-locked pulse train. The laser spectrum of mode-locking operation.

4.4. Picosecond Nd:GdVO$_4$ at 912nm for three levels operation

As the most efficient laser host medium, the Nd:YVO$_4$ crystal has been widely used for diode-pumped mode-locking laser research [4.15-20]. Compare to Nd:YVO$_4$ crystal, Nd:GdVO$_4$ possesses many significant advantages such as higher thermal conductivity, larger absorption cross-section and larger stimulated emission cross-section [4.21]. In addition, Nd:GdVO$_4$ emits the shortest wavelength radiation on the $^4F_{3/2}$ - $^4I_{9/2}$ transition for its smallest splitting (409 cm^{-1}) of the lower laser level. However, the quasi-three levels laser is more difficult to operate than the four level laser and would cause the increase in ground-state absorption loss and laser threshold, and then reducing the output energy of the laser.

In the past, much more attention had been paid on passively Q-switched and mode-locked lasers with Nd:GdVO$_4$ crystal operating at 1064 nm and 1342 nm [4.22-26]. However, passively mode-locked Nd:GdVO$_4$ laser at quasi-three-level were less reported. H. W. Yang et al have reported a passively Q-switched Nd:GdVO$_4$ laser at 912 nm with V^{3+}:YAG as the saturable absorber [4.27]. Fei Chen *et al* have obtained passively Q-switched mode-locked Nd:GdVO$_4$ laser at 912 nm [4.28]. In 2006, our group successfully demonstrated the CW passively mode-locked Nd:GdVO$_4$ laser for three level laser operation [4.29]. In this section, we present a stable CW mode-locked Nd:GdVO$_4$ laser at 912 nm with the total output of 128 mw and the pulse width of 6.5 ps [4.30].

We employed a four-mirror folded cavity, shown in Fig. 9. The pump source is a commercially available fiber-coupled diode-laser which could emit a rated maximum power of 30W at 808 nm, with a core diameter of 200 μm and a N.A. of 0.22. The dimensions of the crystal are 3x3x4 mm^3 with a concentration of 0.2 at%. The crystal was coated and wrapped with indium foil and then mounted in a water-cooled copper block. The water temperature was maintained at 10 ℃. The SESAM (Batop GmbH, Germany) used for mode-locking is with a modulation depth of 2% at 912 nm and a saturation fluence of 70 μJ/cm^2. A plane mirror coated HR at 912 nm replaced the SESAM, and then the laser was operating at 912 nm. The maximum output power of 1.45 W was obtained with an incident pump power of 20.3 W. Then we put the SESAM instead of the plane mirror in the cavity. At pump power above 13.4 W, a stable CW mode-locking state was observed. The repetition rate was 178 MHz and the maximum output power was 128 mw at a pump power of 19.7 W, as shown in Fig.10.

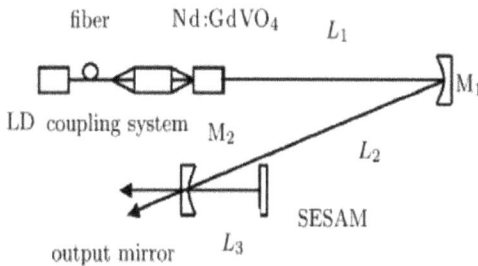

Figure 9. Schematic diagram of the passively mode-locked 912 nm Nd:GdVO$_4$ laser.

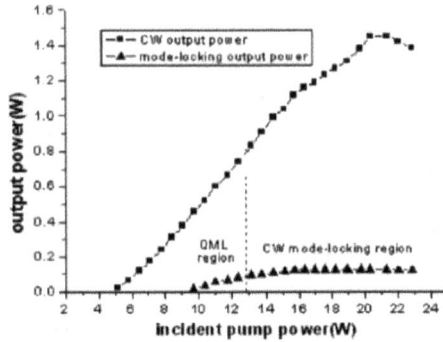

Figure 10. Dependence of the laser CW output and mode-locking output on the incident pump power. QML: Q-switched mode-locking.

The pulse width was measured with a homemade non-collinear second-harmonic-generation autocorrelator. The trace of the SHG autocorrelation is shown in Fig. 11. Assuming a Gaussian pulse profile, we estimated the pulse duration to be approximately 6.5 ps. The corresponding spectrum was measured, it has two peaks at the central wavelengths of 912.3 nm and 912.7 nm, respectively, and the both corresponding bandwidths were 0.25 nm (FWHM).

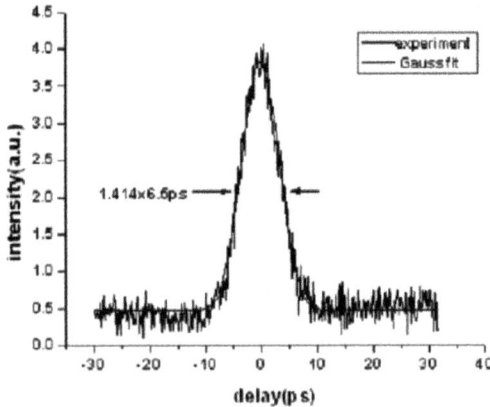

Figure 11. The measured autocorrelation trace of the pulses of the passively mode-locked 912 nm Nd:GdVO₄ laser.

4.5. Nd:LuVO₄ laser at 916 nm for three levels operation

As in vanadate family, Nd:LuVO₄ has larger absorption and emission cross sections than those of Nd:YVO₄ and Nd:GdVO₄, and thus has attracted much attentions. By employing the Nd:LuVO₄ crystal (length of 5.5 mm and doping with 0.1 at.% Nd³⁺ concentration) and using the similar Z-folded cavity design as in passively mode-locked Nd:GdVO₄ laser, we obtained a stable passively cw mode-locked Nd:LuVO₄ laser at 916nm [4.31]. The pulses train has 6.7 ps of pulse duration (shown in Fig. 12) at repetition rate of 133 MHz. The average output power was 88mW under the pump power of 17.1 W.

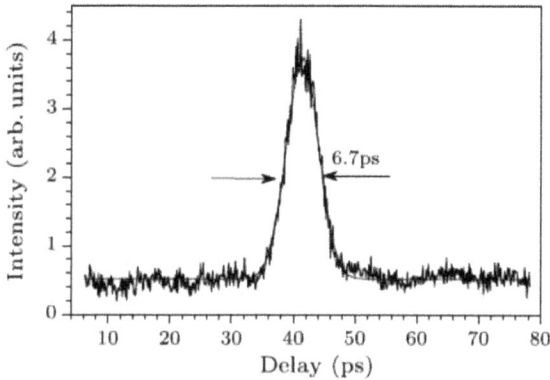

Figure 12. Autocorrelation trace of the mode-locked Nd:LuVO₄ laser.

5. Mode-locked femtosecond and picosecond Yb-doped laser

The Yb³⁺-doped materials have excellent advantages among the available laser hosts, such as high quantum efficiency, absence of excited-state absorption, direct diode laser pumping, as well as large emission bandwidth which can support femtosecond pulses generation, and so on. Up to now, extensive mode-locking research has been reported with various Yb-doped materials. In the following section, we will present some numerical and experimental studies on several passively mode-locked Yb-doped lasers.

5.1. Numerical and experimental investigation of the Yb:YAG laser at a wavelength of 1.05 μm

Among a variety of Yb-doped crystals, the Yb:YAG crystal is one of the most important laser media for several important advantages: excellent thermal-mechanical properties, ease of growth in high-quality crystal, and a high doping concentration without quenching, etc. Remarkable progress has been achieved with ultrashort Yb:YAG lasers because of these favorable properties [5.5, 5.6]. Typically, the Yb:YAG crystal has two main emission wavelengths, 1.03 and 1.05 μm respectively. Ultrafast Yb:YAG laser operating at 1.05 μm has special advantages compared with the one at 1.03 μm. First, the gain around 1.05 μm is flatter and can support femtosecond pulses with broader spectrum and shorter pulse width. Yb:YAG laser pulses as short as 100 fs have been demonstrated at this wavelength. In contrast, limited by the width of the narrow gain peak at 1030 nm, the shortest pulse achieved at this wavelength has the width of 340 fs [5.7], and most femtosecond pulses are longer than 500 fs [5.5, 5.6]. Second, ultrashort pulses at 1.05 μm can be useful in high energy glass laser facilities as the seeding source.

To obtain oscillation at 1.05 μm by the Yb:YAG laser, one must suppress the oscillation at 1030 nm. Some researchers use specially coated mirrors to distinguish these two neighboring wavelengths [5.7,5.8], which, however, inevitably brings additional cavity losses and leads to low laser efficiency. Investigations on the preferred emission wavelength versus the length and the ion concentration of the Yb:YAG crystal will be discussed below, we also described a novel method to obtain efficient 1.05 μm operation based on the Yb:YAG laser[5.9,5.10].

5.1.1. Theoretical investigation

The electronic diagram of the ytterbium ion is shown in Figure 13. It's a typical quasi-three-level system. As the zero phonon line is very narrow and the corresponding absorption cross section is lower, the most efficient pump transition for Yb:YAG is l_1 to u_2 at 940 nm. Emission transitions are from u_1 to l_2 (1024 nm), to l_3 (1030 nm) and to l_4 (1050 nm). For quasi-three-level longitudinally pumped laser system, the thickness of the gain medium is more crucial for the laser oscillation than in a four-level system. In a quasi-three-level system, the terminal level of the laser transition is thermally populated. Thus minimum pump intensity is required for reaching population inversion. As the pump is absorbed when traveling in the gain medium, this minimum intensity is reached after a crystal length which is the so-called optimum length. In the following, the different optimum length at different oscillation wavelength in the Yb:YAG laser will be investigated by the model developed in [5.11] and [5.12].

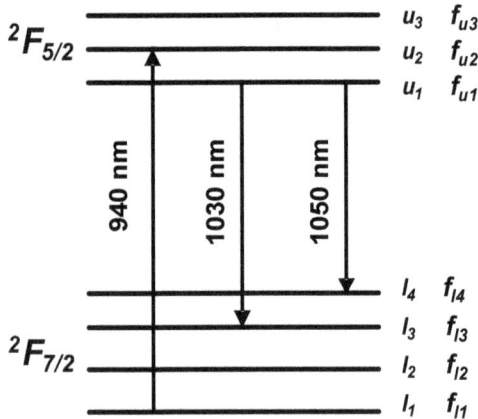

Figure 13. Energy level diagram of Yb³⁺ ion in the Yb:YAG crystal.

The amplification of the laser wave and the absorption of the pump wave are described by equations (5.1a) and (5.1b):

$$\frac{dI_l^\varepsilon}{I_l^\varepsilon} = \varepsilon g_0 \{ X_u - f_l \} dz \tag{3a}$$

$$\frac{dI_p^{\varepsilon'}}{I_p^{\varepsilon'}} = -\varepsilon' \alpha_0 \{ f_p - X_u \} dz \tag{3b}$$

where X_u is the population density of the excited state. $X_u = N_u / N_{Yb}$. N_u is the population of the excited state and N_{Yb} is the Yb^{3+} ions concentration. The linear coefficients of gain g_0 and absorption α_0 are given by equations (4):

$$g_0 = \sigma_l N_{Yb} \left(f_{lk} + f_{u1} \right) \qquad \alpha_0 = \sigma_p N_{Yb} \left(f_{l1} + f_{uj} \right) \tag{4}$$

with $f_l = \dfrac{f_{lk}}{f_{lk} + f_{u1}} \qquad f_p = \dfrac{f_{l1}}{f_{l1} + f_{uj}}$,

where σ_p and σ_l are the absorption and emission cross sections, and f_{jk} are the Boltzmann partition factors of the sublevel k of the manifold j, ε and ε' are ±1 relative to the direction of propagation of the laser and the pump beams respectively. The dynamic equation for the population reads:

$$\tau_u \frac{dX_u}{dt} = I_p \left(f_p - X_u \right) - X_u - I_l \left(X_u - f_l \right) \tag{5}$$

where τ_u is the excited state life time and I_p and I_l are the pump and laser intensity travelling in both directions normalized to the saturation intensity given by (6):

$$I_{sat}^l = \frac{h\nu_l}{\left(f_{lk} + f_{u1} \right) \tau_u \sigma_l} \qquad I_{sat}^p = \frac{h\nu_p}{\left(f_{l1} + f_{uj} \right) \tau_u \sigma_p} \tag{6}$$

In CW regime, equation (5) reads: $I_p \left(f_p - X_u \right) - X_u - I_l \left(X_u - f_l \right) = 0$. Then:

$$X_u = \frac{f_p I_p + f_l I_l}{1 + I_p + I_l} \qquad I_i = I_i^+ + I_i^- \tag{7}$$

With regard to equation (3a), the laser beam is reabsorbed when $X_u < f_l$. Then minimum pump intensity required for bleaching the amplifier medium is given by:

$$X_u = f_l \qquad I_p^{min} = \frac{f_l}{f_p - f_l} \tag{8}$$

For single pass pumping where the pump is travelling following one direction, the pump transmission required for inverting the amplifier medium is given by:

$$I_p(0)\Gamma = I_p^{min} \qquad \Gamma = \beta = \frac{I_p^{min}}{I_p(0)} \tag{9}$$

As the pump is absorbed when traveling in the amplifier, this minimum intensity required for bleaching the amplifier medium at the laser wavelength is reached after an amplifier length we call optimum length.

For single pass pumping and CW laser, the output intensity has been derived in [5.12]:

$$I_{out}(L) = (1 - R_s)\sqrt{R_m} \frac{g_0\left(\frac{I_p(0)(1-\Gamma)}{\alpha_0} - f_l L\right) + \ln\sqrt{R_m R_s}}{\left(1 - \sqrt{R_m R_s}\right)\left(\sqrt{R_m} + \sqrt{R_s}\right)} \tag{10}$$

Where R_s and R_m are the reflectivity of the output and rear mirror respectively (assuming all the losses of the cavity are taken into account in the loss of the rear mirror). By (3a) and (3b), it is possible to find a relation connecting the pump and the laser intensities. Then, Γ reads:

$$\Gamma = \sqrt{R_m R_s}^{-\frac{\alpha_0}{g_0}} \exp\left\{-\alpha_0\left(f_p - f_l\right)L\right\} \tag{11}$$

The optimum length for which the laser wave is not reabsorbed leads to maximum output power. Then, the pump transmission must be equal to β leading to:

$$L_{opt} = \frac{-1}{f_p - f_l}\ln\left\{\sqrt{R_m R_s}^{\frac{1}{g_0}}\beta^{\frac{1}{\alpha_0}}\right\} \tag{12}$$

This result can also be obtained by computing the length for which $I_{out}(L)$ is maximum using formula (10). In CW laser operation of a three-level system, inversion density of gain media remains constant above laser threshold. Then, for a given laser medium, pump transmission is fixed and determined by cavity loss level and spectroscopic parameters of gain media. It is worth to mention that L_{opt} as well as β depends on the pump intensity.

	Absorption	Emission	
λ (nm)	940	1030	1050
$f_u + f_l$	1.04E+00	0.75	7.00E-01
σ (cm²)	7.60E-21	3.30E-20	4.80E-21
τ_u (ms)	0.95	0.95	0.95
f_l	0.838	0.0626	0.0205

Table 2. Values of the crystal parameters

Now the length and the ion concentration of the crystal can be investigated to find out that under what kind of condition the laser oscillation at 1050 nm can be preferred. The emission transitions of Yb: YAG are from upper level u_1 to lower level l_3 (1030 nm) and l_4 (1050 nm). First, as the l_4 energy level is higher than the l_3 one, its thermal population is lower and the f_l parameter at room temperature is 0.0284 against 0.0646. Then, the corresponding optimum crystal length is larger and more pump energy is absorbed. Using (10), the optimum crystal length versus the rear mirror reflectivity for various pump intensities can be calculated. The values of the crystal parameters used for the computation are reported on Table 2. The cross sections are spectroscopic cross sections [5.13].

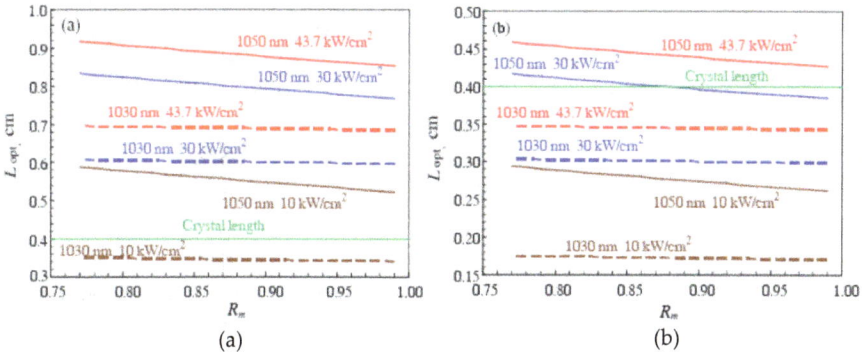

Figure 14. Optimum length versus reflectivity of the rear mirror for various pump intensities with (a) 5% and (b) 10% doping.

Figure 14 shows the calculated optimum lengths for the two wavelengths of 1030 and 1050 nm and for 5 at. % (a) and 10 at. % (b) doping concentration. It can be noticed that, for a crystal length of 4 mm with 5 at. % doping, the crystal length is much lower than the optimum lengths for 1050 nm oscillation at different pump intensities, but close to the 1030 nm optimum length. As a result, oscillation at 1030 nm is favored. It shows good agreement with the results reported in [5.7]. In that paper, a 5%-doped, 3.5-mm-long Yb:YAG crystal was used, and under the pump intensity of about 36 kW/cm², laser oscillation was achieved at the wavelength of 1030 nm. But for the 10 at. % doped crystal, one can notice that the situation is different. The preferred crystal length is much shorter than that of 5 at. % doping crystal. Only when the crystal is very short, oscillation at 1030 nm will be preferred. If the crystal length reaches a proper value, such as 4 mm, the 1030 nm laser is more likely to be re-absorbed and suppressed. This indicates a new way to suppress 1030 nm and get the oscillation at 1050 nm only by choosing the Yb:YAG crystal with proper ion concentration and optimized length.

5.1.2. Mode-locked Yb:YAG laser at the wavelength of 1.05 μm

The experiment is performed with a 10 at. % doped Yb:YAG crystal with the length of 4 mm. A typical Z-shape cavity was used (Figure 15(a)). All the coatings of the intra-cavity mirrors are identical for 1030 and 1050 nm. A CW Ti:sapphire laser at 940 nm with 2 W power was used as the pump. The pump intensity at the front face of the crystal is calculated to be about 43.7 kW/cm² at 2 W pump power. The measured absorption of the crystal to the pump light is 91 %, which is in good agreement with the theoretically calculated value of 92.6% from equation (9) by the model.

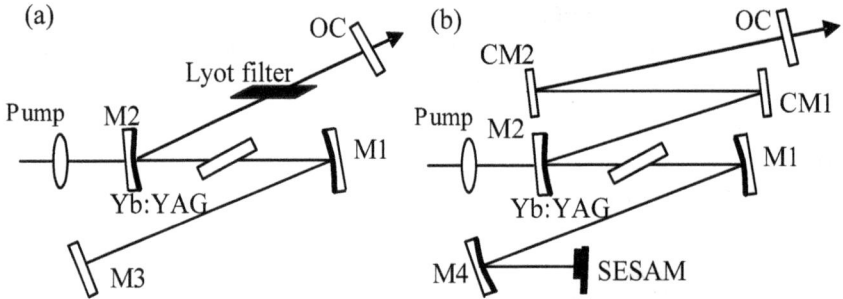

Figure 15. Cavities used to study the (a) CW and (b) mode-locked laser performance of a Yb:YAG laser.

For CW operation, two kinds of output couplers were used, with the transmittivity of 1% and 2.5% respectively. The wavelength of the free-running laser was measured with a scanning spectrometer. When the pump power was increased from the threshold power to 2 W, the emitting wavelength was keeping at 1050 nm. This can be well explained by Figure 14(b). When the pump intensity was increased from zero to 43.7 kW/cm², the length of the crystal is near the optimum length of 1050 nm oscillation and much longer than that of 1030 nm oscillation. 1030 nm lasing is more likely to be reabsorbed and oscillation at the wavelength of 1050 nm can be preferred. Under the pump power of 2 W, the maximum output power was as high as 650 mW, leading to a slope efficiency as high as 45.8%.

Based on the CW 1050 nm operation, the passively mode-locking operation of this laser was also investigated. The cavity layout is shown in Figure 15(b). An output coupler with 0.5% transmittivity was used in this setup. Without intracavity dispersion compensation, stable mode-locked pulses in picosecond regime were obtained. A typical intensity autocorrelation trace (obtained by an FR-103MN autocorrelator, Femtochrome Research, Inc.) of the output pulses is shown in Figure 16(a). Assuming a sech² pulse shape, one can obtain the FWHM pulse duration of 1.8 ps. A simultaneous measurement of the pulse spectrum is illustrated by the insertion.

A pair of Gires-Tournois interferometer (GTI) mirrors, which introduce group delay dispersion (GDD) of -1200 fs² from 1020 to 1080 nm by single pass, were used to obtain femtosecond laser operation. It's enough to compensate the positive GDD caused by the Yb:YAG crystal and the net intracavity GDD remain at a minus value. With this alignment, stable self-starting soliton-like pulses were obtained. Figure 16(b) shows the measured autocorrelation trace and the spectrum of the pulses. The pulse width is 170 fs assuming a sech²-shape pulse and the FWHM spectral bandwidth reaches 7 nm. The central wavelength redshifted from 1050 to 1053 nm. It's worth to mention that the central wavelength of this femtosecond Yb:YAG laser is exactly the working wavelength of high energy Nd:glass based ultrafast amplifier system. It indicates that the femtosecond Yb:YAG laser has the potential to be an excellent seed source for the above system. Under the full pump power of 2 W, the average power of the femtosecond pulses is 180 mW at a repetition rate of 80 MHz, corresponding to the peak power of 13.3 kW.

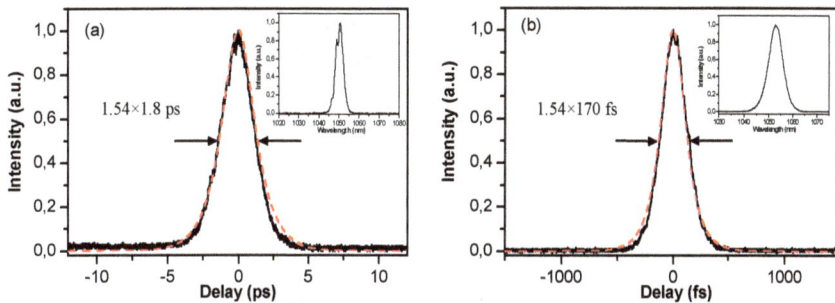

Figure 16. (a) Intensity autocorrelation trace of the picosecond pulses; inset, the corresponding laser spectrum. (b) Intensity autocorrelation trace of the femtosecond pulses; inset, the corresponding laser spectrum.

5.2. Diode-pumped efficient femtosecond ceramic Yb:YAG laser and picosecond Yb:LSO laser

Highly transparent Yb:YAG ceramics for solid-state laser gain medium have been developed in recent years, by the vacuum sintering technique and nanocrystalline technology [5.14]. This Yb:YAG ceramics have such advantages as favorable mechanical properties [5.15], high doping concentration, low cost, and easy fabrication of large-size samples. All of these advantages promise Yb:YAG ceramic extensive potential for the development of high-efficiency and high-power ultrafast laser.

In the first Yb:YAG ceramic laser experiment [5.14], the cw output power was 345 mW with a slope efficiency of 26%. With the improvement on the Yb^{3+} doping concentration, excellent cw laser performances by Yb:YAG ceramics have been demonstrated. Dong et al. realized a 1.73 W cw output power with the absorbed pump power of 2.87 W, corresponding to a slope efficiency as high as 79% [5.16]. A cw output power as high as 6.8 W with the slope efficiency of 72% was further demonstrated by Nakamura et al [5.17]. In the pulsed operation, Yoshioka et al. demonstrated the first mode-locked operation of a Yb:YAG ceramic laser [5.18]. The reported efficiency for the femtosecond operation was, however, relatively low; the output power was 250 mW at the pump power of more than 26 W.

In this section, a high-power and high-efficiency operation of a diode-end-pumped femtosecond Yb:YAG ceramic laser will be described [5.19]. For self-starting femtosecond laser operation, more complicated cavity configuration is needed than for CW laser operation. Lots of additional cavity losses are thereby introduced, such as the insertion losses by the prisms, the non-saturable loss by the SESAM, and the broadband reflection losses by intra-cavity reflectors, etc. To obtain efficient mode-locking operation, several major improvements were made compared with Ref. 5.17. First, other than prisms, highly reflective negative-dispersion mirrors were adopted for dispersion compensation. This avoided the insertion losses by the prisms, which usually caused low efficiency in mode-locked Yb-doped solid-state lasers. Experimental results on this kind of lasers have shown the significant efficiency increase by negative-dispersion mirrors replacing prisms [5.20, 5.21]. Second, a piece of SESAM with the non-saturable loss parameter as low as 0.3% was used to further minimize the intra-cavity losses. Third, the transmission of the output coupler mirror was also optimized for femtosecond operation. Finally, high pumping intensity is much helpful to obtain efficient laser operation. For this purpose, a 7-W high brightness fiber coupled diode laser (Jenoptik laserdiode GmbH) at 968 nm was adopted as the pump. The fiber core diameter is 50 μm, with a numerical aperture of 0.22.

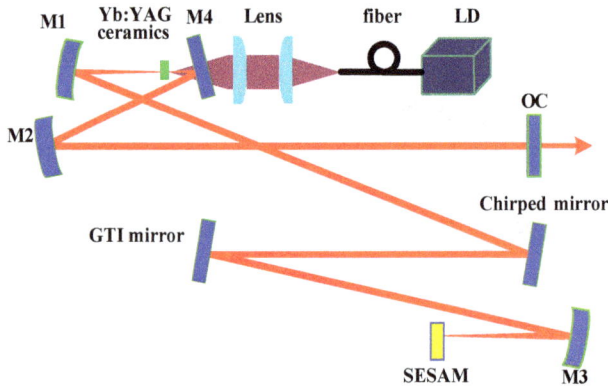

Figure 17. Schematic of the mode-locked Yb:YAG ceramic laser.

A schematic of the laser cavity and pump geometry is shown in Figure 17. The pump laser was reimaged into the Yb:YAG ceramic through a coupling system. The cavity was a standard Z-shape structure, with a piece of high reflective plane mirror folding one arm of the cavity to fit the short focusing length of the pump light. M1 and M3 were curved folding mirrors with ROC of 200 mm, and M2 with ROC of 300 mm. A 2-mm-long 10 at. % doped Yb:YAG ceramic was used for the gain medium. The absorption of the ceramic to the pump laser was around 50% in the experiment due to the lower and narrower absorption peak at 968 nm than at 940 nm. Negative dispersion was introduced by a Gires-Tournois interferometer (G-TI) mirror and a chirped mirror. The G-TI mirror provides second-order dispersion compensation of about -1000 ± 300 fs^2 per rebound in the spectral range from 1030 to 1050 nm and the chirped mirror provides -120 ± 20 fs^2 from 1000 to 1100 nm. The cavity was designed to sustain a fundamental mode with a beam waist of 65 μm × 67 μm in the crystal, and a waist of 58 μm × 65 μm on the SESAM. The total cavity length corresponds to a repetition rate of 64.27 MHz.

Different mirrors were used as output couplers with transmissions of 1%, 2.5%, and 4% from 1020 to 1080nm. Stable soliton mode-locking regime was observed with 1% and 2.5% output couplings. With 4% coupling, the output power under the maximum pump was slightly higher than the output power by the 2.5% coupler, however, cw mode-locking couldn't be realized at this coupling condition. Figure 18 shows the output power as a function of the absorbed pump power for these two output couplers. With the decreasing of mirror reflectivity from 99% to 97.5%, the threshold incident pump power for cw mode-locking operation increased from 1.9 to 2.5 W. The maximum CW mode-locked output power was achieved with the 97.5% reflectivity output coupler. Under the incident pump power of 7 W (the measured absorbed pump power was 3.5 W), stable mode-locked output power of 1.9 W was obtained, corresponding to a slope efficiency of 76% with respect to the absorbed pump power. With the 1% output coupler, the maximum mode-locked power was 1.2 W with the slope efficiency of 44.7%.

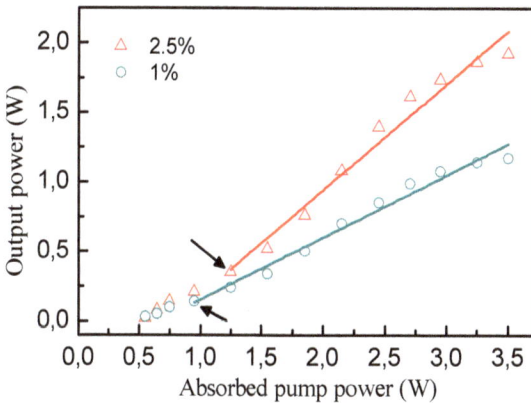

Figure 18. Average output power versus absorbed pump power of the mode-locked Yb:YAG ceramic laser with output couplings of 2.5% (triangles) and 1% (circles). The cw mode-locking thresholds are indicated by arrows.

At the highest output power, the pulse duration was measured to be 418 fs (sech^2 assumption). The FWHM width of the spectrum was 3.4 nm at the central wavelength of 1048 nm (Figure 19 left). The radio-frequency spectrum was also measured by an electrical spectrum analyzer to characterize the performance of femtosecond pulse train at the high power operation. As depicted by Figure 19, the spectrum shows a clean peak at a repetition rate of 64.27 MHz without side peaks, implying that the Q-switching instabilities have been fully depressed. A wider acquisition frequency span (from 0 to 300 MHz with 100 kHz resolution bandwidth) was also performed (inset in Figure 19 right), which was a clear indication for the single-pulse operation in high output power level. The results presented above definitely affirmed that the Yb:YAG ceramic is an excellent laser medium for high-power and high-efficiency diode-pumped ultrafast lasers and amplifiers.

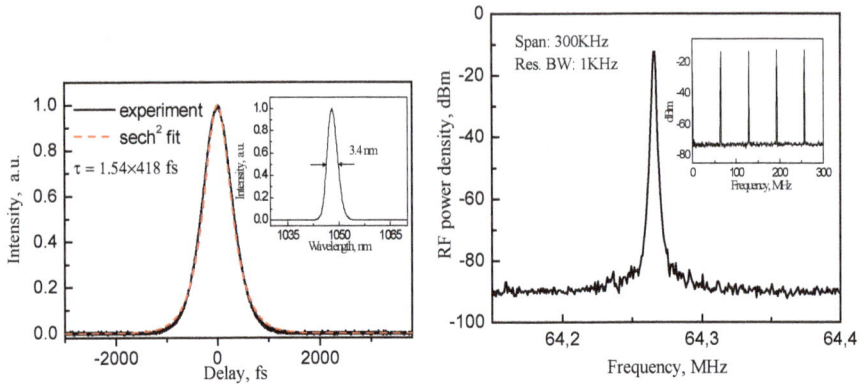

Figure 19. Left, intensity autocorrelation trace of the mode-locking laser pulses with 1.9 W average power; inset, the corresponding spectrum. Right, the rf spectrum at fundamental repetition frequency of the 418 fs pulse train; inset, resolution bandwidth of 100 kHz and span of 300 MHz.

With a similar experimental layout as depicted in Figure 17, a cw mode-locked picosecond Yb:LSO laser was also realized with a 5% doped 3-mm-thick $Yb^{3+}:Lu_2SiO_5$ crystal as the laser medium [5.22]. At the pump power of 5.4 W from the fiber coupled diode laser, 0.72 W stable mode-locked pulses were obtained with an output coupler of 1% transmittivity. However, the thermal loading prevented higher output power when the pump power increased. The central wavelength of the mode-locked pulses lay on 1058 nm with a FWHM bandwidth of 3.5nm. The pulse duration was 5.1 ps, as shown in Figure 20.

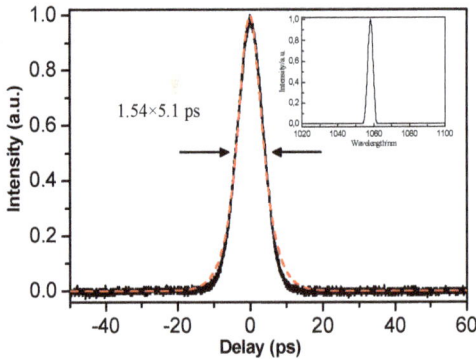

Figure 20. Intensity autocorrelation trace of the mode-locking laser pulses; inset, the corresponding spectrum.

5.3. Femtosecond Yb:GYSO laser

Yb^{3+}-doped $GdYSiO_5$ (Yb:GYSO), a promising ytterbium-doped crystal, has shown several attractive advantages compared to many recent Yb-doped materials. It exhibits $^2F_{7/2}$ ground state splitting as high as 995 cm^{-1}, which makes the population of the transition lower level much less sensitive to the temperature. Yb:GYSO crystal also possesses a comparatively high fluorescence life time of 1.92 ms and good mechanical properties. Du et al. demonstrated the first efficient tunable CW Yb:GYSO laser operation [5.1], and the picosecond Yb:GYSO laser was reported by another group later [5.2]. We describe here the realization of a femtosecond Yb:GYSO laser [5.3]. The setup for the laser is shown in Fig.21. A 3-mm-long, 5%-doped Yb:GYSO crystal is pumped by a CW Ti:sapphire laser at the wavelength of 976 nm. Three pieces of curved folding mirrors with the radius of curvature (ROC) of 100mm (M1-M3) are used to reduce the beam waist inside the active medium and on the SESAM. The SESAM was designed for 0.4% modulation depth and saturation fluence of 120 μJ/cm^2. Its nonsaturable loss was specified to be 0.3%, and the relaxation time less than 500 fs. A pair of chirped mirrors (CM1, CM2) were used in another arm of the cavity for group velocity dispersion compensation. The chirped mirrors were designed with GDD of -120 fs^2 per bounce within the wavelength range from 1000 nm to 1200 nm. Considering the amount of positive GDD introduced by the 3-mm-long 5%-doped Yb:GYSO crystal [5.4], the net intra-cavity GDD was remained at minus value by introducing two bounces onto the chirped mirrors.

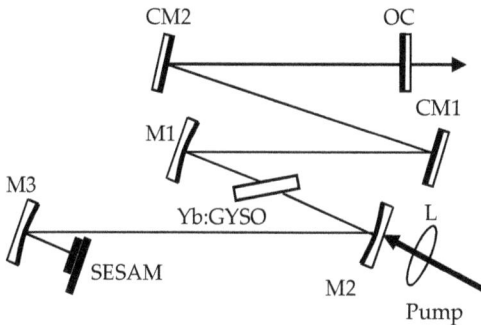

Figure 21. Cavity used to study the femtosecond laser performance of Yb:GYSO. M1, M2, M3, ROC 10 cm; CM1, CM2, flat chirped mirrors; L, lens f=100 mm; OC, 1% output coupler.

With a piece of high reflective mirror in one arm replacing the M3 and SESAM, and removing the chirped mirrors, the CW lasing performance was tested first. At CW lasing, the measured threshold incident pump power was as low as 180 mW, which can be thanks to the large ground state splitting of Yb:GYSO crystal. Output power of 920 mW with diffraction-limited beam pattern was obtained at the wavelength of 1091 nm under 2 W pump power. A tunable range of about 77 nm (from 1033 to 1110nm) was obtained by inserting a Lyot filter in the CW cavity.

After optimization of the cavity alignment shown in Fig. 21, stable and self-starting femtosecond pulses were realized. The autocorrelation trace of the mode-locked pulses is shown in Fig 22(a), the FWHM width of the autocorrelation trace is about 324 fs. If a sech^2-pulse shape is assumed, the mode-locked pulse duration is 210 fs.

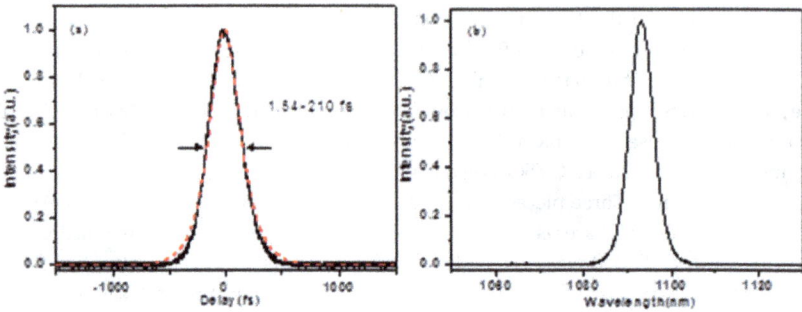

Figure 22. (a) Typical intensity autocorrelation trace of the pulses. The experimental data are shown by the solid curve and the sech2-fitting curve by the dashed curve. (b) The laser spectrum of mode-locking operation.

The spectral width (FWHM) of the pulses was measured as 6.4 nm at the central wavelength of 1093 nm. Fig. 22(b) shows the typical spectrum, compared with the one in CW mode, the central wavelength red shifted slightly from the peak wavelength. The time-bandwidth product is 0.34, which is close to the value of 0.315 for the transform-limited $sech^2$-pulse. This indicates that almost transform-limited pulses were directly obtained from the cavity. However, the gain bandwidth was not fully covered by the obtained spectrum, especially in the shorter wavelength. That means the gain of Yb:GYSO crystal is not fully exploited. Considering the 77 nm tunability achieved in the CW mode, even shorter femtosecond pulses should be possible. Under the full pump power of 2 W from the CW Ti:sapphire laser, the average mode-locking power output from the Yb:GYSO laser was 300 mW at a repetition rate of 80 MHz, corresponding to an energy per pulse of 3.75 nJ and a peak power of 17.9 KW.

A higher mode-locking laser power can be possible by replacing the pump laser by a high power diode laser at 976 nm. In view of the excellent mechanical properties of the Yb:GYSO crystal, a new kind of femtosecond laser with high output power and compactness is very promising.

5.4. Picosecond and femtosecond Yb:YGG laser

Yttrium gallium garnet (YGG) is another garnet crystal. Similar to YAG and GGG, $Y_3Ga_5O_{12}$ (YGG) has many desirable advantages for laser materials—stable, hard, optically isotropic, having good thermal conductivity (9 W/mK), and accepting substitutionally trivalent ions of both rare-earth and iron groups [4.8]. Yb^{3+} doped yttrium gallium garnet (Yb:YGG) was first reported as a scintillator [5.23-24]. The most interesting property is that the bandwidth of its emission spectrum is nearly four times boarder than Yb:YAG 's [5.23]. The high-quality Yb:YGG crystal suitable for laser operation had been grown though optical floating zone method by H. Yu et al for the first time [5.25], and the special thermal properties, including the specific heat, thermal expansion coefficient, thermal diffusion coefficient, and thermal conductivity had been investigated. Tab.3 shows properties comparisons of Yb:YGG, Yb:YAG and Yb:GGG [5.25]. In this section, we reviewed the diode pumped passively mode-locked Yb:YGG laser [5.26-27].

Crystals	Yb:YGG (10at%)	Yb:YAG	Yb:GGG
Symmetry	Cubic	Cubic	Cubic
Thermal expansion coefficient (10^{-6} K^{-1})	3.8	8.18 (10at%)	~7 (Pure)
Specific Heat (J/gK)	0.43	~0.63 (10at%)	~0.37 (Pure)
Thermal diffusion coefficient (mm^{2}s^{-1})	1.33	1.62 (10at%)	~3 (Pure)
Thermal Conductivity (W/m·K)	3.47	~4.8 (10at%)	7.5 (Pure)
Absorption Cross-sections (10^{-20}cm^2)	2.7 (970 nm)	0.77 (941 nm)	0.6 (945 nm)
Emission Cross-sections (10^{-20}cm^2)	2.6 (1025 nm)	2.03 (1031 nm)	1.1 (1031 nm)
FWHM at emission peak	22 nm	10 nm	~10 nm
Positive band at effective gain cross-sections (β=0.5)	120 nm	~100 nm	~80 (nm)

Table 3. Comparisons of Yb:YGG, Yb:YAG and Yb:GGG

The Yb:YGG single crystal employed in the experiment was grown by the optical floating zone method, which was fine polished and antireflection coated at a broad spectrum range around 1 μm with cross section of 3 mm×3 mm and length of 3.2 mm. A high brightness fiber-coupled diode laser emitting at 970 nm (Jenoptik, JOLD-7.5-BAFC-105) was used to end-pump the laser medium. The pump laser output from the fiber (with 50 μm core diameter and 0.22 numerical aperture) was coupled into the laser crystal where the laser spot radius was about 30 μm. Fig. 23 (a) is the schematic of the pumping geometry and laser cavity. A Z-fold cavity was employed for mode-locking experiment. M1 was a plane dichroic mirror with high transmission at 970 nm and high reflection at 1020-1100 nm; M2, M3 and M4 were concave mirrors, with radii of curvature of 300 mm, 200 mm and 200 mm, respectively. Passive mode-locking was realized by using a semiconductor saturable absorber mirror (SESAM) (BATOP), which has a saturable absorption of 0.4 % at 1040 nm, a saturation fluence of 120 μJ/cm^2, relaxation time less than 500 fs, and non-saturable loss of less than 0.3 %. A plane-wedged mirror with transmission rate of 2.5 % was used as the output coupler (OC). According to the ABCD matrix, the cavity was very stable and had large parameter space. After optimization of the cavity alignment, laser pulses as short as 2.1 ps were generated with an average power more than 1 W under incident pumping power of 5.5 W, where multi-pulse will interrupt the stable mode-locking operation when pumping power was higher than 5.5 W.

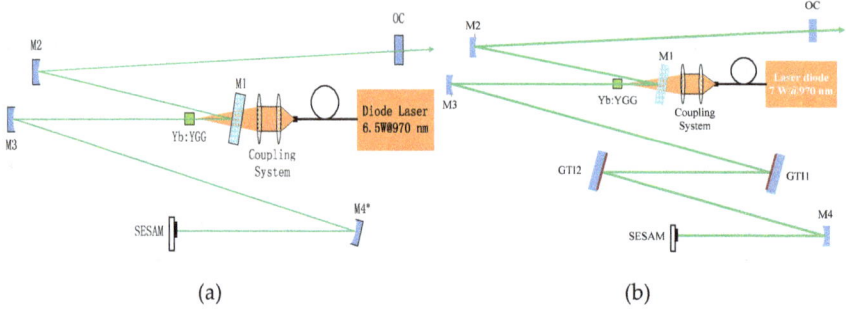

Figure 23. (a) experiment layout of the picoseconds Yb:YGG laser; (b) experiment layout of the femtoseconds Yb:YGG laser

In order to obtain very short pulses, dispersion conversation is an important issue. A pair of Gires–Tournois interferometer (GTI) mirrors, with group-velocity dispersion of -1400 fs^2 per bounce within the wavelength range from 1025 to 1045 nm, was used for dispersion compensation, as shown in Fig. 23 (b). Limited by the available pump power, the maximum output power was 570 mW under 6.9 W of incident pump power. We measured the intensity autocorrelation trace by using the commercial noncollinear autocorrelator (Femtochrome, FR-103MN). As shown in Fig. 24 (a), the FWHM width of the autocorrelation trace was about 360 fs. If a sech^2-pulse shape was assumed, the mode-locked pulse duration was 245 fs. The Fig. 24 (b) depicts the corresponding spectrum of the stable mode locking, which had a FWHM bandwidth of 5.8 nm at the central wavelength of 1045 nm. The time-bandwidth product was calculated to be 0.39.

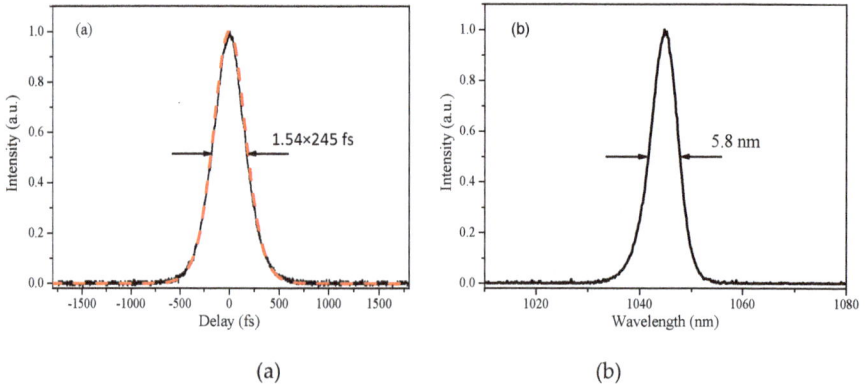

Figure 24. (a) Intensity autocorrelation trace of the pulse. The experimental data are shown by the solid curve and the sech^2-fitting curve by the dashed curve. (b) The laser spectrum of mode-locking operation.

6. Mode-locked femtosecond Cr⁴⁺-doped laser

After several decades of development, there are many kinds of gain media which can operate in near infrared region, but only a few of them can be used in mode-locked lasers. In chapters 4 and 5, Nd-doped and Yb-doped mode-locked lasers have been described. Most Nd-doped lasers work in picosecond region, so do some Yb-doped lasers. Although some Yb-doped lasers can deliver femtosecond pulses, the generated spectral bandwith is relatively narrow and can't support very short pulses.

There are two chromium doped media, namely chromium doped forsterite ($Cr^{4+}:Mg_2SiO_4$) and chromium doped yttrium aluminium garnet ($Cr^{4+}:YAG$), are very attractive for femtoseond laser application. They have very broad emission range which can support few tens of femtoseond pulse generation. Combined with the interesting wavelength ranges they cover (1.3 and 1.5 μm, respectively), these lasers have find wide applications in optical communication [6.1], optical coherence tomography [6.2], biophotonics [6.3] and so on. In this section, we will present our work on the development of these two kind of ultrafast lasers.

6.1. Efficient femtoseond Cr:forsterite laser

Forsterite is an anisotropic crystal belonging to the orthorhombic space group. Cr^{4+}:forsterite crystal shows many excellent characteristics as a laser gain medium. Some important parameters are listed in Table 4 and compared with those of Ti:sapphire crystal. Except for low thermal conductivity and relatively small emission cross section, other physical parameters of Cr^{4+}:forsterite crystal are comparable with those of Ti:sapphire crystals. The Cr^{4+}:forsterite crystal can support the generation of very short femtosecond pulses.

Crystal name	Ti:Sapphire	Cr:Forsterite
Density (g/cm³)	3.98	3.217
Melt point (°C)	2050	1895
Moh's hardness	9	7
Thermal conductivity (W/(m·k))	33	8
Thermal expansion (K⁻¹)	5×10^{-6}	9.5×10^{-6}
Tuning range (nm)	660~1050	1130~1367
Upper level lifetime (us)	3.2	2.7
Emission cross section (cm²)	4×10^{-19}	1.44×10^{-19}
Center wavelength (nm)	795	1244(cw),1235(pulsed)
Shortest pulse width ever achieved (fs)	<5 fs[6.4]	14 fs[6.5]

Table 4. Parameter comparison of Ti:sapphire and Cr:forsterite

Cr^{4+}:forsterite has very wide absorption and emission spectrum, as shown in Figure 25. The use of absorption peaks at 470 nm and 730 nm are typically excluded due to lack of pump sources. The relatively lower but much wider peak near 1 μm is preferred because of the vast availability of the pump laser. Besides, the quantum defect in the crystal will be much smaller when pumped at this wavelength. Although the emission spectrum extends from 600 nm to 1400 nm, wavelengths below 1100 nm overlaps with the absorption spectrum and suffers significant reabsorption, which prevents stimulated emission. Only the part above 1100 nm is suitable for laser operation. Up to now, the shortest pulse duration with Cr^{4+}: forsterite laser of 14 fs was achieved with 80 mW output power [3.49]. We'll present our work on the development of highly efficient self-starting femtosecond Cr:forsterite laser here. With a 7.9 W Yb doped fiber laser as the pump, we obtained stable femtosecond laser pulses with an average power of 760 mW, yielding a pump power slope efficiency of 12.3% [6.6].

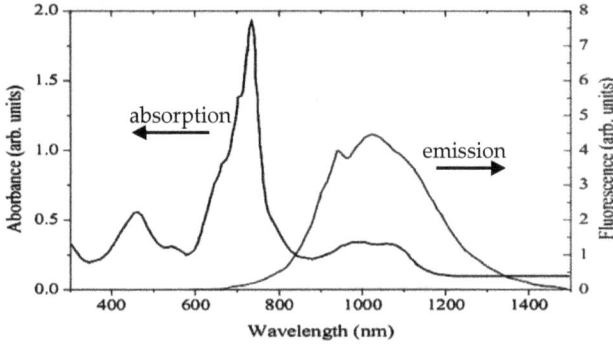

Figure 25. Absorption and emission spectrum of Cr^{4+}:forsterite

The schematic diagram of an efficient Cr^{4+}:forsterite laser is shown in Figure 26 A 1064 nm linearly polarized cw fiber laser is used as the pump (AYDLS-PM-10, Amonics), which deliver 7.9 W average power at maximum. The dimension of Cr^{4+}:forsterite crystal is 4 mm× 2 mm×9 mm and the crystal is cut for propagation of light along the a axis and emitting beam polarization along the c axis (P_{mnb} notation). Both ends are cut at Brewster's angle and polished in order to reduce the reflective loss to the minimum. Due to the low thermal conductivity, the crystal is wrapped in thin indium foil and tightly held in a copper holder. A thermoelectric cooler cools the holder down to 5°C. Although lower temperature can bring higher output power, water condensation on crystal surface will induce unstable operation.

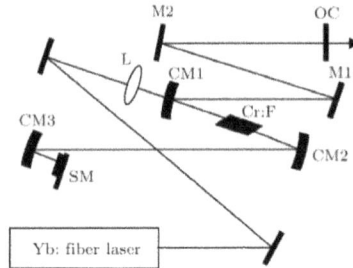

Figure 26. Schematic of the femtosecond Cr:forsterite laser. CM1–CM3: Chirped mirrors, ROC 100 mm; M1 and M2: plane chirped mirrors.

In order to obtain femtosecond pulses, dispersion compensation is important, especially when the used laser crystal is relatively long. By using the formula 13[6.8], the group-delay dispersion of Cr^{4+}:forsterite is calculated and shown in Figure 27.

$$GDD = \frac{1}{2\times5.17}\left[2\times84.42+6\times116.70(\omega-\omega_0)+12\times(-101.21)(\omega-\omega_0)^2+20\times125.08(\omega-\omega_0)^3\right] \quad (13)$$

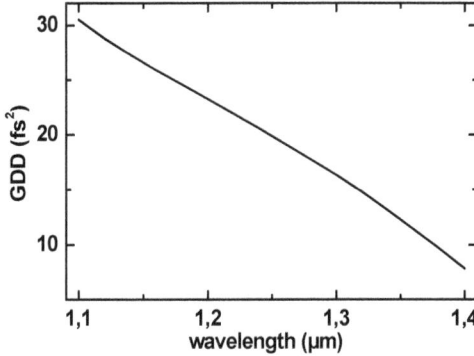

Figure 27. Group-delay dispersion from 1 mm Cr^{4+}:forsterite crystal

The 9 mm long Cr^{4+}:forsterite crystal introduces about 150 fs^2 positive GDD. In order to compensate these dispersion, all the reflective mirrors in cavity are chirped mirrors except the 3% output coupler (OC) and the saturable mirror (SM). Each bounce on CM1, CM2 and CM3 brings in -60±20 fs^2, and -70±20 fs^2 on M1 and M2, respectively. Considering the positive GDD introduced by 1.8-m-long air, the final net intra-cavity GDD after one single trip is about -130 fs^2. Due to the strong oscillations in negative GDD introduced by all the chiped mirrors, a certain amount of negative GDD is necessary to insure the stable mode locking. The semiconductor saturable absorber mirror (SESAM), which is highly reflectively coated from 1240 nm to 1340 nm, has a modulation depth of 2.5% and saturable fluence of 70 $\mu J/cm^2$. In order to reduce the inserting losses, we chose a nonsaturable loss of the SESAM less than 0.5%.

A lens with the focal length of 100 mm is used to focus the pump beam into the crystal. In order to achieve good matching and overlapping between pump and resonant beam, the radius of curvature (ROC) of CM1 and CM2 are 100 mm. Besides, because stable mode locking requires sufficient energy fluence on SESAM, CM3 with ROC of 100 mm is used to focus the resonant beam on it.

The relationship between pump power and output power is shown in Figure 28. The pump power threshold for stable mode locking is 1.8 W, with corresponding output power of 49 mW. The maximum output power of 760 mW is obtained when pump power increases to 7.9 W, which indicates record high slope efficiency as 12.3%. When optimizing the output power of mode-locking operation, we found that a slightly adjustment of the concave mirror

CM2 was necessary when the laser cavity was firstly optimized at other pump powers. We contribute this to the thermal loading effect in the laser crystal. The mode-locking power of the Cr:forsterite laser was almost linear till the pump power up to 7.9 W. Hence we believe that even higher output power can be expected if we use a pump laser with higher power.

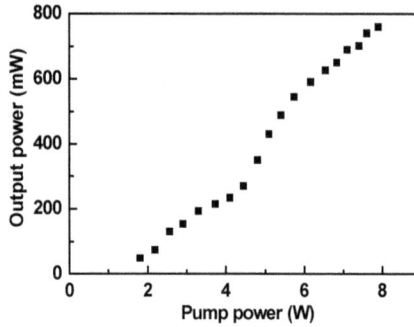

Figure 28. Variation of the mode locking output power as a function of the pump power.

The intensity autocorrelation trace (measured by a FR-103MN autocorrelator, Femtochrome Research, Inc.) and spectrum (measured by AQ-6315A, ANDO) are shown in Figure 29. The autocorrelation trace captured by an oscilloscope shows a FWHM of 71 fs, which indicates the pulse duration of 46 fs by assuming a $sech^2$ pulse shape. The spectrum of mode locking operation is centered at 1277 nm, with a width of 45 nm. A time-bandwidth product of 0.38 is obtained, indicating that the output pulses are nearly transform limited.

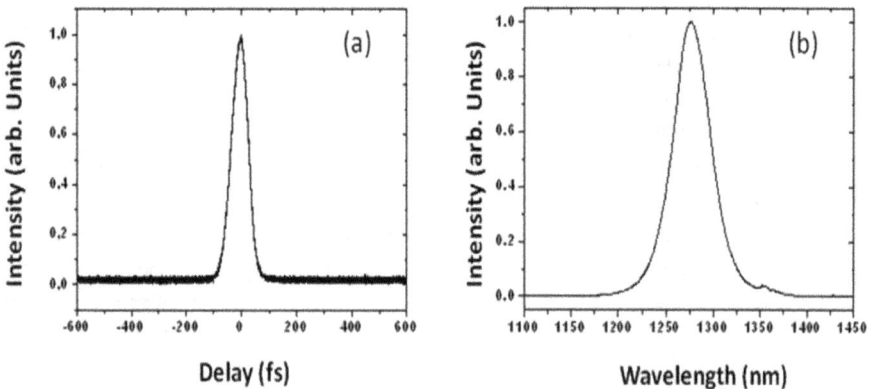

Figure 29. (a) Intensity autocorrelation trace of the output pulses and (b) spectrum of mode locking operation.

The mode-locked Cr^{4+}: forsterite laser can stably work for hours if the environment is dry and stable. The most probable factor which disturbs the mode locking is the water condensation on crystal surfaces. If the environment is quite humid, nitrogen blowing or dehumidifier is required. Based on this efficient femtosecond Cr:forsterite laser, we also realized red light femtosecond laser output with a intra-cavity frequency-doubled Cr:forsterite laser configuration [6.7].

6.2. Self-starting mode-locked femtosecond Cr^{4+}:YAG laser

Chromium doped yttrium aluminium garnet (Cr^{4+}: YAG) is a remarkable gain medium which generates near infrared laser beam. Moreover, it can support femtosecond laser operation for its wide emission spectrum. The first laser operation in Cr^{4+}: YAG was reported in 1988 [6.9]. In 1993, 26 ps pulses were obtained by using active mode locking [6.10]. So far, femtosecond pulses from Cr^{4+}:YAG can be obtained by active mode locking [6.11], kerr-lens mode locking [6.12-13] and SESAM mode locking [6.14-15]. The record of shortest pulses was reported by Ripin et al. in 2002 [6.16].

The absorption spectrum of Cr^{4+}: YAG extends from 950 nm to 1100 nm. It's convenient to use all-solid-state Nd:YAG laser or Yb doped fiber laser as pump source. The emission spectrum extends from 1250 nm to 1650 nm, as shown in Figure 30. Although the peak is at 1.39 μm, the most reported output wavelengths are near 1.5 μm.

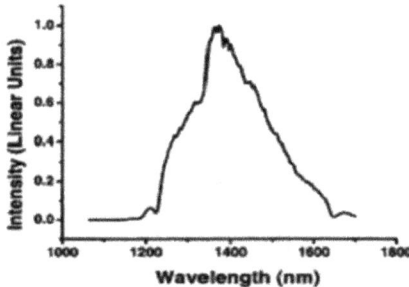

Figure 30. Emission spectrum of Cr^{4+}:YAG

Cr^{4+}:YAG is suitable to generate femtosecond pulses near 1.5 μm for its wide emission spectrum and some other characteristics. There are still some obvious drawbacks though. First, the thermal conductivity is relatively low. The severe thermal lens effect is an annoying problem which causes the unstability of the intra-cavity laser modes. Second, the pump beam and resonant beam both suffer re-absorption effect, which decreases the pump efficiency and increases intra-cavity loss. However, The Cr^{4+}:YAG crystal is still a good gain medium in near infrared range near 1.5 μm. It can support femtosecond pulses generation and has the potential to be used in optical fiber communication, femtobiology, femtochemistry and so on. In this section, a very compact and self-starting femtosecond Cr^{4+}:YAG laser with a pulse width of 65 fs was described [6.17].

The schematic diagram of self-starting mode locking Cr^{4+}:YAG laser is shown in Figure 31.

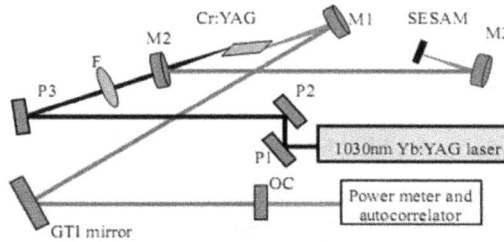

Figure 31. The schematic diagram of the self-starting mode locking Cr^{4+}: YAG laser.

M1, M2 and M3 are concave mirrors with ROC=100 mm, anti-reflective coated (T>98%) at 870~1050 nm and highly-reflective coated (R>99.9%) at 1420~1720 nm. A Gires-Tournois interferometer (GTI) mirror compensates the group-delay dispersion, which can introduce -500±50 fs^2 GDD for each bounce. In order to avoid high intra-cavity loss, an output coupler (OC) with a transmission rate of 1% is used. The pump laser is a commercial Yb:YAG laser which delivers horizontally polarized beam at 1030 nm. The Brewster's angle cut Cr^{4+}:YAG crystal rod is 20 mm long and has a diameter of 5 mm. The laser medium absorbs 90% of the pump energy in this experiment. To reduce the thermal lens effect, the crystal is wrapped in thin indium foil and tightly held in a copper holder. Recirculated cooling water keeps the temperature at 10°C.

For a 20-mm-long Cr^{4+}:YAG, the GDD can be calculated by the following formula [6.12]:

$$GDD = -15296 + 119.83v - 0.2054v^2 \tag{14}$$

The crystal introduces positive GDD of 226.8 fs^2 at 1500 nm for each single pass. Considering the GDD introduced by air, after compensated by GTI mirror, the net GDD is about -230 fs^2 per single pass in the cavity, as shown in Figure 32. The main drawback by using GTI mirror is that the dispersion compensation isn't precisely adjustable as by using prisms. However, using GTI mirror avoids additional insertion loss and makes the cavity structure more compact and robust.

Under pump power of 9 W, stable mode locked pulses are obtained by fine adjusting the positons of M1, M2, Cr^{4+}:YAG crystal and the SESAM. The highest output power is 95 mW, indicating low slope efficiency. As mentioned before, the re-absorption effect causes large loss for resonant laser. Besides, the OC with low transmission rate also limits the output power. OCs with higher transmission rate are tested, but stable mode locking is unable to achieve.

The mode locking spectra and interferometric autocorrelation trace of output pulses are shown in Figure 33. In Figure a, the solid curve presents the results when the net intra-cavity dispersion is about -230 fs^2. The corresponding spectrum width (FWHM) is 45 nm, centered at 1508 nm. The dashed curve corresponds to the mode-locking results with a net intra-cavity dispersion of -720 fs² by using another mirror to reflect the beam on GTI mirror once more. The spectrum width (FWHM) is 34.5 nm and a small blue shift can be observed. An interferometric autocorrelation trace corresponding to the solid spectrum curve is shown in Figure b. By assuming a sech² pulse shape, the pulse duration is 65 fs, indicating a time-bandwith product of $\Delta\nu\Delta\tau = 0.372$.

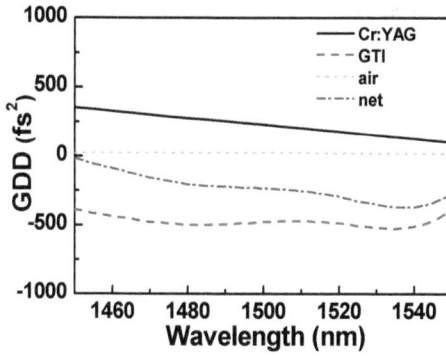

Figure 32. Intra-cavity second-order dispersion contributions.

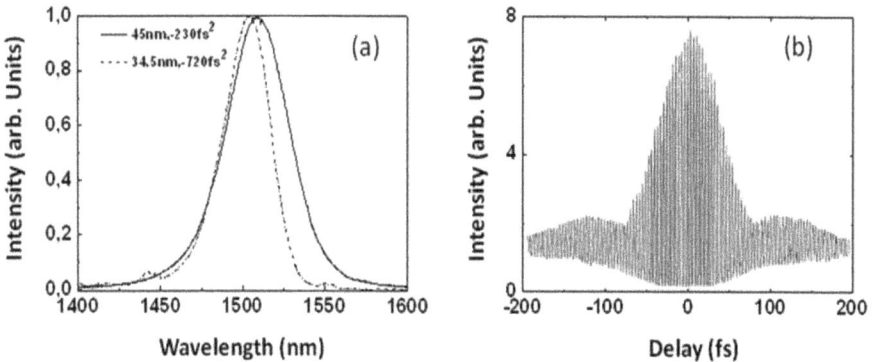

Figure 33. (a) Spectra of mode locking Cr⁴⁺:YAG laser, solid curve corresponds to net GDD of -230 fs², dashed curve corresponds to -720 fs²; (b) Interferometric autocorrelation trace of mode locking pulses, the pulse duration is 65 fs.

7. Conclusion

We have described the generation of ultrafast laser pulses with a series of Nd, Yb or Cr-doped gain media. Stable laser pulses were obtained around central wavelengths from 900 nm to 1500 nm and in the picosecond to femtosecond regimes. The pulse duration for Nd-doped mode-locked lasers is in picosecond regime, while for Yb-doped media it can be as short as several tens of femtosecond. The experimental results are summarized in the Table 5. It contains a wide range of very interesting and practically useful all solid-state passively mode-locked ultrafast laser sources in near infrared. Further investigations toward even shorter pulses and higher average power are underway. The direct amplification of all-solid-state ultrafast laser is also in progress toward higher pulse energy.

Gain media	P_{in}(W)/λ_P(nm)	P_{out}(mW)	λ(nm)	$\Delta\lambda$(nm)	Δt
Nd:YVO₄	9W@808	1700	1064		2.3ps
Nd:GdVO₄	19.7W@808	128	912		6.5ps
Nd:LuVO₄	17W@808	88	916		6.7ps
Nd:GSAG	16.7W@808	510	942.6	0.65	8.7ps
Nd:GGG	24W@808	3200	1062.5	0.25	15ps
	20W@808	320	937.4	0.3	22.7ps
Yb:YAG	2W@940 Ti:S	180	1053	7	170fs
Yb:YAG ceramics	7W@968	1900	1048	3.4	418fs
Yb:GYSO	2W@976 Ti:S	300	1093	6.4	210fs
Yb:LSO	2W@976 Ti:S	70	1047/1066		3.6ps
Yb:YGG	7W@970	570	1045	5.8	245fs
Cr:forsterite	7.9W@1064	760	1277	45	46fs
Cr:YAG	9W@1030	95	1508	45	65fs

Table 5. The experimental results of ultrafast lasers in our lab

Author details

Zhiyi Wei, Binbin Zhou, Yongdong Zhang, Yuwan Zou, Xin Zhong, Changwen Xu and Zhiguo Zhang
Beijing National Laboratory for Condensed Matter Physics and Institute of Physics, Chinese Academy of Sciences, Beijing, China

Acknowledgments

We thank the helpful technical discussions and support of the laser media by G. L. Bourdet, J.Xu, H.J. Zhang

8. References

[1]   U. Keller, K. J. Weingarten, F. X. Kartner, et al, "Semiconductor saturable absorber mirrors (SESAM's) for femtosecond to nanosecond pulse generation in solid-state lasers," IEEE J. Selected Quantum Electron. 2, 435 (1996).

[2]   G. J. Spühler, T. Südmeyer, R. Paschotta, M. Moser, K. J. Weingarten and U. Keller, "Passively mode-locked high-power Nd:YAG lasers with multiple laser heads," Appl. Phys. B 71, 19 (2000).

[3]   D. Burns, M. Hetterich, A. I. Ferguson, E. Bente, M. D. Dawson, J. I. Davies and S. W. Bland, "High-average-power (>20 W) Nd:YVO4 lasers mode locked by stain compensated saturable Bragg reflectors," J. Opt. Soc. Am. B 17, 919 (2000).

[4]   G. P. A. Malcolm, P. F. Curley and A. I. Ferguson, "Additive pulse modelocking of a diode pumped Nd:YLF laser," Opt. Lett. 15, 1303 (1990).

[5]   B. Ileri, C. Czeranowsky, K. Petermann, and G. Huber, "Mixed garnet laser crystals for water vapor detection," in Proceedings of IEEE Conference on Lasers and Electro-Optics Europe, 2005 (IEEE, 2005), p. 10.

[6]   T. Kellner, F. Heine, G. Huber, C. Honninger, B. Braun, F. Morier-Genoud, M. Moser, and U. Keller, "Soliton mode-locked Nd:YAlO3 laser at 930 nm," J. Opt. Soc. Am. B 15, 1663 (1998).

[7]   A. Schlatter, L. Krainer, M. Golling, and R. Paschotta, D. Ebling, and U. Keller "Passively mode-locked 914-nm Nd:YVO4 laser," Opt. Lett. 30, 44 (2005).

[8]   P. Blandin, F. Druon, F. Balembois, P. Georges, S. Lévêque-Fort and M. P. Fontaine-Aupart, "Diode-pumped passively mode-locked Nd:YVO4 laser at 914 nm," Opt. Lett. 31, 214 (2006).

[9]   J. Aus der Au, D. Kopf, F. Morier-Genoud, M. Moser, and U. Keller, "60-fs pulses from a diode-pumped Nd:glass laser," Opt. Lett. 22, 307 (1997).

[10] C. Hönninger, G. Zhang, U. Keller and A. Giesen, "Femtosecond Yb:YAG laser using semiconductor saturable absorbers," Opt. Lett. 20, 2402 (1995).

[11] M. Weitz, S. Reuter, R. Knappe and R. Wallenstein, "Passive mode-locked 21 W femtosecond Yb:YAG laser with 124 MHz repetition-rate," In Technical Digest of Conference on Lasers and Electro-Optics (optical society of America, Washington, DC, 2004), paper CTucc.

[12] H. Luo, D. Tang, G. Xie, H. Zhang, L. Qin, H. Yu, L. Ng, and L. Qian, "High-power passive mode-locking of a diode pumped Yb:GdVO4 laser," Optics Communications 281, 5382–5384 (2008).

[13] W. Li, Q. Hao, H. Zhai, H. Zeng, W. Lu, G. Zhao, L. Zheng, L. Su, and J. Xu, "Diode-pumped Yb:GSO femtosecond laser," Optics Express 15, 2354–2359 (2007).

[14] F. Thibault, D. Pelenc, F. Druon, Y. Zaouter, M. Jacquemet, and P. Georges, "Efficient diode-pumped Yb^{3+}: Y2SiO5 and Yb^{3+}: Lu2SiO5 high-power femtosecond laser operation," Opt. Lett. 31, 1555 (2006).

[15] F. Brunner, G. Spühler, J. Au, L. Krainer, F. Morier-Genoud, R. Paschotta, N. Lichtenstein, S. Weiss, C. Harder, A. Lagatsky, and others, "Diode-pumped femtosecond Yb:KGd(WO4)2 laser with 1.1-W average power," Opt. Lett. 15, 1119–1121 (2000).

[16] F. Brunner, T. Südmeyer, E. Innerhofer, F. Morier-Genoud, R. Paschotta, V. Kisel, V. Shcherbitsky, N. Kuleshov, J. Gao, K. Contag, and others, "240-fs pulses with 22-W average power from a mode-locked thin-disk Yb:KY(WO$_4$)$_2$ laser," Opt. Lett. 27, 1162–1164 (2002).

[17] U. Griebner, S. Rivier, V. Petrov, M. Zorn, G. Erbert, M. Weyers, X. Mateos, M. Aguiló, J. Massons, and F. Díaz, "Passively mode-locked Yb:KLu(WO$_4$)$_2$ oscillators," Opt. Express 13, 3465–3470 (2005).

[18] A. Garcia-Cortes, J. M. Cano-Torres, M. Serrano, C. Cascales, C. Zaldo, S. Rivier, X. Mateos, U. Griebner, and V. Petrov, "Spectroscopy and Lasing of Yb-Doped NaY(WO$_4$)$_2$: Tunable and Femtosecond Mode-Locked Laser Operation," IEEE Journal of Quantum Electronics 43, 758–764 (2007).

[19] F. Druon, S. Chénais, P. Raybaut, F. Balembois, P. Georges, R. Gaum'e, S. Mohr, D. Kopf, and others, "Diode-pumped Yb:Sr$_3$Y(BO$_3$)$_3$ femtosecond laser," Opt. Lett. 27, 197–199 (2002).

[20] F. Druon, F. Balembois, P. Georges, A. Brun, A. Courjaud, C. Hönninger, F. Salin, A. Aron, F. Mougel, and others, "Generation of 90-fs pulses from a mode-locked diode-pumped Yb^{3+}:Ca$_4$GdO(BO$_3$)$_3$ laser," Opt. Lett. 25, 423–425 (2000).

[21] A. Yoshida, A. Schmidt, V. Petrov, C. Fiebig, G. Erbert, J. Liu, H. Zhang, J. Wang, and U. Griebner, "Diode-pumped mode-locked Yb:YCOB laser generating 35 fs pulses," Opt. Lett. 36, 4425-4427 (2011).

[22] M. Delaigue, V. Jubera, J. Sablayrolles, J. P. Chaminade, A. Garcia, and I. Manek-Hönninger, "Mode-locked and Q-switched laser operation of the Yb-doped Li$_6$Y(BO$_3$)$_3$ crystal," Appl. Phys. B 87, 693–696 (2007).

[23] S. Rivier, A. Schmidt, C. Kränkel, R. Peters, K. Petermann, G. Huber, M. Zorn, M. Weyers, A. Klehr, G. Erbert, and others, "Ultrashort pulse Yb:LaSc$_3$(BO$_3$)$_4$ mode-locked oscillator," Optics Express 15, 15539–15544 (2007).

[24] M. Lederer, M. Hildebrandt, V. Kolev, B. Luther-Davies, B. Taylor, J. Dawes, P. Dekker, J. Piper, H. Tan, and C. Jagadish, "Passive mode locking of a self-frequency-doubling Yb:YAl$_3$(BO$_3$)$_4$ laser," Opt. let. 27, 436–438 (2002).

[25] N. Coluccelli, G. Galzerano, L. Bonelli, A. Di Lieto, M. Tonelli, and P. Laporta, "Diode-pumped passively mode-locked Yb:YLF laser," Optics Express 16, 2922–2927 (2008).

[26] P. Klopp, V. Petrov, U. Griebner, K. Petermann, V. Peters, and G. Erbert, "Highly efficient mode-locked Yb:Sc$_2$O$_3$ laser," Opt. let. 29, 391–393 (2004).

[27] U. Griebner, V. Petrov, K. Petermann, and V. Peters, "Passively mode-locked Yb:Lu$_2$O$_3$ laser," Opt. Express 12, 3125–3130 (2004).

[28] A. Schmidt, V. Petrov, U. Griebner, R. Peters, K. Petermann, G. Huber, C. Fiebig, K. Paschke, and G. Erbert, "Diode-pumped mode-locked Yb:LuScO$_3$ single crystal laser with 74 fs pulse duration," Opt. let. 35, 511–513 (2010).

[29] F. Druon, F. Balembois, and P. Georges, "Ultra-short-pulsed and highly-efficient diode-pumped Yb:SYS mode-locked oscillators," Opt. Express 12, 5005–5012 (2004).

[30] D. Li, X. Xu, C. Xu, J. Zhang, D. Tang, Y. Cheng, and J. Xu, "Diode-pumped femtosecond Yb: CaNb$_2$O$_6$ laser," Opt. Lett. 36, 3888–3890 (2011).

[31] Y. Zaouter, J. Didierjean, F. Balembois, G. Lucas Leclin, F. Druon, P. Georges, J. Petit, P. Goldner and B. Viana, "47-fs diode-pumped Yb^{3+}:CaGdAlO$_4$ laser," Opt. Lett. 31,119 (2006).

[32] J. Saikawa, Y. Sato, T. Taira, and A. Ikesue, "Passive mode locking of a mixed garnet Yb: YScAlO ceramic laser," Appl. Phys. Lett. 85, 5845 (2004).

[33] A. Shirakawa, K. Takaichi, H. Yagi, M. Tanisho, J. Bisson, J. Lu, K. Ueda, T. Yanagitani, and A. Kaminskii, "First mode-locked ceramic laser: Femtosecond Yb^{3+}: Y_2O_3 ceramic laser," Laser Phys. 14, 1375–1381 (2004).

[34] M. Tokurakawa, K. Takaichi, A. Shirakawa, K. Ueda, H. Yagi, S. Hosokawa, T. Yanagitani, and A. Kaminskii, "Diode-pumped mode-locked Yb^{3+}:Lu_2O_3 ceramic laser," Opt. Express 14, 12832–12838 (2006).

[35] M. Tokurakawa, K. Takaichi, A. Shirakawa, K. Ueda, H. Yagi, T. Yanagitani, and A. A. Kaminskii, "Diode-pumped 188 fs mode-locked Yb:YO ceramic laser," Appl. Phys. Let. 90, 071101 (2007).

[36] H. Yoshioka, S. Nakamura, T. Ogawa, and S. Wada, "Diode-pumped mode-locked Yb:YAG ceramic laser," Opt. Express 17, 8919–8925 (2009).

[37] H. Yoshioka, S. Nakamura, T. Ogawa, and S. Wada, "Dual-wavelength mode-locked Yb:YAG ceramic laser in single cavity," Opt. Express 18, 1479–1486 (2010).

[38] A. Giesen, H. Hügel, A. Voss, K. Wittig, U. Brauch, and H. Opower, "Scalable concept for diode-pumped high-power solid-state lasers," Appl. Phys. B 58, 365–372 (1994).

[39] A. Giesen and J. Speiser, "Fifteen Years of Work on Thin-Disk Lasers: Results and Scaling Laws," IEEE Journal of Selected Topics in Quantum Electronics 13, 598–609 (2007).

[40] T. Südmeyer, C. Kränkel, C. R. E. Baer, O. H. Heckl, C. J. Saraceno, M. Golling, R. Peters, K. Petermann, G. Huber, and U. Keller, "High-power ultrafast thin disk laser oscillators and their potential for sub-100-femtosecond pulse generation," Appl. Phys. B 97, 281–295 (2009).

[41] C. Kränkel, D. J. H. C. Maas, T. Südmeyer, and U. Keller, "Ultrafast Lasers in Thin-Disk Geometry," High Power Laser Handbook, p. 327, 2011.

[42] R. Peters, C. Kränkel, K. Petermann, and G. Huber, "Power scaling potential of Yb: NGW in thin disk laser configuration," Appl. Phys. B 91, 25–28 (2008).

[43] K. Petermann, D. Fagundes-Peters, J. Johannsen, M. Mond, V. Peters, J. Romero, S. Kutovoi, J. Speiser, and A. Giesen, "Highly Yb-doped oxides for thin-disc lasers," Journal of crystal growth 275, 135–140 (2005).

[44] C. Baer, C. Kränkel, C. Saraceno, O. Heckl, M. Golling, T. Südmeyer, R. Peters, K. Petermann, G. Huber, and U. Keller, "Femtosecond Yb:Lu_2O_3 thin disk laser with 63 W of average power," Opt. lett. 34, 2823–2825 (2009).

[45] C. R. E. Baer, C. Kränkel, C. J. Saraceno, O. H. Heckl, M. Golling, R. Peters, K. Petermann, T. Südmeyer, G. Huber, and U. Keller, "Femtosecond thin-disk laser with 141 W of average power," Opt. Lett. 35, 2302–2304 (2010).

[46] C. Kränkel, J. Johannsen, R. Peters, K. Petermann, and G. Huber, "Continuous-wave high power laser operation and tunability of Yb:$LaSc_3(BO_3)_4$ in thin disk configuration," Appl. Phys. B 87, 217–220 (2007).

[47] O. Heckl, C. Kränkel, C. Baer, C. Saraceno, T. Südmeyer, K. Petermann, G. Huber, and U. Keller, "Continuous-wave and modelocked Yb: YCOB thin disk laser: first demonstration and future prospects," Opt. Express 18, 19201–19208 (2010).

[48] S. Ricaud, A. Jaffres, P. Loiseau, B. Viana, B. Weichelt, M. Abdou-Ahmed, A. Voss, T. Graf, D. Rytz, M. Delaigue, E. Mottay, P. Georges, and F. Druon, "Yb:CaGdAlO₄ thin-disk laser," Opt. Lett. 36, 4134–4136 (2011).

[49] N.V. Kuleshov, V.G. Shcherbitsky, V.P. Mikhailov, S. Hartung, T. Danger, S. Kück, K. Petermann, G. Huber, "Excited-state absorption and stimulated emission measurements in Cr⁴⁺:forsterite, " J. Lumin. 75, 319 (1997).

[50] C. Chudoba, J. G. Fujimoto, E. P. Ippen, H. A. Haus, U. Morgner, F. X. Kärtner, V. Scheuer, G. Angelow and T. Tschudi, "All-solid-state Cr:forsterite laser generating 14-fs pulses at 1.3 μm," Opt. Lett. 26, 292 (2001).

[51] D. J. Ripin, C. Chudoba, J. T. Gopinath, J. G. Fujimoto, E. P. Ippen, U. Morgner, F. X. Kärtner, V. Scheuer, G. Angelow and T. Tschudi, "Generation of 20-fs pulses by a prismless Cr⁴⁺:YAG laser," Opt. Lett. 27, 61 (2002).

[52] http://www.coherent.com.

[53] F. Kallmeyer, M. Dziedzina, X. Wang, H. J. Eichler, C. Czeranowsky, B. Ileri, K. Petermann, and G. Huber, "Nd:GSAG-pulsed laser operation at 943 nm and crystal growth," Appl. Phys. B 89, 305-310 (2007).

[54] C. D. Brandle, Jr. and C. Vanderleeden, "Growth, Optical Properties, and CW Laser Action of Neodymium-Doped Gadolinium Scandium Aluminum Garnet," IEEE J. Quantum Electron. 10, 67 (1974).

[55] S. Wang, X. Wang, F. Kallmeyer, J. Chen, and H. J. Eichler, "Model of pulsed Nd:GSAG laser at 942 nm considering rate equations with cavity structure," Appl. Phys. B 92, 43-48 (2008).

[56] F. Kallmeyer, M. Dziedzina, D. Schmidt, H.-J. Eichler, R. Treichel, and S. Nikolov, "Nd:GSAG laser for water vapor detection by lidar near 942 nm," Proc. SPIE 6451, 64510J (2007).

[57] Changwen Xu, Zhiyi Wei, Yongdong Zhang, Dehua Li, Zhiguo Zhang, X. Wang, S. Wang, H.J.Eichler, Chunyu Zhang and Chunqing Gao, "Diode-pumped passively mode-locked Nd: GSAG laser at 942 nm," Opt. Lett. 34, 2324(2009).

[58] J. E. Geusic, H. M. Marcos, and L. G. Van Uitert, "Laser oscillation in Nd-doped yttrium aluminum, yttrium gallium and gadolinium garnet," Appl. Phys. Lett. 4, 182-184 (1964).

[59] G. F. Albrecht, S. B. Sutton, E. V. George et al, "Solid state heat capacity disk laser ," Laser and Particle Beams 16, 605-625 (1998).

[60] http://www.zlxtech.com.cn/home.asp

[61] M.D. Rotter, C.B. Dane, S. Fochs, K.L. Fortune, R. Merrill, B. Yamamoto, "Solid-state heat-capacity lasers: Good candidates for the marketplace," Photon. Spectra 38, 44-56 (2004).

[62] L. Qin, D. Tang, G. Xie, H. Luo, C. Dong, Z. Jia, H. Yu, and X. Tao, "Diode-end-pumped passively mode-locked Nd:GGG laser with a semiconductor saturable mirror," Opt. Commun. 281, 4762-4764 (2008).

[63] Y. D. Zhang, Z. Y. Wei, C. W. Xu, B. B. Zhou, D. H. Li, Z. G. Zhang, H. H. Jiang, S. T. Yin, Q. L. Zhang, D. L. Sun, "Diode-pumped watt-level mode-locked Nd:GGG laser at 1062 nm," The 7ᵗʰ Asia-Pacitific Laser Symposium, Th-P-78 (2010).

[64] Y. D. Zhang, Z. Y. Wei, C. W. Xu, B. B. Zhou, D. H. Li, Z. G. Zhang, H. H. Jiang, S. T. Yin, Q. L. Zhang, D. L. Sun, "Picoseconds pulse generation with a Nd:GGG laser

operating on quasi-three-level transition," 2009 Lasers & Electro-Optics & the Pacific Rim Conference on Lasers and Electro-Optics, 1-2, 670-671 (2009).

[65] A. Schlatter, L. Krainer, M. Golling, R. Paschotta, D. Ebling and U. Keller, "Passively mode-locked 914-nm Nd:YVO$_4$ laser," Opt. Lett. 30, 44-46 (2005).

[66] G. J. Spu"hler, S. Reffert, M. Haiml, M. Moser and U. Keller, "Output-coupling semiconductor saturable absorber mirror," Appl. Phys. Lett. 78, 2733 (2001).

[67] Y. F. Chen, S. W. Tsai, Y. P. Lan, S. C. Wang and K. F. Huang, "Diode-end-pumped passively mode-locked high-power Nd:YVO4 laser with a relaxed saturable Bragg reflector," Opt. Lett. 26, 199-201 (2001)

[68] Ya-Xian Fan, Jing-Liang He, Yong-Gang Wang, Sheng Liu, Hui-Tian Wang and Xiao-Yu Ma, "2-ps passively mode-locked Nd:YVO4 laser using an output-coupling-type semiconductor saturable absorber mirror," Appl. Phys. Lett. 86, 101103 (2005).

[69] Marie-Christine Nadeau, Stéphane Petit, Philippe Balcou, Romain Czarny, Sébastien Montant, and Christophe Simon-Boisson, "Picosecond pulses of variable duration from a high-power passively mode-locked Nd:YVO4 laser free of spatial hole burning," Opt. Lett.35, 1644-1646 (2010).

[70] Y. L. Jia, Z. Y. Wei, J. A. Zheng, W. J. Ling, Y. G. Wang, X. Y. Ma, Z. G. Zhang, "Diode-pumped self-starting mode-locked Nd:YVO4 laser with semiconductor saturable absorber output coupler," Chin. Phys. Lett. 21, 2209 (2004).

[71] Y. L. Jia, W. J. Ling, Z. Y. Wei, Y. G. Wang, X. Y. Ma, "Self-starting passively mode-locking all-solid-state laser with GaAs absorber grown at low temperature," Chin. Phys. Lett. 22, 2575(2005).

[72] A. I. Zagumennyi, V. A. Mikhailov, V. I. Vlasov, A. A. Sirotkin, V. I. Podreshetnikov, Yu. L. Kalachev, Yu. D. Zavartsev, S. A. Kutovoi and I. A. Shcherbakov, "Diode-Pumped Lasers Based on GdVO$_4$ Crystal," Laser Phys. 13, 311-318 (2003).

[73] Jin-Long Xu, Hai-Tao Huang, Jing-Liang He, Jian-Fei Yang, Bai-Tao Zhang, Chun-Hua Zuo, Xiu-Qin Yang and Shuang Zhao, "The characteristics of passively Q-switched and mode-locked 1.06 um Nd:GdVO$_4$ laser with V:YAG saturable absorber," Optical Materials 32, 522 – 525 (2010).

[74] Hou-Ren Chen, Yong-Gang Wang, Chih-Ya Tsai, Kuei-Huei Lin, Teng-Yao Chang, Jau Tang, and Wen-Feng Hsieh, "High-power, passively mode-locked Nd:GdVO$_4$ laser using single-walled carbon nanotubes as saturable absorber," Opt. Lett. 36, 1284-1286 (2011).

[75] Gang Zhang, Shengzhi Zhao, Yufei Li, Guiqiu Li, Dechun Li, Kejian Yang and Kang Cheng, "A dual-loss-modulated Q-switched and mode-locked Nd:GdVO$_4$ laser with AOM and V^{3+}:YAG saturable absorber at 1.34 μm," J. Opt. 13, 035202 (2011).

[76] Kejian Yang, Shengzhi Zhao, Jingliang He, Baitao Zhang, Chunhua Zuo, Guiqiu Li, Dechun Li and Ming Li, "Diode-pumped passively Q-switched and modelocked Nd:GdVO$_4$ laser at 1.34 μm with V:YAG saturable absorber," Optics Express 16, 20176-20185 (2008).

[77] A. K. Zaytseva, C. L. Wangb, C. H. Linc, and C. L. Pana, "Robust Diode-End-Pumped Nd:GdVO$_4$ Laser Passively Mode-Locked with Saturable Output Coupler," Laser Physics 21, 2029 – 2035 (2011).

[78] H. W. Yang, H. T. Huang, J. L. He, S. D. Liu, F. Q. Liu, X. Q. Yang, J. L. Xu, J. F. Yang and B. T. Zhang, "High Repetition Rate Passive Q-Switching of Diode-Pumped

Nd:GdVO$_4$ Laser at 912 nm with V^{3+}:YAG as the Saturable Absorber," Laser Physics 21, 66 – 69 (2011).

[79] Fei Chen, Xin Yu, Xudong Li, Renpeng Yan, Cheng Wang, Deying Chen, Zhonghua Zhang and Junhua Yu, "High power diode-pumped passively Q-switched and mode-locking Nd:GdVO4 laser at 912 nm," Optics Communications 284, 635-639 (2011).

[80] C. Zhang, Z. Y. Wei, L. Zhang, C. Y. Zhang, and Z. G. Zhang, "Passively mode-locked Nd:GdVO4 laser at 912nm," Chinese Physics 15, 2606 (2006).

[81] C.W. Xu, Z.Y. Wei, K.N. He, D.H. Li, Y.D. Zhang, Zhiguo Zhang, "Diode-pumped passively mode-locked Nd:GdVO4 laser at 912 nm," Opt. Commun. 281, 4398 (2008).

[82] He Kun-Na, Wei Zhi-Yi, Xu Chang-Wen, Li De-Hua, Zhang Zhi-Guo, Zhang Huai-Jin, Wang Ji-Yang, Gao Chun-Qing, "Passively Mode-Locked Quasi-Three-Level Nd:LuVO4 Laser with Semiconductor Saturable Absorber Mirror," Chin. Phys. Lett. 25, 4286 (2008).

[83] [5.1] J. Du, X. Liang, Y. Xu, R. Li, Z. Xu, C. Yan, G. Zhao, L. Su, and J. Xu, "Tunable and efficient diode-pumped Yb^{3+}:GYSO laser," Opt. Express 14, 3333 (2006).

[84] W. Li, Q. Hao, L. Ding, G. Zhao, L. Zheng, J. Xu, and H. Zeng, "Continuous-wave and passively mode-locked Yb:GYSO lasers pumped by diode lasers," IEEE J. Quantum Electron. 44, 567 (2008).

[85] B. Zhou, Z. Wei, Y. Zhang, X. Zhong, H. Teng, L. Zheng, L. Su, and J. Xu, "Generation of 210 fs laser pulses at 1093 nm by a self-starting mode-locked Yb:GYSO laser," Opt. Lett. 34, 31 (2009).

[86] W. Yang, J. Li, F. Zhang, Y. Zhang, Z. Zhang, G. Zhao, L. Zheng, J. Xu, and L. Su, "Group delay dispersion measurement of Yb:Gd$_2$SiO$_5$, Yb:GdYSiO$_5$ and Yb:LuYSiO$_5$ crystal with white-light interferometry," Opt. Express 15, 8486 (2007).

[87] E. Innerhofer, T. Südmeyer, F. Brunner, R. Paschotta, and U. Keller, "Mode-locked high-power lasers and nonlinear optics-a powerful combination," Laser Phys. Lett. 1, 82 (2004).

[88] S. V. Marchese, C. R. E. Baer, A. G. Engqvist, S. Hashimoto, D. J. H. C. Maas, M. Golling, T. Südmeyer, and U. Keller, "Femtosecond thin disk laser oscillator with pulse energy beyond the 10-microjoule level," Opt. Express 16, 6397 (2008).

[89] C. Hönninger, R. Paschotta, M. Graf, F. Morier-Genoud, G. Zhang, M. Moser, S. Biswal, J. Nees, A. Braun, G. A. Mourou, I. Johannsen, A. Giesen, W. Seeber, and U. Keller, "Ultrafast ytterbium-doped bulk lasers and laser amplifiers," Appl. Phys. B 69, 3 (1999).

[90] S. Uemura, and K. Torizuka, "Center-wavelength-shifted passively mode-locked diode-pumped ytterbium(Yb):yttrium aluminum garnet(YAG) laser," Jpn. J. Appl. Phys. 44, L361 (2005).

[91] B. Zhou, Z. Wei, D. Li, H. Teng, and G. Bourdet, "Numerical and experimental investigation of a continuous-wave and passively mode-locked Yb:YAG laser at a wavelength of 1.05 μm," Appl. Opt. 48, 5978 (2009).

[92] Zhou Bin-Bin, Wei Zhi-Yi, LI De-Hua, Teng Hao, Bourdet G. L, "Generation of 170-fs Laser Pulses at 1053nm by a Passively Mode-Locked Yb:YAG Laser," Chin. Phys. Lett. 26, 054208 (2009).

[93] G. L. Bourdet, "Theoretical investigation of quasi-three-level longitudinally pumped continuous wave lasers," Appl. Opt. 39, 966 (2000).

[94] G. L. Bourdet, and E. Bartnicki, "Generalized formula for continuous-wave end-pumped Yb-doped material amplifier gain and laser output power in various pumping configurations," Appl. Opt. 45, 9203 (2006).

[95] R. J. Beach, "CW theory of quasi-three level end-pumped laser oscillators," Opt. Communications 123, 385 (1995).

[96] K. Takaichi, H. Yagi, J. Lu, A. Shirakawa, K. Ueda, T. Yanagitani, and A. A. Kaminskii, "Yb^{3+}-doped $Y_3Al_5O_{12}$ ceramics – a new solid-state laser material," Phys. Status Solidi A 200, R5 (2003).

[97] A. A. Kaminskii, M. Sh. Akchurin, V. I. Alshits, K. Ueda, K. Takaichi, J. Lu, T. Uematsu, M. Musha, A. Shirakawa, V. Gabler, H. J. Eichler, H. Yagi, T. Yanagitani, S. N. Bagayev, J. Fernandez, and R. Balda, "New results in studying physical properties of nanocrystalline laser ceramics," Kristallografiya 48, 562 (2003).

[98] J. Dong, A. Shirakawa, K. Ueda, H. Yagi, T. Yanagitani, and A. A. Kaminskii, "Efficient Yb^{3+}:$Y_3Al_5O_{12}$ ceramic microchip lasers," Appl. Phys. Lett. 89, 091114 (2006).

[99] S. Nakamura, H. Yoshioka, Y. Matsubara, T. Ogawa, and S. Wada, "Efficient tunable Yb:YAG ceramic laser," Opt. Communications 281, 4411 (2008).

[100] H. Yoshioka, S. Nakamura, T. Ogawa, and S. Wada, "Diode-pumped mode-locked Yb:YAG ceramic laser," Opt. Express 17, 8919 (2009).

[101] B. Zhou, Z. Wei, Y. Zou, Y. Zhang, X. Zhong, G. L. Bourdet, and J. Wang, "High-efficiency diode-pumped femtosecond Yb:YAG ceramic laser," Opt. Lett. 35, 288 (2010).

[102] A. Lucca, G. Debourg, M. Jacquemet, F. Druon, F. Balembois, P. Georges, P. Camy, J. L. Doualan, and R. Moncorgé, "High-power diode-pumped Yb^{3+}:CaF_2 femtosecond laser," Opt. Lett. 29, 2767 (2004).

[103] F. Thibault, D. Pelenc, F. Druon, Y. Zaouter, M. Jacquemet, and P. Georges, "Efficient diode-pumped Yb^{3+}:Y_2SiO_5 and Yb^{3+}:Lu_2SiO_5 high-power femtosecond laser operation," Opt. Lett. 31, 1555 (2006).

[104] Zhou Bin-bin, Zou Yu-wan, Li De-hua, Wei Zhi-yi, Zheng Li-he, Su Liang-bi, Xu Jun, "The experimental study of the continuous-wave mode-locked picosecond Yb:LSO laser," Chinese J. Lasers (in Chinese) 36, 1806 (2009).

[105] I. A. Kamenskikh, N. Guerassimova, C. Dujardin, N. Garnier, G. Ledoux, C. Pedrini, M. Kirm, A. Petrosyan, D. Spassky, "Charge transfer fluorescence and f–f luminescence in ytterbium compounds," Opt. Mater. 24, 267–274 (2003).

[106] S. Heer, M. Wermuth, K. Krämer, and H. U. Güdel, "Sharp ^2E upconversion luminescence of Cr^{3+} in $Y_3Ga_5O_{12}$ codoped with Cr^{3+} and Yb^{3+}," Phys. Rev. B 65, 125112 (2002).

[107] H. Yu, K. Wu, B. Yao, H. Zhang, Z. Wang, J. Wang, Y. Zhang, Z. Wei, Z. Zhang, X. Zhang, and M. Jiang, "Growth and characteristics of Yb doped $Y_3Ga_5O_{12}$ laser crystal," IEEE J. Quantum Electron. 46, 1689-1695 (2010).

[108] Y. Zhang, Z. Wei, B. Zhou, C. Xu, Y. Zou, D. Li, Z. Zhang, H. Zhang, J. Wang, H, Yu, K. Wu, B. Yao and J. Wang, "Diode-pumped passively mode-locked Yb:$Y_3Ga_5O_{12}$ laser," Opt. Lett. 34, 3316-3318 (2009).

[109] Y. D. Zhang, Z. Y. Wei, Z. G. Zhang, D. N. Qian, L. Lv, X. D. Zeng, H. J. Zhang, H. H. Yu, J. Y. Wang, "Diode Pumped Efficient Continuous Wave and Picoseconds Yb:YGG Laser," (In Chinese) Chinese Journal of Lases 38, 0202005 (2011).

[110] S. L. Gilbert, W. C. Swann and T. Dennis, "Wavelength standards for optical communications," Proc. SPIE 4269, 184 (2001).

[111] P. R. Herz, Y. Chen, A. D. Aguirre and et al. "Ultrahigh resolution optical biopsy with endoscopic optical coherence tomography," Opt. Express 12, 3532 (2004).

[112] B. E. Bouma, G. J. Tearney, I. P. Bilinsky and et al. "Self-phase-modulated Kerr-lens mode-locked Cr: forsterite laser source for optical coherence tomography," Opt. Lett. 21, 1839 (1996).

[113] H. M. Crespo, J. R. Birge, E. L. Falcão-Filho and et al. "Nonintrusive phase stablization of sub-two-cycle pulses from a prismless octave-spanning Ti:sapphire laser," Opt. Lett. 33, 833 (2008).

[114] C. Chudoba, J. G. Fujimoto, E. P. Ippen and et al. "All-solid-state Cr:forsterite laser generating 14-fs pulses at 1.3 um," Opt. Lett. 26, 292 (2001).

[115] Zhou Bin-Bin, Zhang Yong-Dong, Zhong Xin, Wei Zhi-Yi, "Highly Efficient Self-Starting Femtosecond Cr:Forsterite Laser," Chin. Phys. Lett. 25, 3679 (2008).

[116] Zhong Xin, Zhou Bin-Bin, Zhan Min-Jie, Wei Zhi-Yi, "Generation of Red Light Femtosecond Pulses from an Intra-Cavity Frequency-Doubled Cr^{4+}:Forsterite Laser," Chin. Phys. Lett. 27, 044204 (2010).

[117] I. Thomann, L. Hollberg, S. A. Diddams and R. Equall, "Chromium-doped forsterite: dispersion measurement with white-light interferometry," Appl. Phys. 42, 1661 (2003).

[118] N. B. Angert, N. I. Borodin, V. M. Garmash and et al. "Lasing due to impurity color centers in yttrium aluminum garnet crystals at wavelength in the range 1.35-1.45 um," Sov. J. Quantum Electronics 18, 73 (1988).

[119] P. M. W. French, N. H. Rizvi, J. R. Taylor and A. V. Shestakov, "Continuous-wave mode-locked Cr^{4+}:YAG laser," Opt. Lett. 18, 39 (1993).

[120] A. Sennaroglu, C. R. Pollock and H. Nathel, "Continuous-wave self-mode-locked operation of femtoseond Cr^{4+}:YAG laser," Opt. Lett. 19, 390 (1994).

[121] Y. Ishida and K. Naganuma, "Characteristics of femtosecond pulses near 1.5 um in a self-mode-locked Cr^{4+}:YAG laser," Opt. Lett. 19, 2003 (1994).

[122] Y. P. Tong, J. M. Sutherland, P. M. W. French and et al. "Self-starting Kerr-lens mode-locked femtosecond Cr^{4+}:YAG and picosecond Pr^{3+}:YLF solid-state lasers," Opt. Lett. 21, 644 (1996).

[123] Z. Zhang, T. Nakagawa, K. Torizuka and et al. "Self-starting mode-locked Cr^{4+}:YAG laser with a low loss broadband semiconductor saturable-absorber mirror," Opt. Lett. 24, 1768 (1999).

[124] S. Naumov, E. Sorokin, V. L. Kalashnikov and et al. "Self-staring five optical cycle pulses generation in Cr^{4+}:YAG laser," Appl. Phys. B 76, 1 (2003).

[125] D. J. Ripin, C. Chudoba, J. T. Gopinath and et al. "Generation of 20-fs pulses by a prismless Cr^{4+}:YAG laser," Opt. Lett. 27, 61 (2002).

[126] Zhou Bin-Bin, Zhang Wei, Zhan Min-Jie, Wei Zhi-Yi, "Self-starting mode-locked Cr^{4+}:YAG laser by Gires-Tournois interferometer mirror for dispersion compensation," Acta Physica Sinica (in Chinese), 57, 1742 (2008).

Chapter 3

Femtosecond Laser Cavity Characterization

E. Nava-Palomares, F. Acosta-Barbosa, S. Camacho-López
and M. Fernández-Guasti

Additional information is available at the end of the chapter

1. Introduction

Ultrafast pulses are used now days in physics, engineering, medicine and other research areas such as, materials processing, time resolved spectroscopy, optical coherence tomography, pulse propagation, evolution of fast chemical processes, time resolved interference, etc. It is important to understand the physical processes that govern ultra fast laser pulse generation and shaping. In the present work, we characterize and find optimum operating conditions for a commercial Ti:Sa mode locked laser oscillator that generates pulses around 60 fs duration at 80 MHz repetition rate. Typical output power is 200 mW in mode locked operation.

The sharply peaked time structure at the oscillator output is explained in terms of the superposition of monochromatic waves with integer wavelength multiples of the round trip of the cavity. If a large number of waves with the same absolute phase are added with periodically shifted frequencies, the amplitude envelope becomes a periodically peaked train of temporal pulses. In Ti:Sa oscillators self mode-locking is achieved via Kerr lens mode-locking (KLM). The cavity is tuned so that self focusing of the more intense beam profile is favored.

The ultrafast oscillator is pumped by a Nd:YVO$_4$ CW laser. This laser is intra-cavity frequency doubled, the power output at 532 nm is 2.07 W. The pump beam is steered via two 45° mirrors into the IR cavity. The beam is focused into the Ti:Sa crystal with a 50 mm focal length lens as shown in figure 1. This beam passes through an IR convex mirror (C1) before reaching the active medium. The Ti:Sa emission resonates in a folded cavity comprising 2 concave mirrors and 6 plane mirrors. The group velocity dispersion (GVD) is compensated with a two prisms setup. A variable slit is placed between the back resonator mirror and the second prism. Figure 1 shows the Quantronix Ti-light laser layout and table 1 lists its main components. The oscillator together with an amplifier and a parametric generator were placed on a very stable passively damped holographic table [1].

In section 2, we revise the mathematical description of ultra short pulses. In particular, Gaussian pulses and hyperbolic secant pulses bringing together results that are scattered here

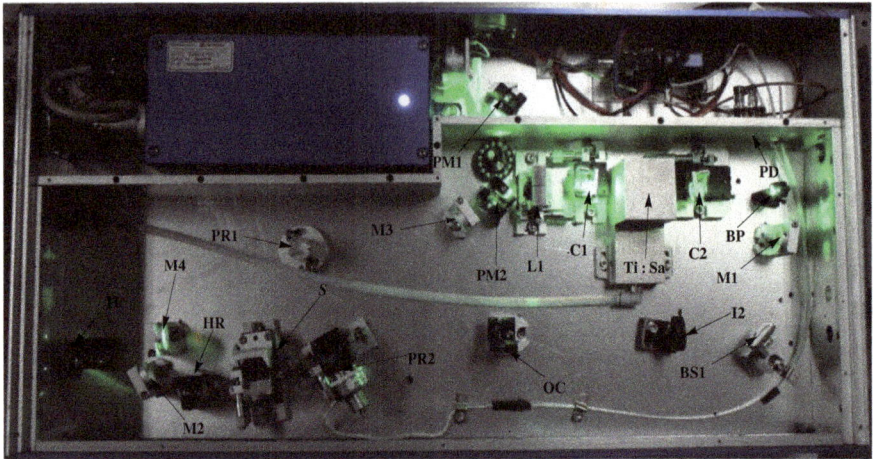

Figure 1. Ti:Sa laser at the Quantum Optics Laboratory, UAM-Iztapalapa

Components	Description
L1	pump beam focusing lens
C1, C2	first and second curved mirrors focused at crystal
M1- M4	IR Cavity folding mirrors
HR	rear highly reflective mirror
OC	output partially reflective mirror
Ti:Sa	Titanium Sapphire crystal
BS1	beam splitter
S	slit
PR1, PR2	first and second GVD compensating prisms
PM1, PM2	mirrors for 532 nm pump, reflection at 45°
I1, I2	iris
BD	beam dump
PD	photo diode

Table 1. Ti:Sa oscillator components.

and there in the literature. In subsection 2.4, the principles of mode-locked laser operation are glossed. Mode-locking is explained in terms of the superposition of monochromatic waves with integer wavelength multiples of the round trip of the cavity. The relevance of a broadband active medium is highlighted. The self mode-locking mechanism is described using the third order Kerr nonlinearity that leads to self focusing of the transverse Gaussian beam profile. The energy content of the output wave is described via a novel approach in subsection 2.5. This formulation avoids the averaging process required by either the Poynting vector or the irradiance functions proposed by Walther, Marchand, Wolf (WMW) and others.

In section 3, power output was measured as a function of mirror C2 position without mode locking. The laser output spectrum was also recorded for different mirror C2 positions within the mode locked regime (ML) with and without CW breakthrough.

In section 4, the pulse repetition rate was carefully measured in order to evaluate the cavity length. The internal photo diode as well as an external fast avalache photo diode were used to measure the pulse train. The difference between the mode locking peak intensity and the continuous CW background light was estimated. These measurements were also performed with much higher resolution using a streak camera.

In section 5, the pulse spectrum was monitored as a function of spectrum intra-cavity cropping. This was achieved through variation of the slit width (S-W) and slit translation (S-T). In section 5.1 the spectrum was evaluated as a function of GVD compensation. For this purpose, the second prism (PR2) position was varied and the spectrum was recorded.

In section 6, a second harmonic generation (SHG) frequency-resolved optical gating (FROG) setup was used to record the temporal features of the pulse: i) the pulse duration, ii) bandwidth, iii) pulse front tilt and iv) spatial chirp. Slit translation, width (S-T) and second prism (PR2) position were adjusted to achieve minimum time bandwidth product (TBP). The pulse measurements were fitted with appropriate curves and correlated to each other. The data obtained from these measurements together with their corresponding spectra from section 5 were fitted to Gaussian and hyperbolic secant time envelopes.

In the last section conclusions are drawn and the main results are abridged.

2. Theory of operation

2.1. Wave description

Ultrafast pulses are usually modeled by a carrier frequency modulated by a temporal envelope. This approach is usually adequate for pulses where the envelope duration is larger than a few oscillation periods. However, for pulses containing only a couple of periods, spatio temporal effects do not permit these type of solutions except for plane waves [2, 3]. It is customary to group the fast oscillating dependence in the phase function whereas the slowly varying features are grouped in a complex amplitude. This approach allows for slowly varying envelope approximations (SVEA) to be readily made [4]. However, a strict amplitude and phase representation requires that both of these functions are real. It is only under these circumstances that appropriate amplitude and phase functions can be used to establish an exact invariant [5].

The representation of a wave using complex variable in its polar representation is

$$\mathbf{E}\left(A, \Phi\right) = \mathbf{A} \exp\left(i\Phi\right), \tag{2.1}$$

where A is the amplitude and Φ is the phase, both of which are real valued functions. To represent a plane wave, the phase is a linear function of time and space, $\Phi = \omega t - \mathbf{k} \cdot \mathbf{r}$, where ω is the angular frequency, t represents time, $\mathbf{k}$ is the wave vector and $\mathbf{r}$ represents position. A plane wave with finite duration, is then described by a linear phase modulated by a time dependent amplitude $A\left(t\right)$

$$E\left(\mathbf{r}, t\right) = A\left(t\right) \exp\left[i\left(\omega t - \mathbf{k} \cdot \mathbf{r}\right)\right]. \tag{2.2}$$

For the sake of simplicity, we shall focus on the temporal domain and neglect polarization effects. Pulsed Nd:YAG lasers usually have a Gaussian envelope. This profile is one of the few functions that can be tackled analitically in the time and frequency domains. On the other hand, ultrafast Ti:Sa lasers usually exhibit a profile closer to a hyperbolic secant curve. However, a hyperbolic secant with linear chirp cannot be solved analitically. For this reason, the problem has been tackled with a hyperbolic tangent chirp that is approximately linear in its central region. Analytical results in terms of Gamma functions can then be obtained. These two procedures are described in subsections 2.2 and 2.3.

2.2. Gaussian envelope

A Gaussian envelope has the form $A(t) = \exp\left(-\Gamma t^2\right)$, where $\Gamma = 1/\tau_G^2$. So the field (2.2) temporal function takes the form

$$E(t) = A_0 \exp\left(-\frac{t^2}{\tau_G^2}\right) \exp\left(i\omega_\ell t\right), \tag{2.3}$$

where ω_ℓ is a constant frequency. Figure 2 shows a plot of a Gaussian pulse. The intensity is defined as $I = |E|^2 = EE^*$, so the intensity of a Gaussian pulse without chirp in the temporal domain is given by the function (2.3) times its complex conjugate

$$I(t) = \frac{1}{2}\varepsilon_0 c \, nE(t) E^*(t) = A_0^2 \exp\left[-\frac{2t^2}{\tau_G^2}\right]. \tag{2.4}$$

Figure 4 shows the intensity in the temporal domain. The full width at half-maximum (FWHM) is obtained by equating at the time when the intensity falls to one half $I = 1/2$,

$$\exp\left[-2\left(\frac{t}{\tau_G}\right)^2\right] = \frac{1}{2},$$

The time half width at half maximum is then

$$\Delta t_{HWHM} = \sqrt{\frac{1}{2}\log(2)}\,\tau_G.$$

In order to get the full width at the half-maximum multiplication by two is required

$$\Delta t_{FWHM} = \sqrt{2\log(2)}\,\tau_G = 1.17741\tau_G. \tag{2.5}$$

To analyze the electric field in the spectral domain, the Fourier transform[1] is evaluated,

$$E(\Omega) = \frac{1}{\sqrt{2\pi}}\int_{-\infty}^{\infty} A_0 \exp\left(-\frac{t^2}{\tau_G^2} + i\omega_\ell t - i\Omega t\right) dt = \frac{A_0 \tau_G}{\sqrt{2}}\exp\left(-\frac{\tau_G^2(\Omega - \omega_\ell)^2}{4}\right). \tag{2.6}$$

---

[1] $E(\Omega) = \frac{1}{\sqrt{2\pi}}\int_{-\infty}^{\infty} \mathcal{E}(t)\exp\left[-i\Omega t\right] dt$

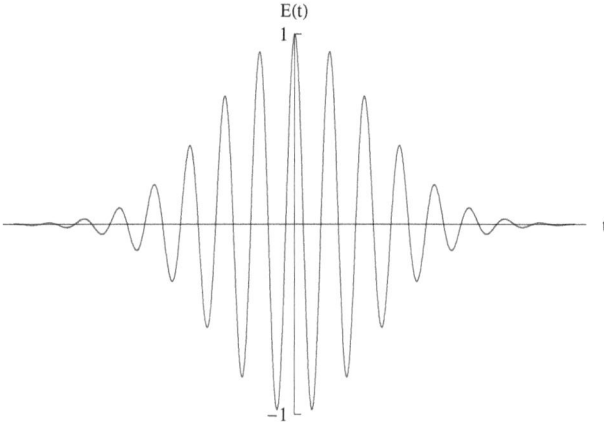

Figure 2. Gaussian pulse.

The intensity in the frequency domain for a Gaussian pulse without chirp is

$$I(\Omega) = \frac{1}{2}\varepsilon_0 c\, n \frac{A_0^2 \tau_G^2}{2} \exp\left[-\frac{\tau_G^2 (\Omega - \omega_\ell)^2}{2}\right]. \tag{2.7}$$

The normalized intensity is obtained upon division by $|E(0)|^2$. The FWHM in the frequency domain is obtained in an analogous fashion as before. The bandwidth measured from the carrier frequency $\triangle\omega_{HWHM} = \triangle\Omega_{HWHM} - \omega_l = \triangle\Omega_{HWHM}$

$$\exp\left[\frac{-\tau_G^2 (\triangle\omega_{HWHM})^2}{2}\right] = \frac{1}{2}.$$

The full width at half maximum is twice the solution to this equation

$$\triangle\omega_{FWHM} = \frac{2\sqrt{2\log(2)}}{\tau_G} = \frac{2.3548}{\tau_G}. \tag{2.8}$$

2.2.1. Gaussian with chirp

Let us now consider a Gaussian pulse with a quadratic phase

$$\Phi(t) = \omega_\ell t - \frac{a}{\tau_G^2} t^2. \tag{2.9}$$

Its instantaneous frequency is given by

$$\omega(t) = \frac{\partial \Phi}{\partial t} = \omega_\ell - \frac{1}{2}\frac{a}{\tau_G^2} t. \tag{2.10}$$

This means that the frequency is time dependent and the pulse is say to be chirped, a is the chirp factor. Figure 3 shows the plot of a chirped Gaussian pulse. With the present sign conventions i.e. eq. (2.10), the chirp factor is positive for down-chirp (frequency decreasing in time) and negative for up-chirp (frequency increasing in time). The electric field for a chirped pulse in the time domain is

$$E(t) = A_0 \exp\left[-(1+ia)\frac{t^2}{\tau_G^2} + i\omega_\ell t\right].\qquad(2.11)$$

The intensity in the time domain of the Gaussian chirped pulse is identical to the Gaussian pulse without chirp (2.4). That is, the time dependent intensity does not depend of the chirp constant.

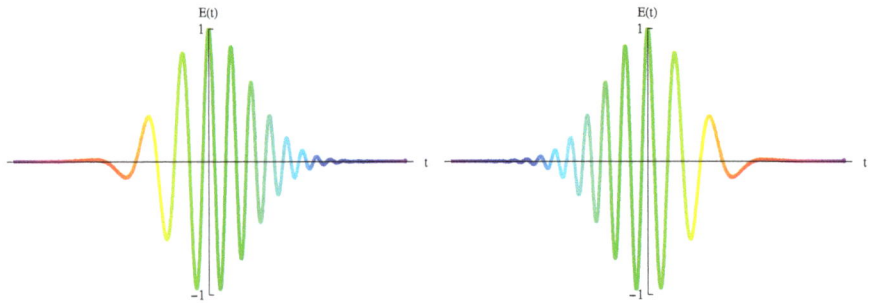

(a) Up-chirp: Negative chirp parameter $a = -3$ (b) Down-chirp: Positive chirp parameter $a = 3$

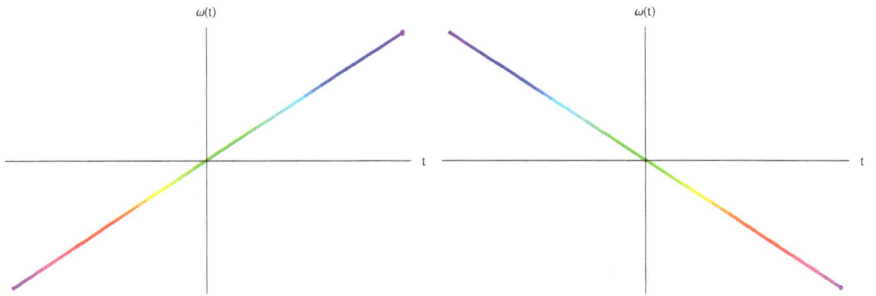

(c) For up-chirp, $a < 0$, the instantaneous frequency has positive slope.   (d) For down-chirp, $a > 0$, the instantaneous frequency has negative slope.

Figure 3. Chirped Gaussian pulse.

The Fourier transform of the quadratic time dependence (2.11) is

$$\tilde{E}(\Omega) = \frac{1}{\sqrt{2\pi}} \int_{-\infty}^{\infty} A_0 \exp\left[-(1+ia)\frac{t^2}{\tau_G^2}\right] \exp(-i\Omega t)\, dt,$$

upon integration

$$\tilde{E}\left(\Omega\right) = \frac{A_0}{\sqrt{2}} \frac{\tau_G}{\sqrt{\left(ia+1\right)}} \exp\left(\frac{i\Omega\tau_G^2}{4\left(a-i\right)}\right).$$
(2.12)

This equation can be written as

$$\tilde{E}\left(\Omega\right) = E\left(\Omega\right) = \frac{A_0}{\sqrt{2}} \frac{\tau_G}{\sqrt[4]{\left(1+a^2\right)}} \exp\left[i\phi - \frac{\omega^2\tau_G^2}{4\left(a^2+1\right)}\right],$$
(2.13)

where $\phi = \left(\frac{1}{2}\arctan\left(a\right) + \frac{a\Omega^2\tau_G^2}{4\left(a^2+1\right)}\right)$. On the other hand, the intensity in the frequency domain is $I\left(\Omega\right) = \varepsilon_0 c\, n\tilde{E}\left(\Omega\right)\tilde{E}^*\left(\Omega\right)$

$$I\left(\Omega\right) = \frac{1}{2}\varepsilon_0 c\, n\varepsilon_0 c\, n\frac{A_0^2}{2} \frac{\tau_G^2}{\sqrt[2]{\left(1+a^2\right)}} \exp\left[i\phi - \frac{\Omega^2\tau_G^2}{4\left(1+a^2\right)}\right] \exp\left[-i\phi - \frac{\Omega^2\tau_G^2}{4\left(1+a^2\right)}\right],$$

$$I\left(\Omega\right) = \varepsilon_0 c\, n\frac{A_0^2}{4} \frac{\tau_G^2}{\sqrt[2]{\left(1+a^2\right)}} \exp\left[-\frac{\Omega^2\tau_G^2}{2\left(1+a^2\right)}\right],$$
(2.14)

and the normalized frequency dependent intensity is

$$I\left(\Omega\right) = \exp\left[-\frac{\Omega^2\tau_G^2}{2\left(1+a^2\right)}\right].$$
(2.15)

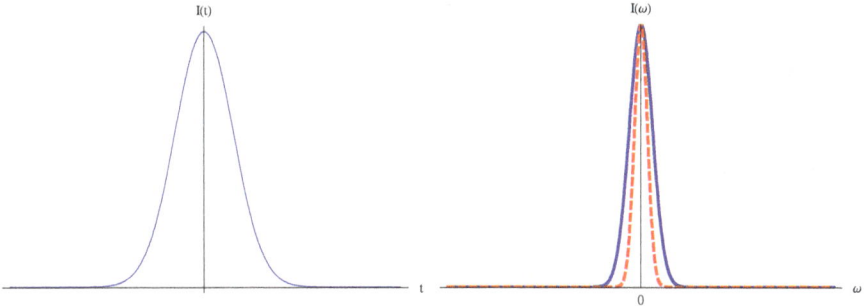

(a) Intensity of a Gaussian pulse with and without chirp in the temporal domain (curves are overlapped). (b) Intensity of a Gaussian pulse with (solid) and without chirp (dashed) in the frequency domain.

Figure 4. Gaussian chirped pulse.

2.2.2. Time bandwidth product

The time bandwidth product of a Gaussian pulse without chirp from (2.5) and (2.8) is

$$\left(\triangle t\triangle\omega\right)_{FWHM} = \sqrt{2\log\left(2\right)}\tau_G\frac{2\sqrt{2\log\left(2\right)}}{\tau_G} \approx 2.2226,$$
(2.16)

or in terms of the frequency

$$(\triangle t \triangle \nu)_{FWHM} = \frac{2 \log (2)}{\pi} \approx 0.4413 . \tag{2.17}$$

For a Gaussian chirped pulse, the FWHM in the temporal domain is the same as the Gaussian pulse without chirp. However, the FWHM in the frequency domain for the chirped pulse is obtained from (2.15)

$$\exp \left[-\frac{\triangle \omega_{HWHM}^2 \tau_G^2}{2 (1 + a^2)} \right] = \frac{1}{2} .$$

The frequency FWHM in the frequency domain for a Gaussian chirped pulse is then

$$\triangle \omega_{FWHM} = \frac{\sqrt{8 (1 + a^2) \log (2)}}{\tau_G} . \tag{2.18}$$

The time bandwidth product from (2.5) and (2.18) is

$$(\triangle t \triangle \omega)_{FWHM} = 4 \log (2) \sqrt{1 + a^2}, \tag{2.19}$$

or in terms of the linear frequency

$$(\triangle t \triangle \nu)_{FWHM} = \frac{2 \log (2)}{\pi} \sqrt{1 + a^2}. \tag{2.20}$$

To abridge, a Gaussian pulse with linear frequency chirp produces a Gaussian spectral distribution broadened by $\sqrt{1 + a^2}$. The time bandwidth product increases by this same amount since the temporal width is unaltered.

2.3. Hyperbolic secant envelope

2.3.1. Phase function without chirp

The electric field with carrier frequency $\frac{d\Phi(t)}{dt} = \omega_\ell$ modulated by a hyperbolic secant envelope $A(t)$ is

$$E(t) = A(t) e^{i\omega_\ell t} = \text{sech} \left(\frac{t}{\tau_s} \right) e^{i\omega_\ell t}.$$

The time dependent intensity is

$$I(t) = \varepsilon_0 c n |A(t)|^2 = \varepsilon_0 c n \, \text{sech}^2 \left(\frac{t}{\tau_s} \right).$$

The time $\triangle t_{HWHM}$ where the intensity is at half maximum is

$$\text{sech}^2 \left(\frac{\triangle t_{HWHM}}{\tau_s} \right) = \frac{1}{2} \quad \Rightarrow \quad \triangle t_{HWHM} = \text{arcsech} \sqrt{\frac{1}{2}} \tau_s = 0.881374 \tau_s.$$

The intensity full width at half maximum is twice this value

$$\Delta t_{FWHM} = 2\text{arcsech}\sqrt{\frac{1}{2}}\tau_s \approx 1.76275\tau_s.$$

The field spectrum is obtained from the Fourier transform of the field

$$E(\Omega) = \mathcal{F}\{E(t)\} = \sqrt{\frac{\pi}{2}}\tau_s\text{sech}\left(\frac{\pi\tau_s(\omega_\ell + \Omega)}{2}\right),$$

the frequency dependent intensity is then

$$I(\Omega) = |E(\Omega)|^2 = \varepsilon_0 c\,n\frac{\pi}{4}\tau_s^2\text{sech}^2\left(\frac{\pi\tau_s(\omega_\ell + \Omega)}{2}\right). \qquad (2.21)$$

The maximum intensity is reached when $\Omega = \omega_\ell$, then $I_M = \varepsilon_0 c\,n\frac{\pi}{4}\tau_s^2$. The frequency $\Delta\Omega_{HM} = \omega_\ell + \Omega_{HM}$ where the intensity is at half maximum is then

$$\frac{\pi}{2}\tau_s^2\text{sech}^2\left(\frac{\pi\tau_s(\Delta\Omega_{HM})}{2}\right) = \frac{\pi}{4}\tau_s^2.$$

The frequency FWHM is then

$$\Delta\Omega_{FWHM} = \frac{4}{\pi\tau_s}\text{arcsech}\left(\sqrt{\frac{1}{2}}\right).$$

The Time-bandwidth product is thus

$$\Delta\Omega_{FWHM}\Delta t_{FWHM} = \left(\frac{4}{\pi\tau_s}\text{arcsech}\left(\sqrt{\frac{1}{2}}\right)\right)\left(2\text{arcsech}\sqrt{\frac{1}{2}}\tau_s\right) \approx 1.979.$$

In terms of the linear frequency

$$TBP = \Delta\nu_{FWHM}\Delta t_{FWHM} = \left(\frac{4}{2\pi^2\tau_s}\text{arcsech}\left(\sqrt{\frac{1}{2}}\right)\right)\left(2\text{arcsech}\sqrt{\frac{1}{2}}\tau_s\right) \approx 0.3148. \qquad (2.22)$$

2.3.2. Phase function with chirp

For a pulse with linear chirp $\frac{d\Phi}{dt} = -\frac{2at}{\tau_s^2}$, the phase is quadratic in time $\Phi(t) = \omega_\ell - \frac{at^2}{\tau_s^2}$. The representation of the chirped pulse with hyperbolic secant envelope is then

$$E(t) = A(t)e^{i\Phi} = \text{sech}\left(\frac{t}{\tau_s}\right)\exp\left[i\left(\omega_\ell t + \frac{at^2}{\tau_s^2}\right)\right].$$

The Fourier transform $\mathcal{F}$ of the function with linear chirp cannot be solved analytically, so a different approach is presented. The frequency rate of change is a measure of the frequency deviation as a function of time $\frac{d\Phi(t)}{dt}$. The spectral width of solid state laser systems can be described via laser rate equations [6]. The refractive index variation arising from the rate equation can describe frequency chirping [7]. This model can be used to describe a nonlinear chirp for hyperbolic secant pulses that is mathematically tractable. The phase equation in terms of laser intensity is [8]

$$\frac{d\Phi(t)}{dt} = \frac{a}{2}\frac{1}{I}\frac{dI}{dt},$$

where a is the phase amplitude coupling factor that will correspond to the chirp parameter as we shall presently see. For $I = \varepsilon_0 c\, n\, \mathrm{sech}^2\left(\frac{t}{\tau_s}\right)$,

$$\frac{d\Phi(t)}{dt} = \frac{a}{2}\left(-\frac{2}{\tau_s}\tanh\left(\frac{t}{\tau_s}\right)\right) = -\frac{a}{\tau_s}\tanh\left(\frac{t}{\tau_s}\right).$$

The Gaussian profile with linear chirp and the hyperbolic secant profile with hyperbolic tangent chirp are similar near the central frequency for time $t \ll \tau_s$, $\frac{d\Phi(t)}{dt} = \left(-\frac{a}{\tau_s^2}t\right)$ as can be seen in figure 5.

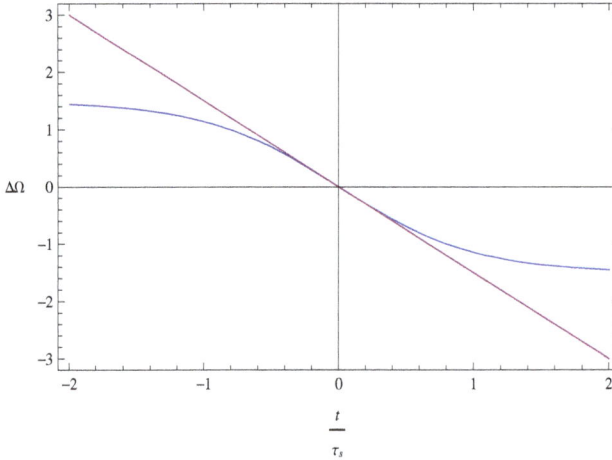

Figure 5. Instantaneous frequency deviation for a hyperbolic tangent chirp and a linear chirp.

The field phase can be obtained by direct integration

$$\frac{d\Phi(t)}{dt} = \frac{a}{2}\frac{1}{I}\frac{dI}{dt} = \frac{a}{2}\frac{d\ln I}{dt} = \frac{d}{dt}\ln I^{\frac{a}{2}}$$

so that $\Phi(t) = \ln I^{\frac{a}{2}}(t)$. Since $I(t) = |E(t)|^2 = \varepsilon_0 c\, n\, \mathrm{sech}^2\left(\frac{t}{\tau_s}\right)$, the phase is then

$$\Phi(t) = \ln\left[\mathrm{sech}^a\left(\frac{t}{\tau_s}\right)\right].$$

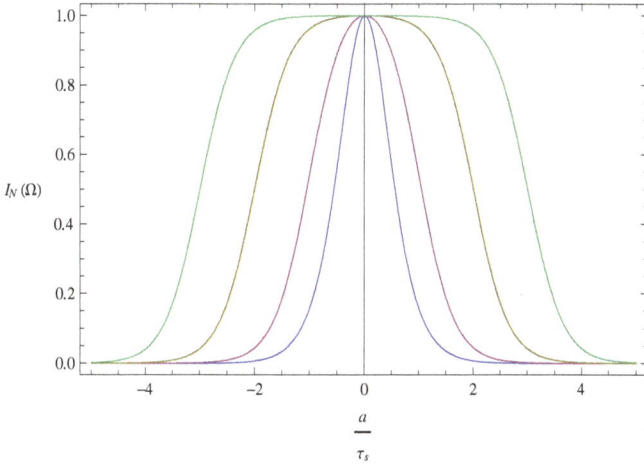

Figure 6. Different profiles as a function of chirp parameter a: 0, 1, 2, 3 with increasing width and flattened top.

The complete electric field is therefore

$$E(t) = A(t)e^{i\Phi(t)} = \text{sech}\left(\frac{t}{\tau_s}\right)\exp\left[i\left\{\ln\left[\text{sech}^a\left(\frac{t}{\tau_s}\right)\right]\right\}\right],$$

that can be economically written as

$$E(t) = \text{sech}^{1+ia}\left(\frac{t}{\tau_s}\right). \tag{2.23}$$

The electric field for a non-linearly chirped pulse is thus described by a hyperbolic secant function raised to a complex power. The imaginary part corresponds to the chirp parameter. The Fourier transform of this function has an analytical solution in terms of Gamma functions with complex argument:

$$E(\Omega) = \mathcal{F}\{E(t)\} = \frac{2^{ia}\tau_s}{\Gamma(1+ia)}\Gamma\left(\frac{1+ia+i\Omega\tau_s}{2}\right)\Gamma\left(\frac{1+ia-i\Omega\tau_s}{2}\right).$$

The normalized power spectrum can then be written as

$$\frac{|E(\Omega)|^2}{|E(0)|^2} = \frac{\text{sech}\left[\frac{\pi}{2}(\Omega\tau_s+a)\right]\text{sech}\left[\frac{\pi}{2}(\Omega\tau_s-a)\right]}{\text{sech}^2\left(\frac{\pi a}{2}\right)}. \tag{2.24}$$

This function is plotted in figure 6 for various chirp parameters. The envelope flattens on the top as the chirp parameter is increased. Notice that when $a = 0$ (unchirped pulse) we obtain simply the previous $\text{sech}^2\left[\frac{\pi}{2}\Omega\tau_s\right]$ power spectrum.

2.4. Mode-locking mechanism

Phase locked laser operation requires a coherent superposition of waves with a well defined phase relationship. The waves have frequencies separated by the cavity round trip time inverse. Each mode has the form

$$E_j = A e^{i(\omega_0 t + j\delta\omega t)},$$

where $\delta\omega$ is the frequency separation between modes and A is the amplitude of each mode. The modes amplitudes may be considered constant and equal in a first approximation within the gain bandwidth. However, they will obviously fall down at the edges of the medium's bandwidth.

2.4.1. Coherent sum of fields

The mode-locked cavity can be explained in terms of the superposition of monochromatic waves with integer wavelength multiples of the round trip of the cavity. The sum of m_{tot} waves traveling in the same direction (labeled with subindex $\Rightarrow$) with constant amplitude A in complex notation is

$$G_{\Rightarrow} e^{i\gamma_{\Rightarrow}} = \sum_{j=m_i}^{m_f} A e^{i(\omega t + j\delta\omega t)}, \tag{2.25}$$

where $\delta\omega = 2\pi\delta\nu$, m_i is the initial mode and m_f is the final mode. Waves do not interact directly between them. However, they are actually added in the presence of charges that respond to the superposition of the fields [9]. In this case, it is within the Ti:Sa crystal where the summation is carried out

$$G_{\Rightarrow} e^{i\gamma_{\Rightarrow}} = A \frac{\sin\left(\frac{(m_f - m_i + 1)\delta\omega t}{2}\right)}{\sin\left(\frac{\delta\omega t}{2}\right)} \exp\left[i\left(\omega t + \frac{(m_f + m_i)}{2}\delta\omega t\right)\right]. \tag{2.26}$$

The amplitude is now time dependent with a periodicity $\frac{\delta\omega t}{2} = 2\pi b$ where b is an integer. The amplitude maxima are obtained from L'hopital rule, the magnitude is

$$G_{\Rightarrow max} = A \left. \frac{\cos\left(\frac{m_{tot}\delta\omega t}{2}\right) \frac{m_{tot}\delta\omega}{2}}{\cos\left(\frac{\delta\omega t}{2}\right) \frac{\delta\omega}{2}} \right|_{\frac{\delta\omega t}{2} = 2\pi} = A m_{tot}, \tag{2.27}$$

where $m_{tot} = m_f - m_i + 1$ is the total number of modes. Therefore, if a large number of waves with the same absolute phase are added with periodically shifted frequencies, the amplitude envelope becomes a periodically peaked train of temporal pulses. For an estimate of the width, consider the first zero of the amplitude function in (2.26), that is $\frac{m_{tot}\delta\omega\Delta t_0}{2} = \pi$. The full width between the first positive and negative zero of the function is

$$\Delta t_{FW0} = \frac{1}{\delta\nu}\frac{2}{m_{tot}} \tag{2.28}$$

Series expansion of the amplitude function is

$$\frac{\sin\left(\frac{m_{tot}\delta\omega\,t}{2}\right)}{\sin\left(\frac{\delta\omega\,t}{2}\right)} = m_{tot} - \frac{\delta\omega^2}{24}\left(m_{tot}^3 - m_{tot}\right)t^2 + \cdots$$

FWHM estimated from this series for large m_{tot} is

$$\triangle t_{FWHM} = \frac{1}{2\pi\delta\nu}\frac{4\sqrt{3}}{m_{tot}} \tag{2.29}$$

Three mechanisms are invoked in laser cavities in order to amplify the mode locked peaks: active feedback - pumping, intra-cavity saturable absorber and self mode-locking via self focusing. In Ti:Sa oscillators self mode-locking is the mechanism responsible for pulse generation. It is achieved via Kerr lens mode-locking (KLM). Kerr refractive index modulation is a third order nonlinear process where the refractive index is proportional to the intensity of the incoming beam $n = n_0 + n_2 I$. The cavity is tuned so that self focusing of the more intense beam profile is favored. In our setup, the second curved mirror (C2) distance from the Ti:Sa crystal is varied in order to achieve the mode-locking regime. The continuous light (CW) pump is a frequency-doubled Nd:YVO$_4$ laser at 532 nm with 2.07 W.

2.5. Energy content

In electromagnetic phenomena, Poynting's theorem is considered to represent the field energy conservation equation. However, the conservation equation obtained either from Maxwell's equations or the wave equation is not unique. The choice of Poynting's conservation equation as representing the electromagnetic energy is to some extent arbitrary and not exempt of counterintuitive predictions [10, p.27-6]. The Wigner function wave description [11, p.287] involves the function times its conjugate; Therefore, it also relies on a complex version of Poynting's energy theorem.

Another conservation equation derived from the wave equation is obtained by invoking the two linearly independent solutions [12]. These two wave solutions are referred to as the complementary fields. The complementary fields approach to wave phenomena requires the existence of two fields that establish a dynamical balance between two forms of energy. The energy content of the wave goes from one field to the other and back such that the total energy is constant for an infinite wave-train [13]. The formalism attempts to address two issues: i) a representation that avoids taking time averages and hence is suitable to describe ultrafast phenomena and ii) a description that gives a clearer picture of the apparent redistribution of energy when waves are superimposed.

The complementary fields energy content of an electromagnetic wave with electric field amplitude $A(\mathbf{r}, t)$ and phase $\Phi(\mathbf{r}, t)$ is equal to

$$\mathcal{E} = \varepsilon A^2(\mathbf{r}, t)\frac{\partial\Phi(\mathbf{r}, t)}{\partial t}. \tag{2.30}$$

The energy density proposed in an earlier communication [12, eqs.(24)-(25)] has been scaled by the permittivity factor ε to establish the equality. This quantity represents the energy exchange between the two complementary fields. Whereas the flow is equal to

$$\mathbf{S}_\perp = \frac{1}{\mu} A^2 (\mathbf{r}, t) \nabla \Phi (\mathbf{r}, t). \tag{2.31}$$

The exchange energy flow can be written in terms of the temporal derivative since $\nabla \Phi = \frac{\partial \Phi}{\partial t} \left(\frac{\hat{k}}{v} \right) = \sqrt{\mu \varepsilon} \frac{\partial \Phi}{\partial t} \hat{k}$, where $\hat{k}$ is a unit vector in the direction of propagation. The flow of the exchange energy is then

$$\mathbf{S}_\perp = \sqrt{\frac{\varepsilon}{\mu}} A^2 (\mathbf{r}, t) \frac{\partial \Phi (\mathbf{r}, t)}{\partial t} \hat{k}. \tag{2.32}$$

The impedance is often written in optics for non magnetic media as $\sqrt{\frac{\varepsilon}{\mu}} = \varepsilon \frac{c}{n} = \varepsilon_0 c\, n$. For a plane monochromatic wave (labeled with subindex j), the amplitude is constant $A_j (\mathbf{r}, t) = A_j$ and the phase is linear in the time and space variables $\Phi_j (\mathbf{r}, t) = \mathbf{k}_j \cdot \mathbf{r} - \omega_j t$, where the wave vector $\mathbf{k}_j$ and the angular frequency ω_j are constant. The assessed quantities (2.30) and (2.31) are then also constant

$$\mathcal{E} = \varepsilon A_j^2 \omega_j, \qquad \mathbf{S}_\perp = \varepsilon_0 c\, n A_j^2 \omega_j \hat{k}_j. \tag{2.33}$$

Two issues should be highlighted, the frequency dependence of the energy density and the fact that no cycle averages have been performed. The linear frequency dependence means that when summing up over the phase locked monochromatic fields there is a frequency weighing factor. On the other hand, since no cycle averages are performed the energy density gives the instantaneous energy content of the wave-field.

2.5.1. Superposition of fields

The energy content of the sum of two or more fields is not equal to the sum of their individual energies but in the case of incoherent superposition. To wit, consider that the fields coexist in a certain region of space-time but there is no medium that allows for the superposition of the fields, that is, the fields do not sum up. The energy content of the jth monochromatic wave from (2.33) is

$$\left| \mathbf{S}_{\perp j} \right| = \varepsilon_0 c\, n\, A^2 (\omega_0 + j\delta\omega),$$

since the modes are equally spaced by $\omega_j = \omega_0 + j\delta\omega$ and $A_j = A$. If the waves are not added, the total energy is equal to the sum of energies of each monochromatic wave is

$$\sum_{j=m_i}^{m_f} \left| \mathbf{S}_{\perp j} \right| = \varepsilon_0 c\, n \sum_{j=m_i}^{m_f} A^2 (\omega + j\delta\omega) = \varepsilon_0 c\, n \left(m_f - m_i + 1 \right) A^2 \left(\omega + \left(\frac{m_f + m_i}{2} \right) \delta\omega \right),$$

that can be written as

$$I_{tot} = |\mathbf{S}_{\perp tot}| = \sum_{j=m_i}^{m_f} \left| \mathbf{S}_{\perp j} \right| = \varepsilon_0 c\, n\, m_{tot} A^2 \bar{\omega} \tag{2.34}$$

where $m_i = 1$, $m_f = m_{tot}$, the initial mode is one and the final mode equals the total number of modes m_{tot} and the mean mode angular frequency is $\overline{\omega} = \omega + \frac{m_{tot}+1}{2}\delta\omega$. The medium bandwidth gain is $\triangle\omega = (m_{tot} + 1)\,\delta\omega$. Notice that the total energy flow as well as the energy flow of each mode are time independent.

In the presence of a material medium, the Ti:Sa crystal in the laser oscillator case, the modes interact via the gain medium. The medium thus responds to the superposition of the fields. In this case, it is the energy of the superimposed fields that needs to be evaluated. The energy content of the coherent sum from (2.30) and (2.26) is

$$I_{coh} = |\mathbf{S}_{\perp coh}| = \varepsilon_0 c\, n\, A^2 \left[\frac{\sin\left(\frac{m_{tot}\delta\omega\, t}{2}\right)}{\sin\left(\frac{\delta\omega\, t}{2}\right)} \right]^2 \overline{\omega}.$$

The energy of the sum of waves is now time dependent and sharply peaked at $\delta\omega\, t = 2\pi b$ for integer b. The maximum intensity is

$$I_{max} = |\mathbf{S}_{\perp coh-max}| = \varepsilon_0 c\, n\, A^2 m_{tot}^2 \overline{\omega}. \tag{2.35}$$

However, this result does not necessarily imply that there has been an energy redistribution in time. It does mean that a device or entity capable of responding to the sum of the fields, such as an electric charge, will detect a large amplitude at the peaks. Outside this region, from the comparison of (2.34) and (2.35), it will detect an intensity approximately $1/m_{tot}$ lower.

3. Medium gain and mode locking operation

3.1. Gain without mode-locking.

The second curved C2 mirror was varied without mode locking while maintaining the slit width (S-W) totally opened (7.73 mm). The IR power output was recorded with a thermopile power detector [2]. The center wavelength was measured with two different monochromators. On the one hand, a scanning Czerny Turner monochromator, 0.5 m focal length[3]. On the other hand, a miniature fiber optic Czerny Turner spectrometer, 0.075 m focal length [4] with 2048 pixels detector array. The C2 mirror distance to the crystal is smaller for larger micrometer readings (see figure 1). The integration time in this latter system was 259 ms. The position was scanned from 9.60 mm up to 10.75 mm. From 9.60 mm to 10.30 mm it was varied in 0.10 mm steps. In the interval between 10.30 mm and 10.75 mm the variation was in steps of 0.05 mm.

Figure 7 shows the overlap of two runs. The output power lies around 200 mW between 9.9 and 10.7 mm C2 mirror position. The gain falls off rapidly before or after this range. However, there is a dip at 10.50 mm. This decrease is due to lasing operation at two CW wavelengths as shown in figure 8.

---

[2] Ophir 3A with Nova 2 meter, 3% accuracy
[3] Pacific MP-1018B
[4] Avantes AvaSpec 2048

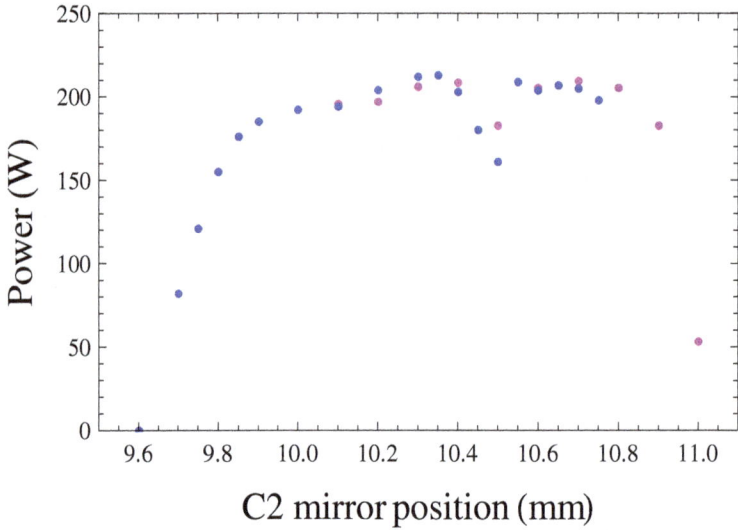

Figure 7. Mirror position vs. output power in non-mode locked operation. Points in pink were recorded directly on the computer whereas the blue ones were taken from the thermopile meter.

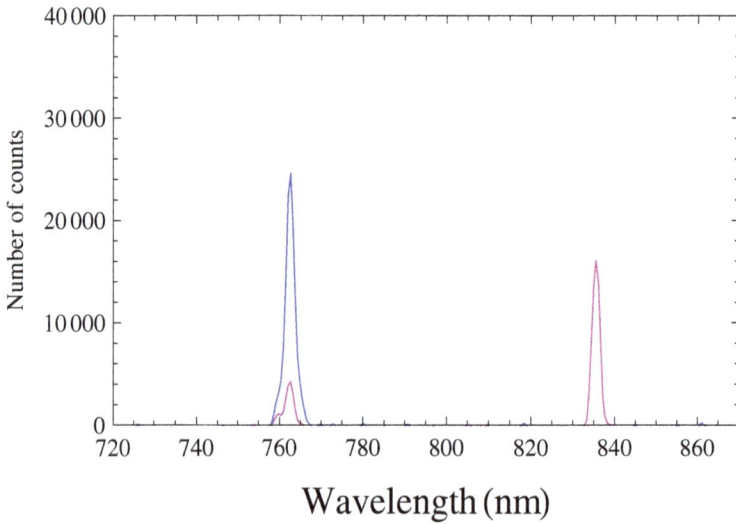

Figure 8. Continuous wave (CW) non-mode locked emission at 9.70 mm (blue) and 10.50 mm (magenta) depending on C2 position. Monochromator integration time is 259 ms.

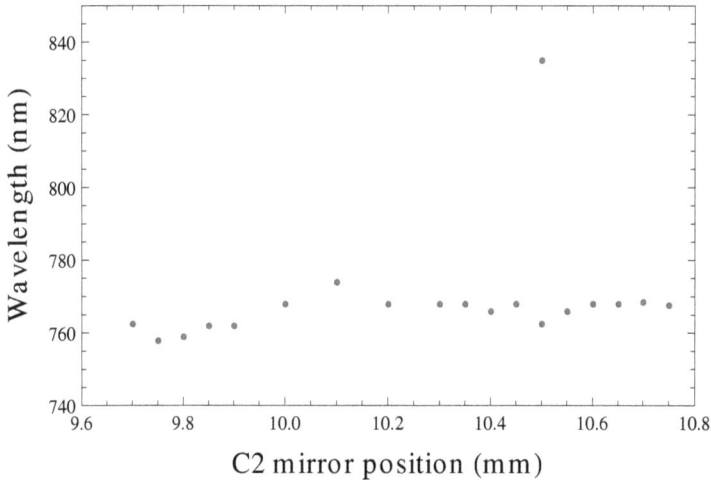

Figure 9. Emission wavelength vs C2 mirror position.

Figure 9 shows the wavelength where the maximum emission took place when C2 mirror was varied. The wavelength remains between 758 nm and 774 nm with the exception of the point at 10.50 mm, that emits at 762.6 nm and 835 nm.

3.2. C2 adjustment in mode-locking operation.

The focusing point within the active medium can be varied with the second curved mirror (C2) in order to obtain mode locking operation [14]. The position when self focussing of the intense beam attains maximum overlap with the pumped region will produce pulsed operation. This position lies on the edge of the gain curve (10.80 mm lies on the right hand border in figure 7). It is necessary to be within the gain curve so that many modes are amplified. However, if some modes experience a very large gain they will prevail over the pulsed many mode superposition. There are two ways to suppress this continuous (CW) light breakthrough: 1) lowering the pump current of the seed laser beam and 2) by moving the second curved mirror C2 [5]. To get the maximum power optimization the second option was chosen. Mode locking operation without CW breakthrough was obtained from 10.81 mm to 10.85 mm micrometer reading as shown in figure 10. Beyond the 10.85 position laser operation was not sustained. The spectrum profile shifts towards longer wavelengths as the beam is focused further into the crystal. From 10.80 mm to 10.81 mm the amplitude decreases due to less overlap between the focused beam and the gain region. Thereafter the amplitude remains constant presumably because self-focusing compensates and focuses the beam within the gain volume. On the other hand, the continuous light CW breakthough vanishes at 10.83 mm position. For C2 greater than 10.83 there is no CW breakthrough because no single mode is amplified above other modes.

---

[5] Quantronix, Ti-Light Operator's manual.

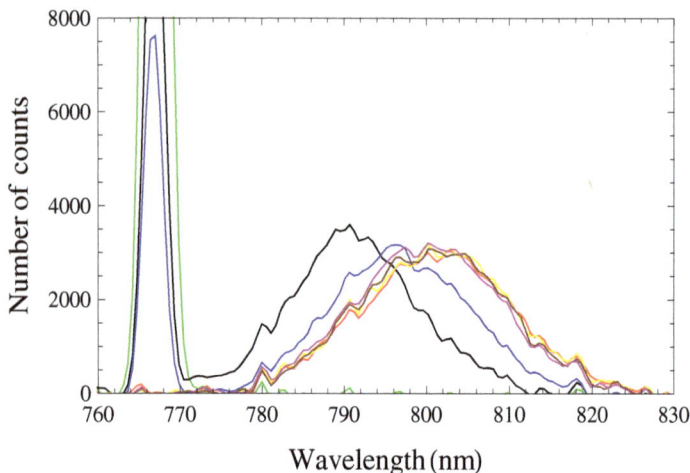

Figure 10. Spectra for different C2 mirror positions. At 10.80 mm two spectra were taken, with (black) and without (green) mode locking . All other spectra were made in mode locked operation; 10.81 mm (blue), 10.82 mm (red), 10.83 (yellow), 10.84 mm (brown) y 10.85 mm (magenta) (Curves from left to right). The continuous sharply peaked contribution vanishes when C2 mirror is located at 10.82 mm and further away. Spectrometer integration time is 259 ms.

4. Pulse train measurement and cavity length

4.1. Repetition rate

The output oscillator pulses were measured with an internal photo diode as well as a fast avalanche external photo diode. The internal photo diode is electronically amplified whereas the external photo diode was not amplified to provide the fastest possible risetime. Both signals gave similar results. The pulse repetition frequency was monitored with a fast oscilloscope[6]. The repetition frequency obtained from the scope trace shown in figure 11 is 77.5 ± 0.1 MHz with a period of $\tau = \frac{1}{\delta v} = 12.90 \pm 0.02$ ns which gives a round trip cavity of $\tau c = \frac{c}{\delta v} = 12.90 \times 10^{-9} \times 3 \times 10^8 = 3.86$ m $= 2L$, thus the cavity optical length is $L = 1.928$ m. The angular frequency mode separation is then $\delta \omega = 2\pi \delta v = 486.9 \pm 0.6$ MHz or in wavelength units @ 800 nm, $\delta \lambda = -\delta v \frac{\lambda^2}{c} = -77.5 \times 10^6 \frac{(800 \times 10^{-9})^2}{3 \times 10^8} = 1.65 \pm 0.02 \times 10^{-13}$ m. For a typical bandwidth of 35 nm (see figure 22b) the number of modes is $m_{tot} = \frac{\Delta \omega}{\delta \omega} = \frac{35}{1.65 \times 10^{-4}} = 2.21 \times 10^5$. The coherent superposition of these modes gives a pulse width, according to (2.28)

$$\triangle t_{FW0} = \frac{1}{\delta v} \frac{2}{m_{tot}} = 12.9 \times 10^{-9} \frac{2}{2.21 \times 10^5} = 116 \, \text{fs}.$$

---

[6] Tektronix 485/R485 analogue oscilloscope with 350 MHz bandwidth or Picoscope 5302 digital scope with 250 MHz bandwidth.

The measured pulse width is in this order of magnitude, although somewhat lower. Recall however, that this estimate is for the first zero rather than width at half maximum. Evaluation from FWHM series estimate (2.29) gives a closer estimate to the values reported in section (6),

$$\triangle t_{FWHM} = \frac{1}{2\pi\delta v}\frac{4\sqrt{3}}{m_{tot}} = 12.9 \times 10^{-9}\frac{1.1027}{2.21 \times 10^5} = 64.36\,\text{fs}.$$

Nonetheless, it should be stressed that pulse width is critically dependent on the pulse envelope. The spectra shown in subsequent sections were integrated over 259 ms as mentioned in section 3.1. Since the repetition rate is 77.5 MHz, the spectra show the average of 2.007×10^7 pulses.

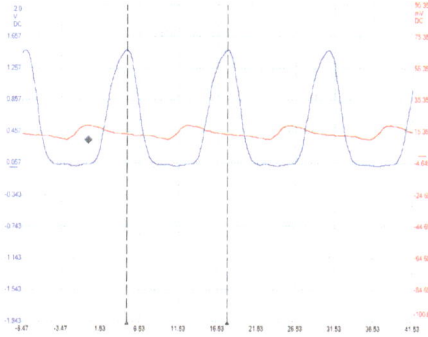

Figure 11. Oscillator output pulses detected with the internal photodiode (blue) and an external avalanche hamamatsu photodiode (red). The horizontal axis represents time in ns and the vertical axis depicts photodiode voltage.

4.2. Mode locking power and CW broadband background

An external fast photo diode was used to evaluate the difference between the mode locking (ML) peak intensity and the background light in between the high intensity peaks. This CW broad spectrum background light should not be confused with the CW narrow bandwidth breakthrough mentioned in subsection 3.2. The broad CW background is always present and is due to the superposition of the cavity modes when they are out of phase. Typical traces are shown in figure 12. The width of the peaks in the order of 1 ns are due to the detection system (photo diode + scope) since the optical pulse duration is four orders of magnitude shorter. The light level stabilizes at $I_{bb} = 10$ a.u. in between pulses. This measurement corresponds to the broad background (bb) emission between pulses in mode-locked operation. There is possibly electrical ringing after some pulses in the blue trace. In all three curves, there are spurious peaks due to analog to digital conversion uncertainty. This assertion is suported by the lack of subsidiary peaks in the streak camera recordings shown in the next section. The average intensity of the narrow bandwidth (nb) in non-mode locked operation is $I_{nb} = 6.3$ a.u. The ratio of these two intensities is $\frac{I_{bb}}{I_{nb}} = 1.59$. This quantity is an estimate of the energy content when a large number of out of phase modes (aprox. 10^5) are present in comparison with flow detected when a few modes are present.

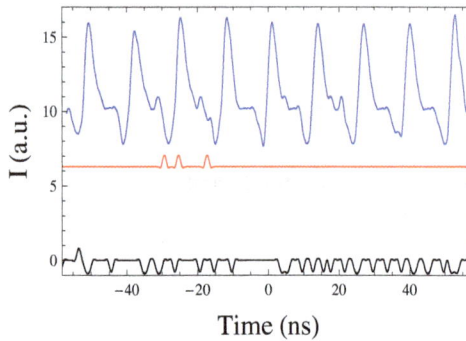

Figure 12. Oscillator pulse train detected with fast avalache photo diode and digital oscilloscope. Mode locked operation (blue trace), non-mode locked (red trace), residual trace with beam blocked in black. Small peaks are most likely due to analog to digital conversion uncertainty.

A streak camera was used to record the oscillator output. Firstly, a fraction of the laser beam output was fed through an optical fiber of 400 μ diameter and two fibers with 5 m length. This way the pulse was spread in time due to dispersion in the multimode fibre to insure that the peak intensity in the streak photocathode was not too high. The streak camera[7] was swept with a 330 ps/mm - 25 ns/mm sweep unit with 100 kHz maximum trigger frequency[8]. This unit was triggered by a pulse generator[9] that was in turn triggered by the laser oscillator electrical signal output. The unit was swept at 6.6, 10, 20, 50 and 100 ns full time bases. A typical intensity plot is shown in figure 13. The time span between pulses is $\tau = 13.12 \pm 0.5$ ns. The error is due to 2.5% non-linearity error in the time sweep. This result is consistent, within error, with the repetition rate obtained from the photo diode measurement.

5. Pulse spectrum and center wavelength

The pulse spectrum can be altered by the position of the second prism or the position and width of the slit placed between this prism and the end mirror.

5.1. Pulse spectrum as a function of prism compensation

In order to keep a narrow pulse, the two prisms are used to compensate the group velocity dispersion (GVD) introduced by the pulse propagation through the titanium sapphire crystal [15, 16]. The second prism (PR2) has a translational motion while the first prism position is fixed. The cavity design would benefit if the first prism (PR1) is also mounted on a traslation stage. The initial operating conditions are shown in table 2.

The spectrum in figure 14 shifts towards longer wavelengths as the second prism (PR2) position decreases. The wing on the right hand side of figure 14 shows that all the profiles

---

[7] Optronis SC-10
[8] Optronis TSU12-10PG sweep unit
[9] Stanford Research DGC535

(a) Streak camera image. The vertical direction shows a transverse beam spatial direction and the horizontal direction represents the time axis.

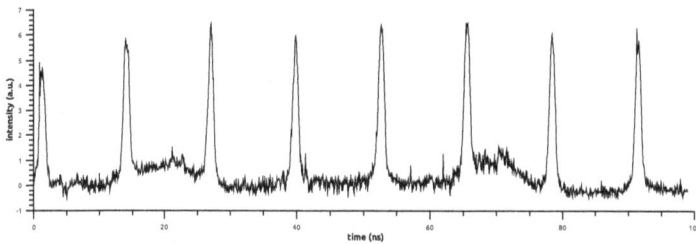

(b) Streak camera intensity time profile.

Figure 13. Streak camera image and graph of Ti:Sa oscillator pulses. The sweep rate is 5 ns/mm, 100 ns full sweep (small background bumps are due to pulses during backtrace).

converge towards 860 nm. However, the wings on the left hand side are displaced as the curve maximum is displaced towards shorter wavelengths. On the other hand, the amplitude increases at longer wavelengths and the bandwidth becomes narrower. In figure 15 a plot of the bandwidth as a function of the prism PR2 position is shown. Another set of spectra under the same operating parameters as shown in table 2 are depicted in figure 16. The same features described in the previous case are observed except for an amplitude decrease rather than increase at longer wavelengths. This discrepancy is possibly due to a slightly different cavity alignment so that shorter or longer wavelengths' gain were favoured in either case.

Element	Position (mm)
Mirror C1 position	3.98
Mirror C2 position	10.85
Lens L1 position	4.12
Prism PR2 initial position	1.30
Slit width (S-W)	7.73
Slit translation (S-T)	3.00

Table 2. Initial conditions when prism PR2 position was varied.

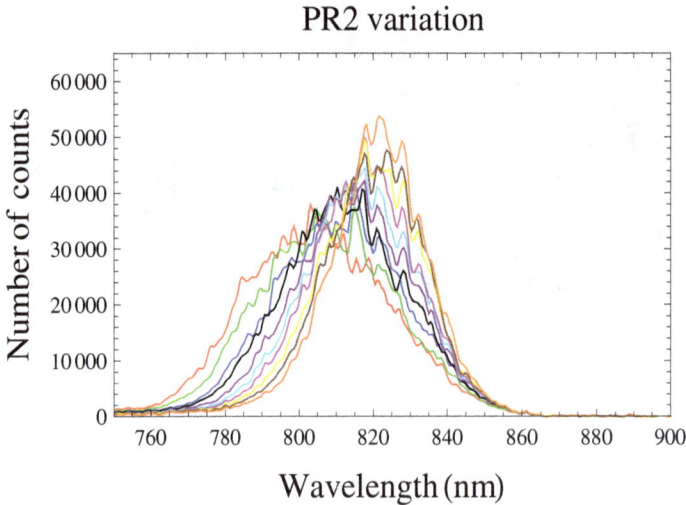

Figure 14. Spectra with mode lock taken at different prism PR2 positions: 0.40 mm(orange), 0.50 mm(brown), 0.60 mm(yellow), 0.70 mm(magenta), 0.80 mm(cyan), 0.90 mm(purple), 1.00 mm(black), 1.10 mm(blue), 1.20 mm(green) y 1.30 mm(red). The PR2 position was varyied by 0.10 mm steps. Spectrum curves shift towards longer wavelengths for larger micrometer readings.

The prism apex moves further into the beam as the micrometer reading increases (see figure 21a). As the beam travels through more prism material, the ML emission shifts towards shorter wavelengths.

5.2. Adjustment of the slit width (S-W)

The slit has two degrees of freedom, translation (S-T) and slit width (S-W) as can be seen in figure 17. Each of these displacements was varied separately.

The slit width micrometer is not calibrated to the real aperture width. The slit aperture was measured with a Vernier and a difference of 0.73 mm was obtained when the slit micrometer was set to zero. Hereafter, this value is added to slit width (S-W) micrometer readings in order to report the actual slit width. It should also be noted that one blade is fixed and only the blade

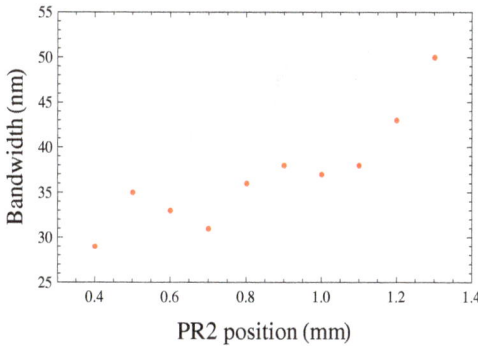

Figure 15. Bandwidth variation as a function of second prism (PR2) position. Micrometer readings between 0.40 and 1.30 mm.

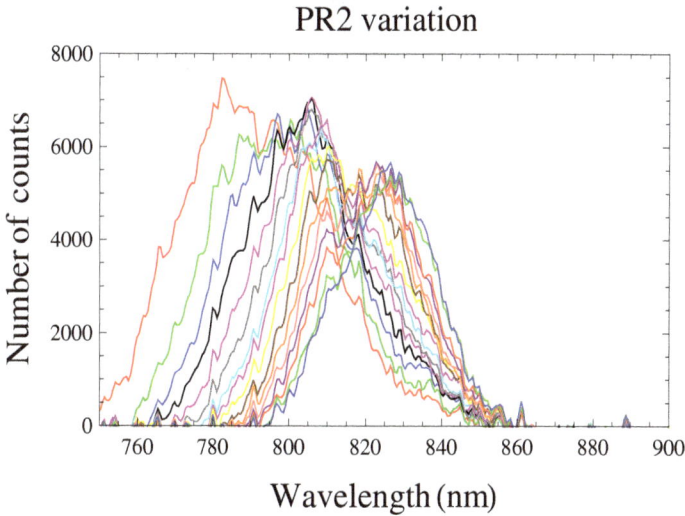

Figure 16. Spectra with mode lock for prism PR2 position between 0.00 and 1.50 mm: 0.00 mm (blue), 0.10 mm (green), 0.20 mm (red), 0.30 mm (purple), 0.40 mm (pink), 0.50 mm (orange), 0.60 mm (brown), 0.70 mm (yellow), 0.80 mm (magenta), 0.90 mm (cyan), 1.00 mm (gray), 1.10 mm (magenta), 1.20 mm (black), 1.30 mm (blue), 1.40 mm (green) and 1.50 mm (red). The position was varied in 0.10 mm steps. Micrometer readings decrease in curves from left to right.

on the blue side of the dispersion actually moves (see figures 21a and 17). The setup would benefit with a slit where both blades are displaced from a central position.

The laser parameters used while the width was changed are abridged in table 3. S-W was set at 3.73 mm and increased in steps of 0.50 mm up to 7.73 mm. Between 3.73mm to 7.73 mm the spectra did not show a noticeable variation because the laser beam is not blocked by the slit. That is, the beam width at the slit plane is smaller than the slit width. This assertion was

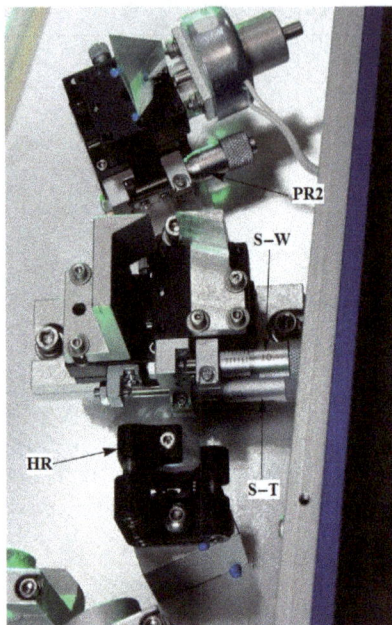

Figure 17. Slit displacement and width: S-T micrometer controls the translation mechanism whereas S-W alters the slit width.

confirmed by direct view with an IR viewer that no appreciable part of the beam was being blocked.

Element	Measurement (mm)
First curved mirror (C1)	3.98
Second curved miror (C2)	10.85
Pump beam focusing lens (L1)	4.12
Second prism position (PR2)	1.10
Initial S-W value	3.23
S-T position	5.50

Table 3. Initial conditions of slit width S-W variation .

The spectra for slit widths ranging from 3.23 to 1.43 mm are shown in figure 18. The S-W was decreased in 0.50 mm steps from 3.23 to 2.23 mm. Thereafter, it was decreased in steps of 0.10 from 1.93 to 1.43 mm. The spectra shifts towards longer wavelengths as the slit is reduced. This is expected because the moving blade crops from the blue side of the dispersion. On the other hand, the bandwidth decreases as the slit is closed although the intensity does not vary in an appreciable way. These features become relevant below the 2 mm width where the beam dispersion width is being considerably clipped. The pulse duration does not necessarily increase as the bandwidth is reduced until the pulse is time-bandwidth limited as we shall see in section 6.3.

S–W variation

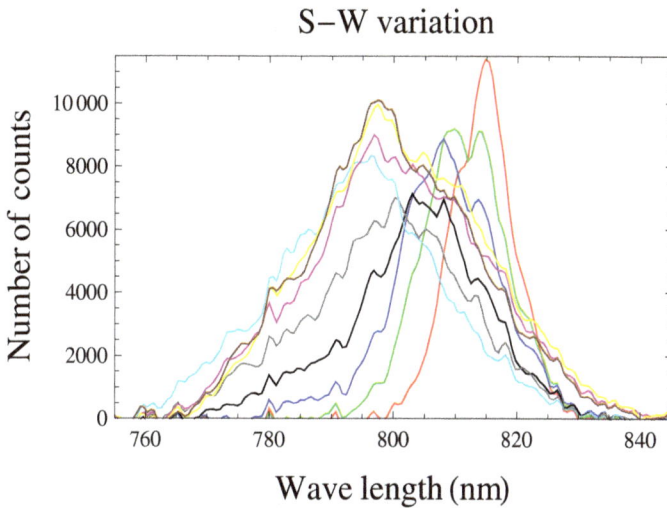

Figure 18. Spectra with mode lock for different slit widths: 1.43 mm (red), 1.53 mm (green), 1.63 mm (blue), 1.73 mm (black), 1.83 mm (gray), 1.93 mm (cyan), 2.23 mm (magenta), 2.73 (yellow) and 3.23 mm (brown). . From 2.23 mm to 3.23 mm the steps were of 0.50 mm. From 1.43 mm to 1.93 mm the S-W the steps were of 0.10 mm. Narrower slit width spectra lie on the right side of the figure.

5.3. Slit translation adjustment

The slit translation was varied from 7.60 mm to 0.40 mm while the slit width (S-W) was totally opened at 7.73 mm. The operating conditions while performing the S-T translation are shown in table 4.

Element.	Measure (mm)
First curved mirror (C1)	3.98
Second curved miror (C2)	10.85
Pump beam focusing lens (L1)	4.12
Second prism position (PR2)	0.5
S-W value	7.73
S-T initial position	7.60

Table 4. Initial conditions of S-T translation.

The spectra are shown in figure 19. ML emission is more intense at shorter wavelengths. The intensity decreases about tenfold for emission centered at 820 nm compared with emission centered at 730 nm. Nonetheless, stable operation is obtained from 730 to 850 nm, that is, in a 110 nm span. The curves suggest that emission can be obtained further to the blue but the micrometer mechanical translation table could not go any further.

In order to analyze these spectra we have separated the contributions in four different groups shown in figure 20. The slit translation (S-T) was decreased in steps of 0.10 mm from 7.60 mm to a 0.40 mm. At 7.60 mm the slit was blocking the redder frecuencies while the blue ones were

S–T variation

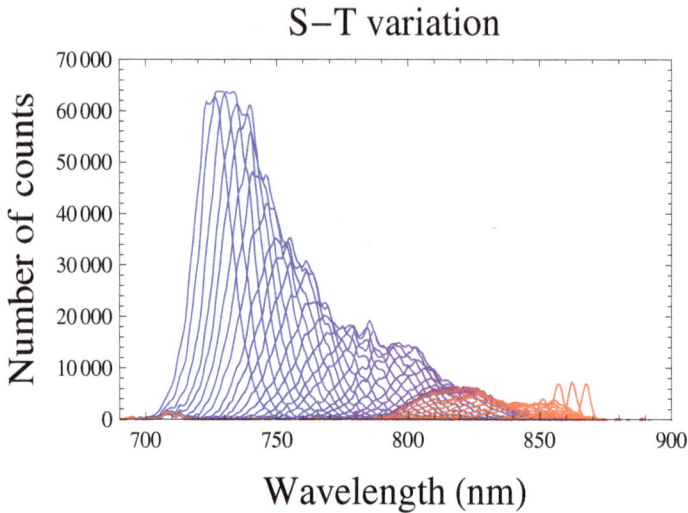

Figure 19. Mode locked spectra as a function of slit translation (S-T) from 7.60 mm to 0.40 mm. The spectrum varies from shorter to longer wavelengths and the intensity decreases for lower micrometer readings.

unblocked. A sketch of the slit together with the beam dispersed light is shown in figure 21a. Blue is dispersed at a larger angle than red in the first prism PR1 shown in photograph 21b. After a two mirror reflection it impringes on the second prism that is inverted with respect to the first one. When the slit was moved towards lower micrometer readings, the longer wavelength frequencies became progressively unveiled. As a result, the center wavelength moved towards longer wavelengths; The bandwidth increased and the spectrum intensity decreased as seen in figure 20a. At 5.20 mm the red side of the spectrum was completely unblocked.

From 5.10 mm to 3.50 mm, , the spectra do not suffer any change and the center wavelength remains constant as well as the bandwidth as shown in figure 20b. The beam is not blocked on neither side by the slit since the beam width is smaller than the slit width. This behaviour continued until the translation micrometer was at 1.60 mm.

Between 1.50 mm to 0.70 mm, the slit started to block the blue side of the spectrum while the red side remained unobstructed. The central emission thus shifts to longer wavelengths as seen in figure 20c. The bandwidth increases while the spectrum intensity continues to decrease.

Finally, from 0.80 mm to 0.40 mm the blue side of the spectrum is further blocked. However, the emission spectra in figure 20d show a different behavior. From left to right, the first two profiles exhibit a bandwidth reduction and an intensity increase on top of a much broader but lower intensity emission. The last three curves exhibit a central wavelength shift but the intensity remains constant while the broader emission is no longer present. This behaviour

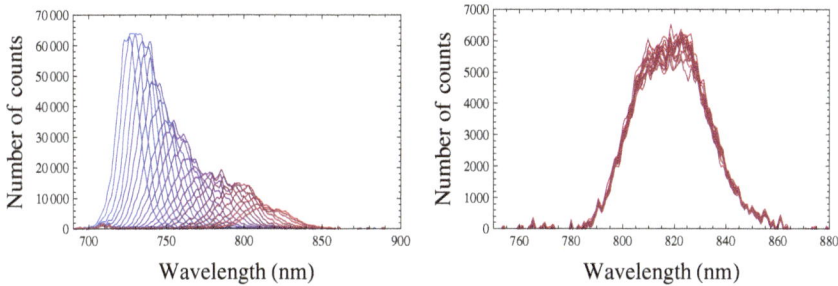

(a) S-T from left to right curves 7.60 mm to 5.20 mm in steps of 0.1 mm

(b) S-T between 5.10 mm to 3.50 mm

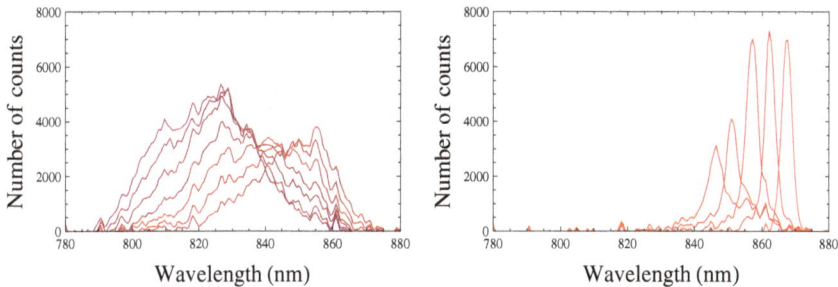

(c) From left to right S-T between 1.50 to 0.70 mm in 0.1 mm steps

(d) S-T between 0.80 mm to 0.40 mm

Figure 20. Oscillator pulse train spectra, with mode lock, as a function of slit translation (S-T). 20a) At higher micrometer readings the slit blocked most part of the beam frequencies on the red side. 20b) Does not exhibit any change in the spectrum since the beam is not blocked by the slit. 20c) For lower micrometer readings most part of the beam blue frequencies are blocked. 20d) Bandwiths are narrower when laser is hardly mode locking.

could suggest continuous light (CW) emission. However, even under these circumstances mode locking (ML) was still being held.

6. Pulse duration and time bandwidth product (TBP)

6.1. Frequency-Resolved Optical Gating (FROG)

FROG is a special kind of nonlinear autocorrelator that measures the pulse duration as well as other features of the pulse. This is possible because the autocorrelator signal beam is spectrally resolved. There are different kinds of FROG depending on the nonlinear medium: Polarization-Gate (PG), Self-Diffraction (SD), Transient-Grating (TG), Second-Harmonic-Generation (SHG) and Third-Harmonic-Generation (THG). Second-Harmonic-Generation is the most sensitive because it involves a second order nonlinearity rather than third. However, being an even order nonlinearity it is time

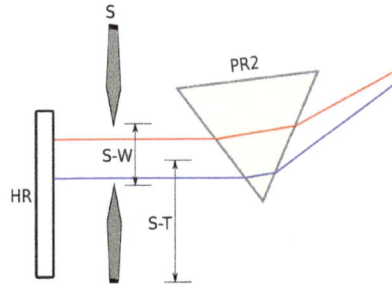

(a) Sketch of the intra-cavity mirror - slit - prism setup.

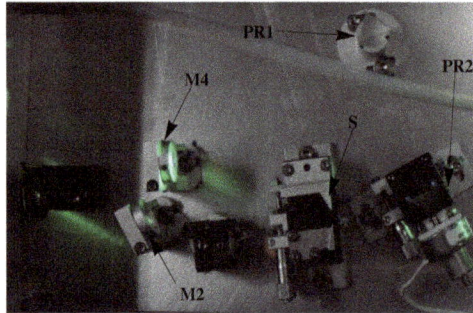

(b) Photograph of the mirror - slit and two prisms setup

Figure 21. Oscillator GVD compensating prisms and spectrum cropping slit. The solenoid that initiates mode locking operation can be seen below the second prism PR2.

symmetric and thus the time direction is ambiguous [17]. In the following experiments we used a simplifed FROG version called grenouille with a SHG crystal, 18 to 180 fs pulse duration measurement capability[10]. Grenouille raw data were processed with Femtosoft QuickFrog software. The pulse retrieval algorithm is reasonably reliable although it sometimes runs wild and it requires resetting. It gives numeric estimates of the pulse duration, TBP, central emission wavelength, pulse front tilt and spatial chirp. However, it does not state a temporal chirp value although the temporal phase curve is presented on screen.

6.2. GVD prism compensation and time bandwidth product (TBP)

The second prism (PR2) position was moved between 1.50 mm and -0.50 mm in steps of 0.10 mm. The initial parameters are summarized in table 5. From the data records, we produced plots of various pulse parameters in figure 22.

The pulse duration is plotted in figure 22a. It decreases as the prism PR2 micrometer reading is reduced. As we mentioned in section 5.1, the beam traverses less prism material as the micrometer reading decreases. The beam is incident on the apex of the first prism. In fact, a little bit of it on one side passes through without hitting the prism. It is expected that the

---

[10] GRENOUILLE 8-20 from Swamp optics

Element	Measurement(mm)
Mirror C1 position	3.98
Mirror C2 position	10.85
Lens L1 position	4.12
Prism PR2 initial position	1.50
Slit width (S-W)	7.23
Slit translation (S-T)	3.50

Table 5. Initial parameters when prism PR2 was moved.

second prism will produce a good GVD compensation when symetrically placed with respect to the first prism. The micrometer was moved until zero was reached and then moved further down as much as the mount allowed. In the innermost position the minimum time seemed to be reached. If the second prism over-compensates or under-compensates the first prism the pulse duration should increase. A minimum is expected when compensation is optimum. For this reason, we modeled the time duration curve with a quadratic fit

$$57.9953 + 12.8317(0.45 + x)^2. \tag{6.1}$$

The minimum time of the fit is located at 57.99 fs whereas the minimum experimental measurement was 62.4 fs. The bandwidth dependence is shown in figure 22b. It also decreases as the prism PR2 position is decreases. Again, a minimum is expected at optimum compensation. A quadratic fit is phenomenologically assigned

$$22.5591 + 2.79782(1.5 + x)^2. \tag{6.2}$$

While the time minimum seems to be reached at -0.5 mm, it is not clear that the bandwidth minimum has been reached yet. Indeed, the fit in equation (6.2) predicts that the minimum bandwidth is located at -1.5 mm.

The time-bandwidth product is plotted in figure 22c. This value was taken from the Frog retrieval algorithm. It should be proportional to the product of the previous two curves. A quadratic fit gives the following parameters

$$0.65104 + 0.381219(0.6 + x)^2. \tag{6.3}$$

The minimum TBP obtained from the fit is 0.65. However, the minimum TBP achieved with this setup is 0.783. This value is somewhat far from a fourier limited pulse as may be seen from comparison with table (6).

On the other hand the center wavelength increases as the prism position moves out of the beam as seen in figure 22d. A linear fit gives the equation

$$827.012 - 22.6234x. \tag{6.4}$$

This frequency dependence is similar to that observed in section (5.1).

The plot of the pulse front tilt is shown in figure 22e. When the two prism compensate each other transversely they should produce minimum pulse front tilt. However, according to the frog measurements we are far from the minimum position. For this reason, although we expect a quadratic fit with a minimum, a linear fit was sufficient for the actual data

$$- 11.8223 + 1.15039x. \tag{6.5}$$

It is not clear whether the pulse front tilt measurement is real or the Grenouille apparatus is slightly miscalibrated. A callibration experiment will be performed to elucidate this point.

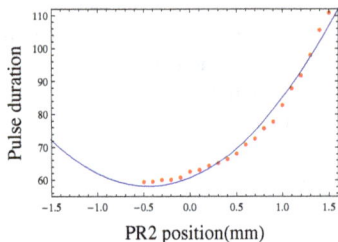

(a) Pulse duration $\triangle t$.

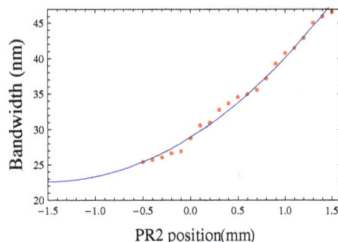

(b) Bandwidth $\triangle \lambda$.

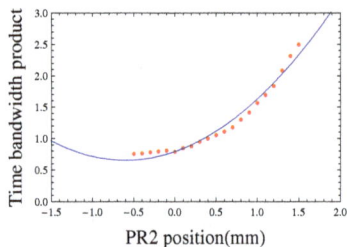

(c) Time bandwidth product (TBP).

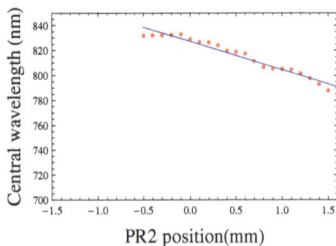

(d) Central emission wavelength λ_c.

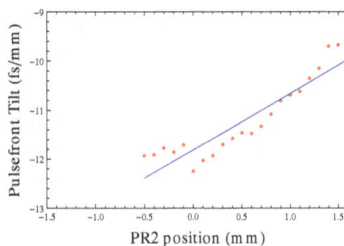

(e) Pulse front tilt (PFT).

Figure 22. $\triangle t$, $\triangle \lambda$, TBP, λ_c, and PFT versus prism PR2 position.

Since we did not achive a Fourier transform pulse, the bandwidth increases while the pulse duration increases too see figures 22a and 22b. Table 6 shows the different Fourier transform limits for different pulse shapes.

Envelope shape	$\varepsilon(t)$	Time bandwidth product
Gaussian function	$\exp\left[-(t/t_0)^2/2\right]$	0.441
Exponential function	$\exp\left[-(t/t_0)/2\right]$	0.140
Hyperbolic secant	$1/\cosh[t/t_0]$	0.315
Rectangle	-	0.892
Cardinal sine	$\sin^2[t/t_0]/(t/t_0)^2$	0.336
Lorentzian function	$\left[1+(t/t_0)^2\right]^{-1}$	0.142

Table 6. Fourier transform pulses for different shapes. Data taken from Rulliere et al. [14].

6.3. Spectrum cropping

The slit width (S-W) was varied between 1.43 mm and 7.23 mm. Table 7 shows the initial conditions for this experiment. It was varied in steps of 0.10 mm between 1.43 mm to 1.93 mm. From 1.93 mm to 2.73 mm the slit translation (S-T) was moved to center the laser beam within the aperture. Two different measurements at 2.73 mm on different days give an estimate of the measurement error in the various parameters shown in figure 23. Larger slit widths had little impact on the plotted parameters.

Element	Measurament(mm)
Mirror C1 position	3.98
Mirror C2 position	10.85
Lens L1 position	4.12
Prism PR2 position	1.50
Slit width initial width (S-W)	1.43
Slit translation (S-T)	3.50

Table 7. Initial conditions when S-W was varying

Let us now begin by describing the bandwidth behaviour that should be directly proportional to the slit width. Indeed, from 1.43 to 2.23 the experimental points lie on a straight line. At 2.23 mm the slit no longer crops the beam and thus becomes insensitive to slit width. The maximum bandwidth is around 35 nm consistent with the plateau shown in figure 15 in section 5.1.

The time bandwidth product dependence on slit width is shown in figure 23b. When the slit begins to crop the beam, the TBP decreases until it reaches the Fourier transform limit. A hyperbolic tangent fit of this curve is

$$0.816 + \frac{0.793}{2}\tanh\left[\frac{x}{0.259}-4.677\right]. \tag{6.6}$$

(a) Bandwidth $\triangle\lambda$.

(b) Time bandwidth product TBP.

(c) Pulse duration $\triangle t$.

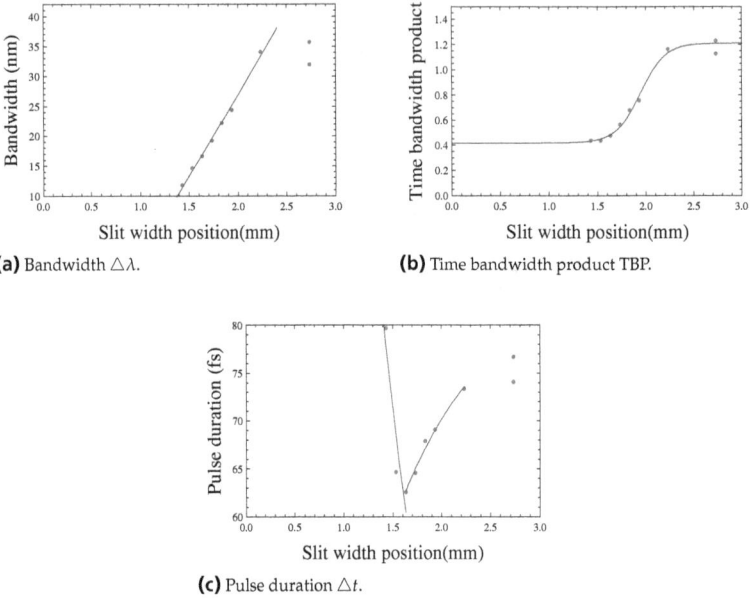

Figure 23. $\triangle\lambda$, TBP and $\triangle t$ versus slit width.

The minimum asymptotic value is $0.816 - \frac{0.793}{2} = 0.419$. This TBP is below the Gaussian envelope limit as seen from table 6, thus suggesting that the pulse envelope must be closer to the hyperbolic secant proposed for solid state lasers [8].

The time duration of the pulse versus slit width is shown in figure 23c. The experimental points show two rather different behaviors. On the one hand, the pulse duration decreases as the slit width is reduced between 2.73 and 1.63 mm. Beyond this value, the pulse duration increases as the slit width is further reduced.

This apparently wild behaviour of the pulse duration can now be nicely explained in terms of the bandwidth and the TBP. The pulse duration is reduced as frequency components far in the wings are cropped because these frequencies have a large spread in time due to dispersion. For large slit widths, the TBP follows the strict inequality product $\triangle t\triangle\nu > c_B$. Thus both quantities may decrease as long as the equality is not reached. However, when the TBP limit is reached, the pulse is Fourier transformed and it should fulfill the $\triangle t\triangle\nu = c_B$ equality. If $\triangle\nu$ (or equivalently $\triangle\lambda$) is further reduced as it happens beyond the 1.63 mm slit width, the time duration $\triangle t$ must increase so that the equality holds. The pulse duration in the Fourier transform limited region should fulfill an inverse relationship with slit width since

$$\triangle t = \frac{c_B}{\triangle\nu} \propto \frac{1}{\text{s-w}}.$$

The fit gives the following parameters

$$-0.818912 + \frac{55.2703}{x}. \tag{6.7}$$

For the non Fourier limited region the pulse width increases with increasing bandwidth. Since the bandwidth only increases up to the active medium - cavity maximum bandwidth it may also be fitted to an inverse function

$$89.8668 - \frac{24.7288}{x}. \tag{6.8}$$

The asymptotic value for this function is 89.86 fs.

A screen shot of the Fourier limited pulse (TBP=0.437) produced by the grenouille femtosoft software is shown in figure 24. The pulse duration is 64.7 fs and its bandwidth is 14.65 nm. The pulse exhibits a negligible temporal chirp as expected. Nonetheless and contrary to our expectations, the pulse front tilt is reported to be still rather large -10.3 fs/mm.

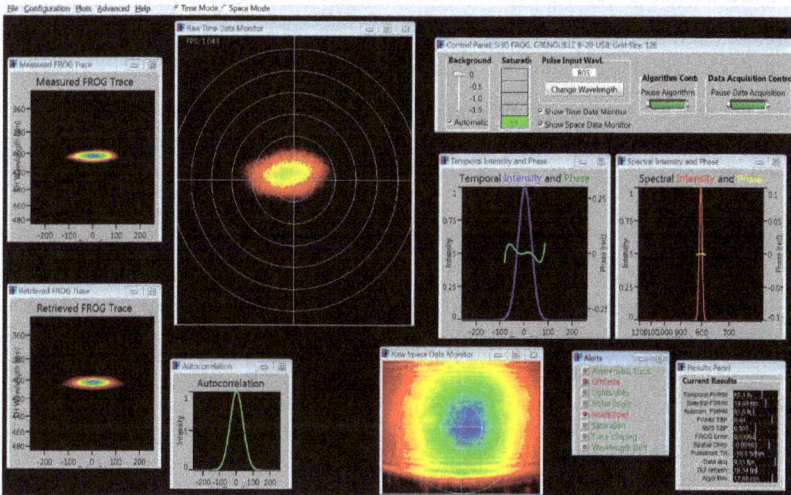

Figure 24. Screen shot for a Fourier limited pulse with slit width 1.53 mm. Grid characteristics 128 points; Delay increment: 4.03 fs; Center Wavelength: 807.3 nm Frog Error was: 0.01039 Temporal FWHM: 64.7 fs Spectral FWHM: 14.65 nm, FWHM Time-Bandwidth Product: 0.437, RMS Time-Bandwidth Product: 0.501.

6.4. Envelope shape

The data obtained from the frequency resolved optical gating measurements as well as the detailed spectra from integrated spectrometer measurements were analyzed in order to obtain the pulse envelope function. Two sets of spectra were analyzed. On the one hand, the spectrum optimized for minimum pulse duration and thereafter minimum TBP while maintaining the pulse duration. On the other hand, typical spectra shown in figure 20b, where the spectrum is rather broad and TBP is large.

6.4.1. Minimum pulse duration and minimum TBP spectrum

A Gaussian function without chirp (2.7) was fitted to the ML spectrum shown in figure 25. The parameters obtained without chirp are: $\tau_G \frac{2\pi c}{\lambda_\ell^2} 10^{-9} = 0.1086$, center wavelength $\lambda_\ell = 2\pi c/\omega_\ell = 809.628$ nm and the quality of the fit is $R^2 = 0.99770$. Once the time duration $\tau_G = 37.78$ fs was established from the previous fit, the Gaussian function with chirp (2.11) was fitted to the data. The fitted linear chirp parameter a was negligible $a = 4.02016 \times 10^{-8}$. The time duration obtained from the fit is $\triangle t_{FWHM} = \tau_G \cdot 1.17741 = 44.48$ fs. This fourier transform limited calculation underestimates the pulse duration given by the FROG retrieval measurement of 56.4 fs. If the TBP obtained from the FROG measurements is used in order to evaluate the Gaussian envevelope, $\tau_G = 53.43$ fs and the pulse width is $\triangle t_{FWHM} = 62.90$ fs; This time overestimates the FROG value by 11.5 %.

The hyperbolic secant function fit to the same data gave the following parameters: $\tau_s \frac{2\pi c}{\lambda_\ell^2} 10^{-9} = 0.06678$, center wavelength $\lambda_\ell = 2\pi c/\omega_\ell = 809.572$ nm and the quality of the fit is $R^2 = 0.99705$. Once the time duration τ_S was established from the previous fit, the hyperbolic secant function with chirp was adjusted. The chirp parameter was $a = 0.105855$. This small chirp parameter value improved slightly the curve fit, yielding $R^2 = 0.99790$.

The best fits with the Gaussian and hyperbolic secant envelopes give similar R^2 results. It is therefore not possible to discern whether the envelope follows one or the other profile under these conditions.

6.4.2. Typical spectra with broad spectrum and large TBP

A typical ML spectrum as shown in figure 20b was fitted to a Gaussian envelope function. Since the functional dependence of the pulse width τ_G and the linear chirp parameter a in the exponential argument are similar, no substantial difference is obtained when chirp is added. The Gaussian fit parameters are: $\tau_G \frac{2\pi c}{\lambda_\ell^2} 10^{-9} = 0.07079$, center wavelength $\lambda_\ell = 2\pi c/\omega_\ell = 817.608$ nm and the quality of the fit is $R^2 = 0.99105$ is rather poor.

The ML spectrum shown in figure 20b was fitted to a hyperbolic secant envelope function with and without chirp. The data are plotted against wavelength rather than frequency; Therefore the scaling $\Omega - \omega_\ell \rightarrow \frac{2\pi c}{\lambda^2}(\lambda - \lambda_c)$ is required. The frequency dependent sech^2 intensity function without chirp given by (2.21) has parameters $\tau_s \frac{2\pi c}{\lambda^2} = 0.0357$, centered at $\lambda_\ell = 2\pi c/\omega_\ell = 817.59$ nm and $R^2 = 0.98442$. On the other hand, the normalized intensity function with chirp function (2.24) was fitted with parameters: $\tau_s \frac{2\pi c}{\lambda^2} = 0.0649$, chirp coefficient $a = 1.122$ centered at $\lambda_\ell = 2\pi c/\omega_\ell = 817.56$ nm and $R^2 = 0.99472$. These fits are shown in figure 27. The function with temporal chirp is better adjusted due to the flat top of the central portion of the spectrum. The comparison between these fits is better visualized in a plot of the residuals shown in figure 28.

The hyperbolic secant function with chirp gives the smaller residuals and is thus the best fit. We can conclude that for oscillator pulses with considerable chirp, the pulse envelope is closer to a hyperbolic secant function.

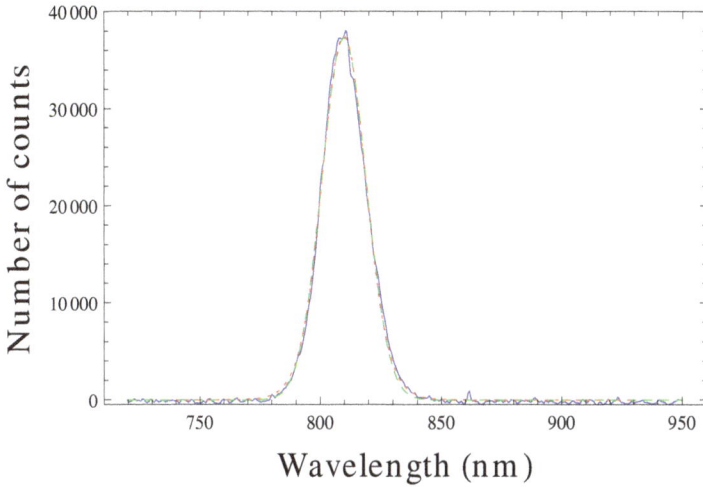

Figure 25. Spectrum of ML femtosecond oscillator taken with miniature spectrometer, 400 ms integration time. The oscillator parameters are optimized for minimum pulse duration (56.4 fs) and minimum TBP (0.624). Fits with Gaussian (dashed green line) and a hyperbolic secant (dotted red line) functions are shown. Fits with or without chirp make little difference.

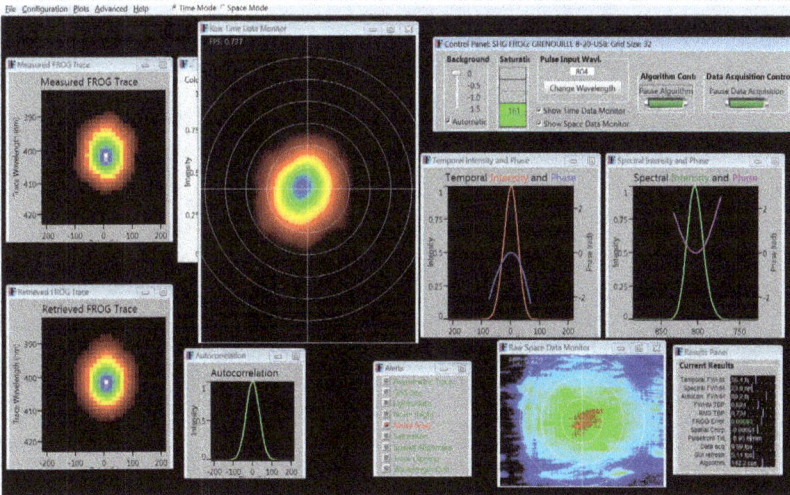

Figure 26. Screen shot for pulse with slit width 1.50 mm and slit translation 5.92mm. Center Wavelength: 809.6 nm Frog Error was: 0.00693 Temporal FWHM: 56.4 fs Spectral FWHM: 23.8 nm, FWHM Time-Bandwidth Product: 0.624.

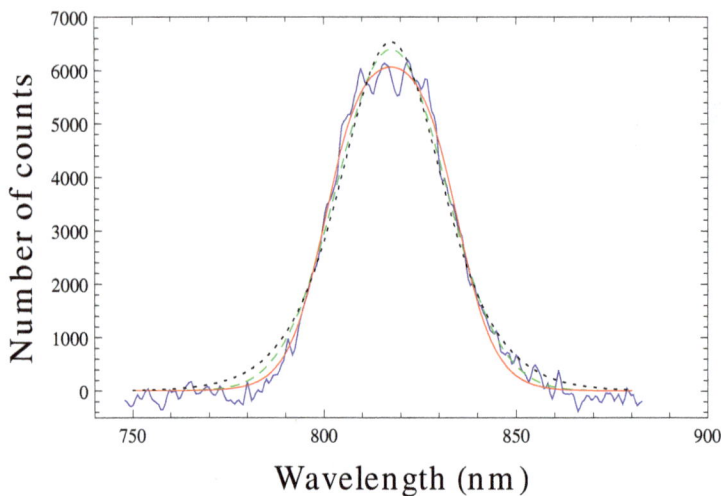

Figure 27. Gaussian (green dotted line) and Hyperbolic secant (HS) envelope fits to a typical spectrum (blue solid line) shown in figure 20b. The black dashed curve is a HS fit without chirp whereas the red continuous curve has a chirp factor of $a = 1.122$.

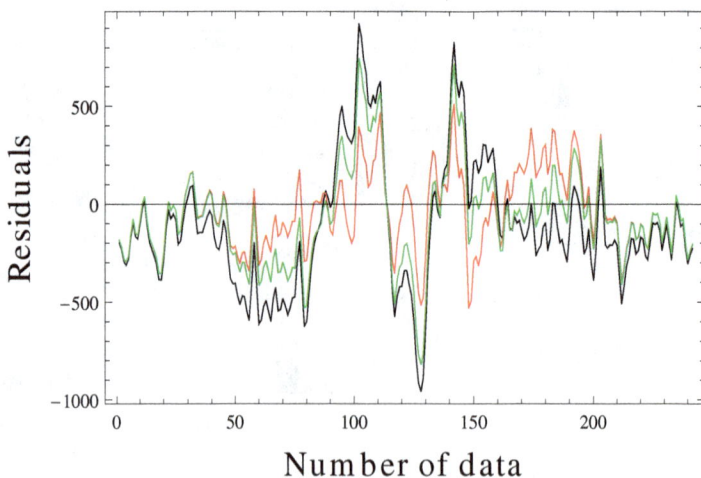

Figure 28. Residuals for the fits shown in figure 27. The Gaussian (green), Hyperbolic secant (HS) with (red) and without (black) chirp show that the best fit corresponds to the HS with chirp.

7. Conclusions

A Ti:Sa femtosecond laser oscillator was optimized for different operation purposes:

- A minimum pulse duration of 56.4 fs was obtained. The prims PR2 and the slit were adjusted to produce the minimum TBP of 0.624 while maintaining the shortest pulse duration.

- A time-bandwidth product of 0.783 with 62.4 fs pulse duration, 28.7 nm bandwidth and central wavelength of 828.6 nm was obtained when the prism PR2 was adjusted and the spectrum was not cropped.

- A Fourier limited pulse with time-bandwidth product of 0.437 with pulse duration between 63 to 79.7 fs, 11.82 to 18 nm bandwidth and central wavelength around 804.1 nm was obtained with appropriate bandwidth cropping with the intra-cavity slit.

- Oscillator emission was tuned from 730 to 850 nm with output power above 200 mW.

- The cavity adjustments would be more flexible if the first prism (PR1) is also mounted on a translation stage. A slit with two movable blades would also be welcome.

The wave description for Gaussian and hyperbolic secant envelopes was presented in detail. Results scattered in the literature were brought together in particular for the hyperbolic profile. The energy content of the wave field has been outlined with a novel continuity equation described in terms of complementary fields. These linearly independent fields do not require a temporal averaging over the pulse cycle. Therefore, they are well suited to describe ultrafast pulses or even few or single cycle pulses.

Acknowledgements

We are grateful to Raúl Rangel CISESE and his collaborators Jacob Licea Rodríguez and Alexandro Ruiz de la Cruz for useful comments and much needed advice.

Author details

E. Nava-Palomares, F. Acosta-Barbosa, S. Camacho-López and M. Fernández-Guasti.
Lab. de Óptica Cuántica, Depto. de Física, Universidad A. Metropolitana - Iztapalapa, 09340 México D.F., Ap. postal. 55-534, Mexico, url: http://luz.izt.uam.mx

8. References

[1] G. Muñoz, C. García, M. Muñoz, and M. Fernández-Guasti. Mesa holográfica pasiva ultra-estable: análisis de vibraciones. In *XXIII reunión anual de óptica*, sept 2010.
[2] Qiang Lin, Jian Zheng, Jianming Dai, I-Chen Ho, and X.-C. Zhang. Intrinsic chirp of single-cycle pulses. *Phys. Rev. A*, 81(4):043821, Apr 2010.
[3] Qiang Lin, Jian Zheng, and Wilhelm Becker. Subcycle Pulsed Focused Vector Beams. *Phys. Rev. Lett.*, 97(25):253902, 2006.

[4] J. C. Diels and W. Rudolph. *Ultrashort Laser Pulse Phenomena: Fundamentals, Techniques, and Applications on a Femtosecond Time Scale, Academic Press, 2nd edition, Elsevier 2006.* Elsevier Science Publishers, 2nd edition edition, 2006.

[5] M. Fernández-Guasti. The Wronskian and the Ermakov - Lewis invariant. *International Mathematical Forum*, 4(16):795 – 804, 2009.

[6] C. H. Henry. Theory of the linewidth of semiconductor lasers. *IEEE J. Quant. Elec.*, 18(2):259–264, 1982.

[7] T. L. Koch and J. E. Bowers. Nature of wavelength chirping in directly modulated semiconductor lasers. *Electr. Lett.*, 20(25/26):1038–1040, Dec 1984.

[8] P. Lazaridis, G.Debarge, and P. Gallion. Time-bandwidth product of a chirped sech2 pulses. *Optics Lett.*, 20(10):1160–1163, 1995.

[9] M. Fernández-Guasti, E. Nava, F. Acosta, and R. Chandrasekar. Physical processes behind a Ti:Sa femtosecond oscillator. In *Optics and Photonics 2011*, volume 8121 of *The nature of light: What are photons? IV*, pages 812118–1–10. SPIE, 2011.

[10] R. P. Feynman, M. Sands, and R. Leighton. *Lectures on physics*, volume II. Addison Weseley, 1972.

[11] M. A. Alonso. Wigner functions in optics: describing beams as ray bundles and pulses as particle ensembles . *Adv. Opt.Phot.*, 3:272–365, 2011.

[12] M. Fernández-Guasti. Complementary fields conservation equation derived from the scalar wave equation. *J. Phys. A: Math. Gen.*, 37:4107–4121, 2004.

[13] M. Fernández-Guasti. The necessity of two fields in wave phenomena. In *Optics and Photonics 2011*, volume 8121 of *The nature of light: What are photons? IV*, pages 81210R–1–12. SPIE, 2011.

[14] Claude Rulliere, editor. *Femtosecond laser pulses.* Advanced texts in physics. Springer, 2nd edition, 2005.

[15] O. E. Martinez, J. P. Gordon, and R. L. Fork. Negative group-velocity dispersion using refraction. *J. Opt. Soc. Am. B*, 1(10):1003–1006, October 1984.

[16] O. E. Martinez. Grating and prism compresors in the case of a finite beam size. *J. Opt. Soc. Am. B*, 3(7):929–934, July 1986.

[17] R. Trebino. *Frequency resolved optical gating.* Kluwer Academic Publ., 2000.

Chapter 4

Longitudinally Excited CO_2 Laser

Kazuyuki Uno

Additional information is available at the end of the chapter

1. Introduction

In the mid-infrared region (3 – 30 μm), there are only several kinds of commercial laser as shown in Fig. 1. A CO_2 laser is very important because the CO_2 laser emits high output energy and various pulse shapes at the wavelength region between 9.2 μm and 11.4 μm with a high absorptivity in many substances. The first CO_2 laser was developed by C. K. N. Patel in 1964 and had a continuous output power of a few milliwatts (Patel, 1964). In 1968, the pulsed CO_2 laser was developed by A. E. Hill and had a few joules (Hill, 1968). Nowadays, various types of CO_2 lasers have been developed. The characteristics of CO_2 laser pulses are determined by the discharge tube structure (a longitudinal excitation scheme like Fig. 2 (a) or a transversal excitation scheme like Fig. 2 (b)), the discharge types (DC discharge, RF discharge, or pulsed discharge) and the oscillation system (CW oscillation, Q-switched oscillation, or pulsed oscillation).

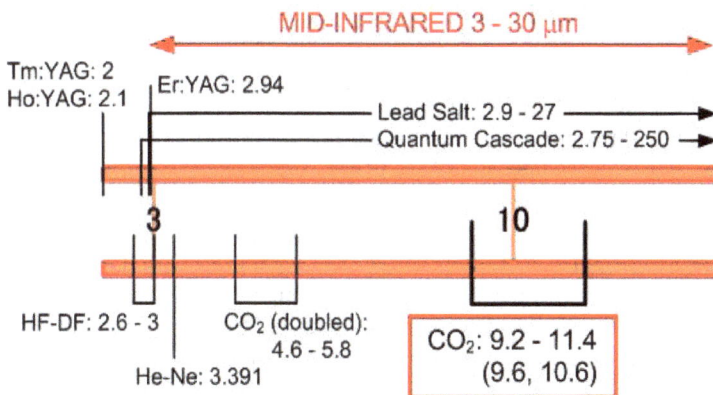

Figure 1. Mid-infrared commercial lasers (Photonics.com). Unit is μm.

(a)

(b)

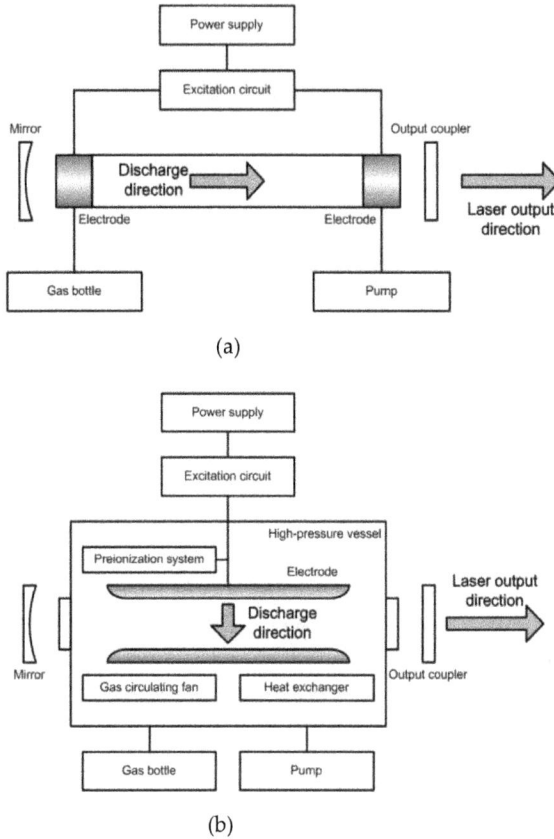

Figure 2. Excitation system. (a) Longitudinal excitation system. (b) Transversal excitation system.

In the CW oscillation and the long pulse oscillation (a pulse width of 10 μs – 10 ms), a low power CO_2 laser (< 10W) is used as a medical laser like tooth and skin treatments. A CO_2 laser emitted 20 – 100 W is used for cutting a nonmetallic substance like a film, cloth, leather, paper, rubber, a plastic, acrylics, etc. A high power laser emitted 100 – 400 W is used for cutting metal, glass, and ceramics. A short pulse CO_2 laser that emits an output energy of a few joules and a pulse width of about 100 ns is used for puncturing and patterning of a nonmetallic substance, removal of surface polymer material which does not give a damage to a metal board, and etc. Specially, a large and high output CO_2 laser called "Lekko VIII" that emitted an output energy of 10 kJ and a pulse width of 1 ns was developed in Institute of Laser Engineering, Osaka University, Japan, 1981 (Yamanaka et al, 1981). Recently, the CO_2 laser attracts attention as a driver laser for a Laser-Produced-Plasma Sn-EUV source (13.5 nm) and a excitation laser for a Terahertz laser (CH_3F, C_2H_2Cl, CH_3OH, D_2O, and etc., 30 – 300 μm) (Ueno et al, 2007; He et al, 2010).

2. Principle of CO$_2$ laser

The CO$_2$ laser oscillates in the vibrational level of CO$_2$ molecular. The CO$_2$ laser has about 100 oscillation lines at the center of 9.6 μm and 10.6 μm between 9.2 μm and 11.4 μm. Fig. 3 shows the energy diagram of CO$_2$ laser. Generally, the CO$_2$ laser is pumped by glow discharge in the low mixed gas (CO$_2$, N$_2$, He, and etc.) pressure. The N$_2$ serves to improve the output energy and the efficiency due to the energy transfer from the vibrational level of N$_2$ to the upper laser level of CO$_2$. However, in the short pulse oscillation, N$_2$ causes a laser pulse tail because of the long lifetime of the vibrational level of N$_2$. The He serves to make the discharge uniform. The CO$_2$ molecular is efficiently pumped by very low electron temperature because the energy of the laser upper level 001 is 2349.2 cm^{-1} and the energy required for excitation of CO$_2$ molecule is about 0.28 eV. However, generally, the CO$_2$ molecular pumped by the higher electron temperature than 0.28 eV because of an electrical gradient to sustain the discharge. The CO$_2$ laser oscillates at 10.6 μm band (001 – 100 transition) and 9.6 μm band (001 – 020 transition) and has about 100 oscillation lines because of the P branch and the R branch related to the rotational level. The laser lower levels of 100 and 020 are relaxed by the collision with the CO$_2$ molecule of ground state. The 010 level serves as a bottleneck. But the thermal relaxation of the 010 level is accelerated by the collision with other molecular, He atom and the laser tube wall. Then, the 010 level returns to the ground state. Therefore, in the efficiently laser oscillation, the N$_2$ molecular that supplies the upper laser level of CO$_2$ and media like He that eases the lower laser level of CO$_2$ are essential. However, in the low-pressure pulse-discharge at the low repetitive operation, He may be unnecessary.

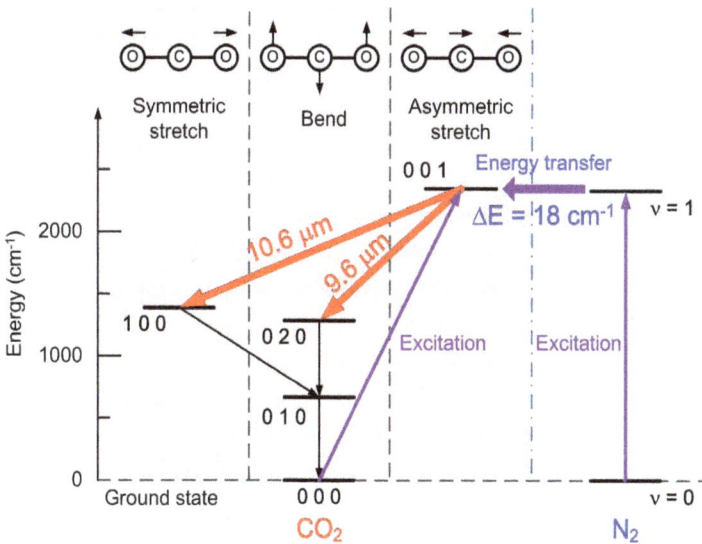

Figure 3. Energy diagram of CO$_2$ laser.

3. Longitudinal excitation scheme

In the longitudinal excitation scheme, the excitation discharge is in the direction of the laser axis. This system was used in the early stage of development of lasers such He-Ne lasers (633 nm), N_2 lasers (337 nm) and CO_2 laser (9.2 – 11.4 µm). In this system, a dielectric tube with the small inner diameter (1 – 2 cm in CO_2 lasers) and the long length (> 30 cm in CO_2 laser) is used as a discharge tube. Metallic electrodes are attached to each end of the tube. The long discharge length provides a high breakdown voltage (> 20 kV) at a low gas pressure (< 10 kPa). Therefore, this system does not require a high gas-pressure vessel. Moreover, this laser may oscillate at narrow spectral width because the low-pressure operation produces the small pressure broadening. A uniform discharge can be obtained easily because of the fast electron drift velocity (the long mean free path) of the low pressure operation; additionally, the discharge uniformity is not affected by residual charges in the longitudinal excitation because a discharge takes place in a discharge tube with a long length and small inner diameter. In a RF discharge, an electronic trapping occurs. A uniform discharge takes place because electrons vibrate by the high-frequency electric field and are trapped between the discharge gaps. In a pulsed discharge, a uniform discharge takes place by a diffuse streamer discharge is formed by diffusing uniformly and progressing the minute spark discharge with which an avalanche and optical ionization are combined (Fig. 4).

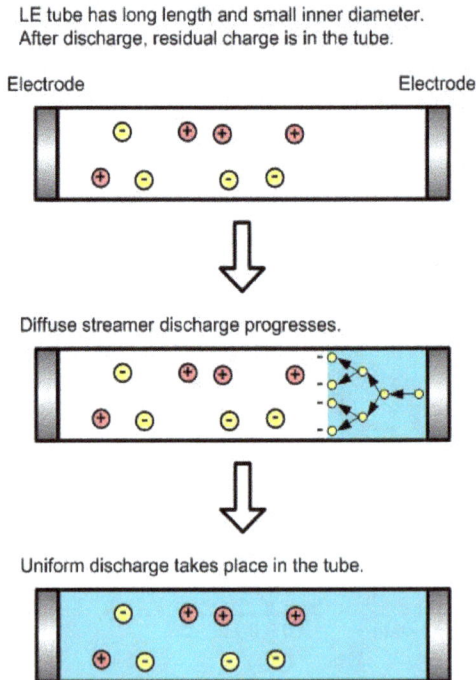

Figure 4. Longitudinally pulsed discharge.

The tube is covered with metal foil.

Electrode   Metal foil   Electrode

Excitation circuit

⇩

Corona discharge takes place along the tube wall.

Excitation circuit

⇩

Uniform discharge takes place in the tube.

Excitation circuit

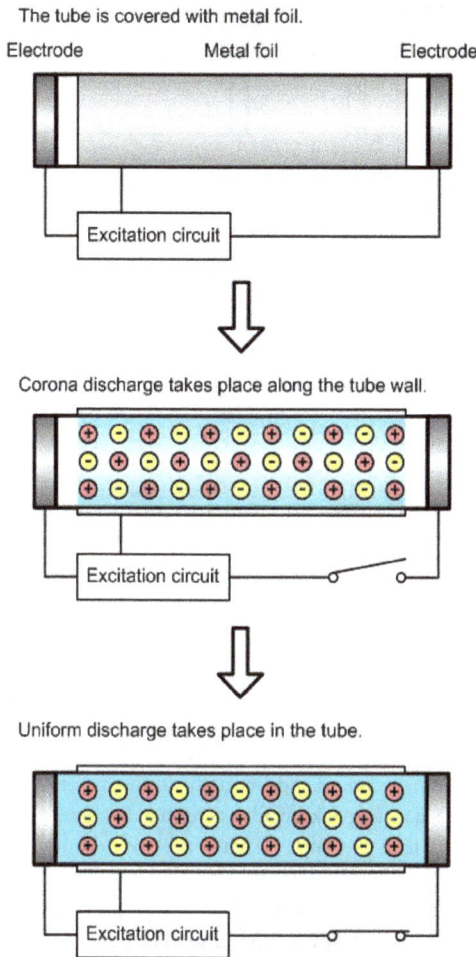

Figure 5. Longitudinally pulsed discharge with strong preionization.

Therefore, the longitudinal excitation scheme does not require a pre-ionization device like UV and X-ray source. On the other hand, when the pre-ionization is required in the high-pressure operation (> 1 atm), the discharge tube is covered with Al foil (Fig. 5, 6) (El-Osealy et al, 2002b; Uno et al, 2009; Uno et al, 2012a). A corona discharge occurs along the tube wall before the main discharge to serve as pre-ionization. Therefore, nothing is put in the discharge tube. In the longitudinal excitation scheme without pre-ionization, a gas lifetime is long. The gas lifetime is a phenomenon in which the laser output energy decreases with operation time in a sealed-off operation. Impurities increases with time in the discharge tube, UV and X-ray for the pre-ionization cannot reach the center part of the discharge space, the uniform discharge cannot take place, the localization of the discharge occurs, and

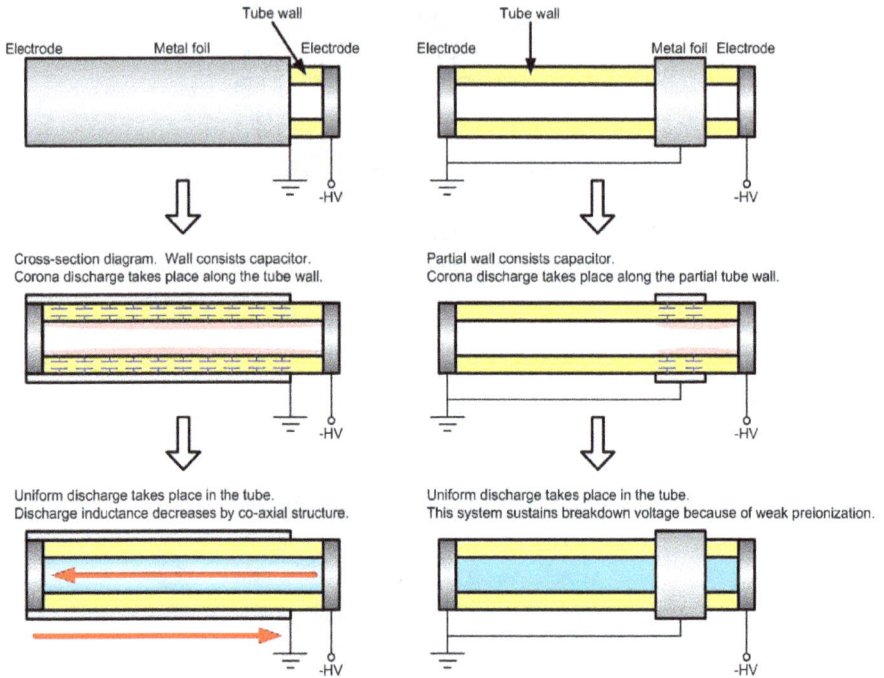

Figure 6. Longitudinally pulsed discharge with simple preionization.

the laser output energy decreases. Therefore, this scheme without the pre-ionization is suitable for the long lifetime operation. Additionally, in this scheme, the perfect hard-sealed discharge tube without a gasket packing like a rubber O-ring is producible because the structure of the discharge tube is simple. Therefore, this scheme is suitable for the long lifetime gas laser. In fact, a low-output sealed-off CO_2 laser has a CW oscillation or a high repetition oscillation without a gas-flowing system and water-cooling system effectively. Additionally, in this scheme, the beam section of the laser is circular because the discharge tube is used a dielectric tube like a glass tube and a ceramics tube. All the optical elements may be circular. This scheme produces laser beam with good beam quality because the length of optical cavity is long enough to the diameter of aperture (the inner diameter of the discharge tube). Therefore, in the longitudinal excitation scheme, a high performance laser can be developed with a simple, portable, and low-cost device.

In the longitudinal excitation scheme, as shown in Table 1, various laser oscillation from mid-infrared to vacuum ultraviolet has been reported. In recent years, especially, new gas lasers, for example, a longitudinally excited CO_2 laser with short laser pulse like TEA and Q-switched CO_2 laser (Uno et al, 2009), a longitudinally excited N_2 laser with the same excitation system as an excimer lamp (Uno et al, 2006), and a longitudinally excited F_2 laser at low-pressure (total pressure of 40 Torr) (El-Osealy et al, 2002b) have been developed.

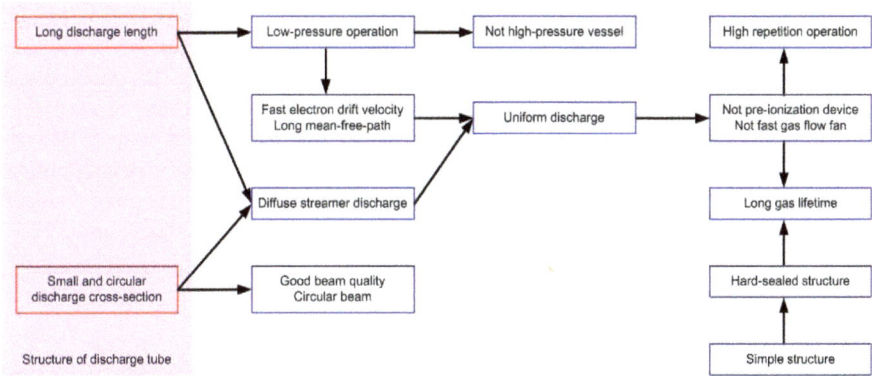

Figure 7. Characteristics of longitudinal excitation system.

Laser	Wavelength		Example of reference
CO$_2$ laser	Mid-infrared	9.2 – 11.4 µm	Patel, 1964; Hill, 1968; Chung et al, 2002; Uno et al, 2009
Xe laser	Near-infrared	1.73 – 3.51 µm	Komatsu et al, 1991
F laser	Visible	630 – 780 nm	Hocker & Phi, 1976; Uno et al, 2008
He-Ne laser		544, 594, 612, 633 nm	Javan et al, 1961
XeF laser	Ultraviolet	351 nm	Burkhard et al, 1981; Cleeschinsky et al, 1981
N$_2$ laser		337 nm	El-Osealy et al, 2001; Uno et al, 2006; Uno et al, 2008
XeCl laser		308 nm	Zhou et al, 1983; Furuhashi et al, 1987
KrF laser		248 nm	Newman, 1978; Eichler et al, 1985
ArF laser	Vacuum-ultraviolet	193 nm	Rosa et al, 1986
F$_2$ laser		157 nm	El-Osealy et al, 2002a; El-Osealy et al, 2002b

Table 1. Longitudinally excited gas lasers.

4. Excitation system

The shape of laser pulse depends on the discharge and oscillation types. The discharge type can be divided into the DC (direct current) discharge, the RF (radio frequency) discharge and the pulsed discharge. The oscillation type can be grouped into the CW (continuous wave) oscillation and the pulsed oscillation including Q-switched oscillation. The DC and RF discharge provides the CW oscillation basically and can provide the long-pulsed oscillation (the pulse width from several µs to several ms) by switching of excitation discharge and the short-pulsed (giant-pulsed) oscillation (the pulse width of about 100 ns) by using Q switch. Although it is well known the pulsed discharge provides the long-pulsed oscillation, the pulsed discharge can provide the short-pulsed oscillation (the pulse width of about 100 ns and the pulse tail of several tens µs) like the Q-switched CO_2 laser and a TEA-CO_2 laser.

4.1. DC discharge

Figure 8 shows a general longitudinally excited CO_2 laser pumped by DC discharge. The DC-CO_2 laser commonly is used a high-voltage (a few tens kV) DC power supply and emits the CW power of several tens kW. The output power can control delicately by adjustment of DC voltage.

A discharge tube is made of a heat-resistant hard glass pipe with an inner diameter of several mm and is the co-axial dual structure for cooling by water. Cooling improves the efficiency and output power of the laser oscillation by decreasing the plasma temperature. Additionally, high output power can be provided by a multi-path system folding back the optical path in the resonator and a multi-beam system which uses multi-tubes. High power DC-CO_2 laser emitting CW power of 1 kW using 20 discharge tubes with the inner diameter of 9 mm and the length of 195 cm has been reported (Deshmukh & Rajagopalan, 2003).

A low power CO_2 laser can operate in a sealed-off mode. In a long-life sealed-off CO_2 laser, a gold catalyst is used to regenerate CO_2 from the dissociation products formed during discharge (Tripathi et al, 1994). In the sealed-off CO_2 laser, the laser beam (output power and beam quality) is stable because discharge and the distribution of particles are uniform. In a high power CO_2 laser, a gas-flowing system circulating fresh gas is used for preventing the decrease of the oscillation efficiency by the deterioration of gas. The longitudinal excitation system has only an axially flowing type.

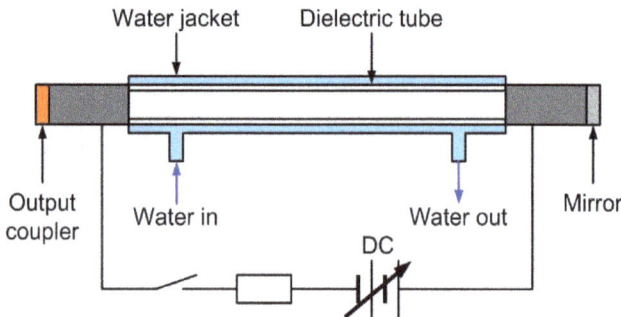

Figure 8. Discharge tube with water jacket.

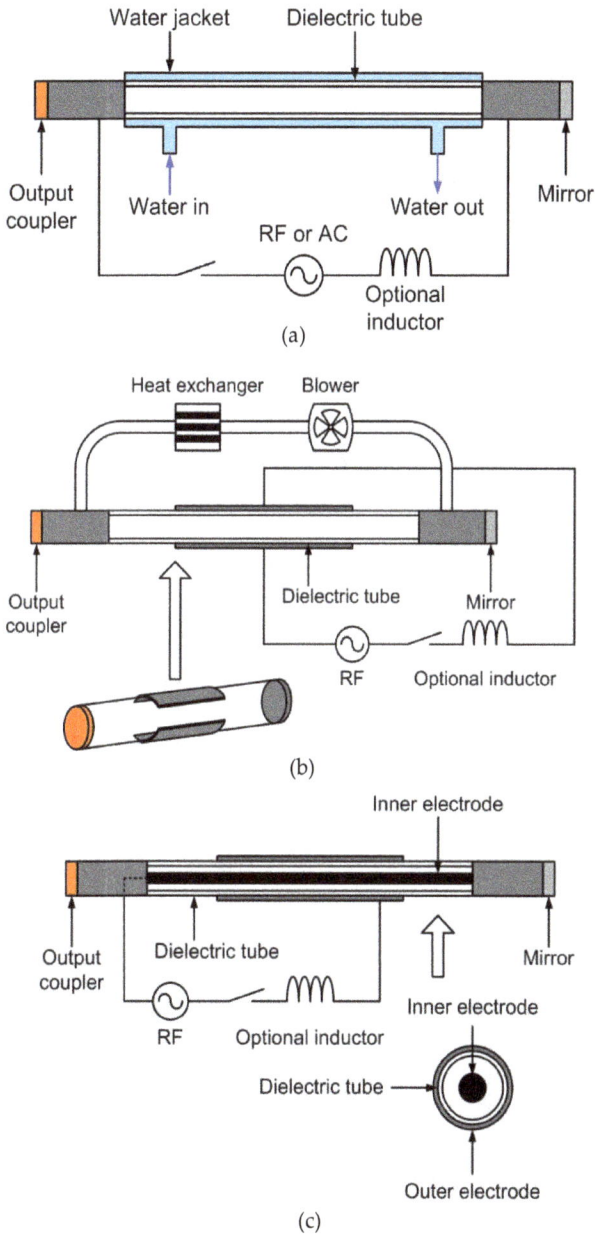

Figure 9. RF discharge tube. (a) Longitudinal excitation (Lee et al, 2000). (b) Transverse excitation (Terai et al, 1993). (c) Coaxial RF-CO$_2$ laser (Bethel, 1998).

4.2. RF discharge

The first CO_2 laser by C. K. N. Patel used the RF discharge excitation, 27 MC/s (Patel, 1964). In the RF discharge excitation, the medium gas is pumped by a high-frequency voltage (typically 13.56 MHz). In this system, electrodes are able to be put on the outer wall of discharge tube. Thus, the electrodes are not sputtered and there are little degradation of gas and contamination of optical components. A compact high-output CO_2 laser can be developed by high discharge input-energy per unit volume. An electronic trapping provides good pulse characteristics.

However, generally, the RF discharge excitation scheme uses the discharge tube like Fig. 9. Fig. 9 (b) looks like the longitudinal excitation scheme, but is the transversal excitation scheme because the direction of discharge is perpendicular to the direction of a laser axis.

In the low pressure region less than 10 kPa, the laser pumped by RF wave more than 0.1 MHz emits CW and the laser pumped by RF wave less than 10 kHz emits pulse shapes. Generally, the pulse oscillation of RF-CO_2 lasers is realized by switching of excitation discharge and the duty ratio of ON and OFF (Nagai, 2000). Fig. 10 (a) is a example of normal pulses by chopping the CW output. In the axial flowing system, the gas existence time in the discharge tube is long and the rise of gas temperature is large. It is difficult to keep the peak value of a pulse constant for a long time. Therefore, a narrow pulse shape called an enhanced pulse like Fig. 10 (b) or a super pulse like Fig. 10 (c) takes place.

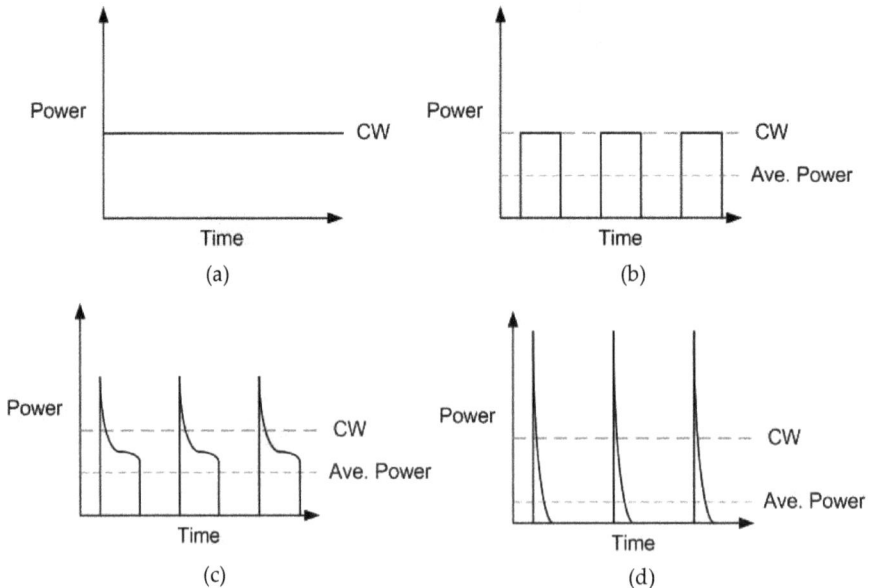

Figure 10. Images of RF-CO_2 laser output (Nagai, 2000). (a) CW. (b) Normal pulses. (c) Enhanced pulses. (d) Super pulses.

In the slab RF-CO$_2$ laser, commercial sealed-off CO$_2$ lasers produces the broad output range from CW 20 W to CW 1 kW and the high peak pulse oscillation of 2.5 kW at the pulse width 50 μs at 1 kHz (Coherent, Inc., DIAMOND E-1000). In a compact fast axial flow CO$_2$ laser, the output power of CW 2.1 kW has been reported (Biswas et al, 2010). Additionally, a low-power CO$_2$ laser pumped by AC discharge of 60 Hz whose frequency is lower than RF discharge has also been reported (Lee et al, 2000).

4.3. Q-switch

A Q switch is a system to produce a giant pulse with a short pulse and a high peak power. A Q value is a standard of the resonator performance and is a ratio of a loss energy to a storage energy in the cavity. It is hard to carry out a laser oscillation when a Q value is low, and it is easy when a Q value is high. The Q switch rapidly decreases the high resonator losses, that is, rapidly increases the Q value.

A CW-CO$_2$ laser using a Q switch produces a giant pulse with a pulse width from a few tens ns to a few μs and a peak power of several kW. In a CO$_2$ laser, the use of a mechanical chopper, rotating mirror (Battou et al, 2008), electrooptical (CdTe) (Tian et al, 2005) or acoustooptical modulators (Xie et al, 2010), or saturable absorber (SF$_6$) (Soukieh et al, 1999) has been reported.

4.4. Pulsed discharge

A CO$_2$ laser is easily oscillated by applying a high voltage pulse of about 20 kV to a longitudinal discharge tube filled by CO$_2$ gas. Fig. 11 shows a longitudinally excited pulsed CO$_2$ laser with a general and easy excitation circuit (Chung et al, 2002; Loy & Roland, 1977). A voltage pulse of several hundreds V is generated by a power supply and is fed to a step-up transformer. The high-voltage pulse of about 20 kV is directly applied to the discharge tube. The high-voltage pulse has a long rise time of about a few hundred μs because the transformer with a large inductance is directly connected to the discharge tube. When the applied voltage reaches the breakdown threshold, a discharge starts and the laser oscillates. The laser output energy depends on the discharge volume and the input energy. The laser pulse has a long width from several tens μs to several ms because the excitation circuit produces a long and high voltage pulse and the discharge formation time is long. For example, the discharge tube with the length of 80 cm applied the high voltage pulse of 25 kV and produced the output power of 35 W and the width of 3 ms at 60 Hz (Chung et al, 2002).

The longitudinally excited CO$_2$ laser produces a short laser pulse like TEA-CO$_2$ laser and Q-switched CO$_2$ laser by a fast discharge. Fig. 12 shows a CO$_2$ laser with a capacitor transfer circuit that is used for transversal excited excimer lasers and is known as a fast discharge circuit (Miyazaki et al, 1986). A low-inductance storage capacitor C$_s$ was charged up to DC voltage of about 20 kV. A spark gap was switched by a trigger pulse from a trigger circuit, and the high voltage was transferred to a buffer capacitance C$_b$. When the voltage reached

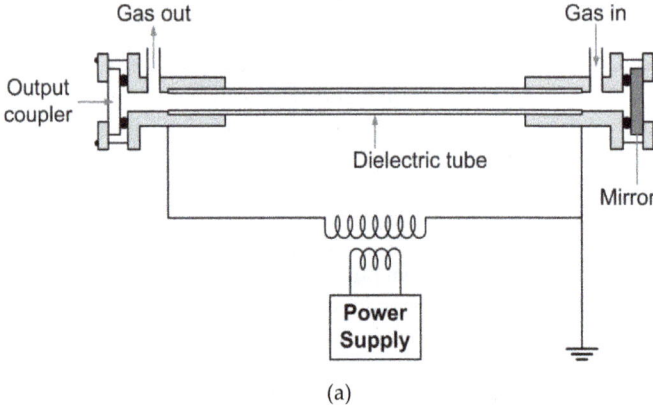

(a)

(b)

Figure 11. Image of longitudinally excited long-pulse CO_2 laser. (a) Schematic diagram. (b) Photograph.

the breakdown threshold, a rapid main discharge took place in the discharge tube. The high-voltage pulse of a rise time less than 100 ns is applied to the discharge tube. Additionally, a high-voltage resistor is connected in parallel with the discharge tube. A low resistance acts as a shunt resistance and provides a rapid discharge.

For example, in our study (Uno et al, 2009), the discharge tube was a ceramic pipe made of alumina, with an inner diameter of 13 mm, an outer diameter of 17 mm, and a length of 45 cm. The discharge tube was covered with an Al sheet. Therefore, the laser tube had a coaxial structure for reducing the discharge impedance. An optical cavity was formed by a ZnSe output coupler with a reflectivity of 80% and a high-reflection mirror with a radius of curvature of 20 m. The distance between the output coupler and the mirror (i.e., the cavity length) was 54 cm. In the excitation circuit, the storage capacitance C_s was 5.4 nF and charged up to DC 20 kV, and the buffer capacitance C_b was 2.8 nF. The fall time of the discharge voltage decreased from 32.8 µs (10 MΩ) to 0.97 µs (100 Ω). In the fast discharge, the gain increases gradually from the start of discharge. When the gain becomes enough

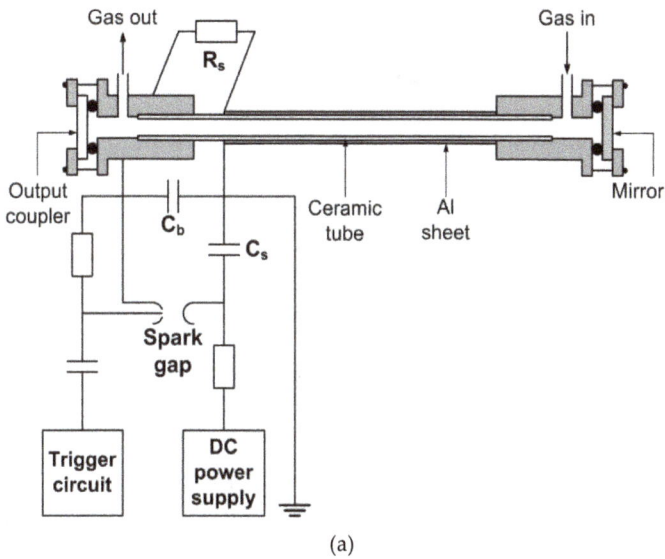

(a)

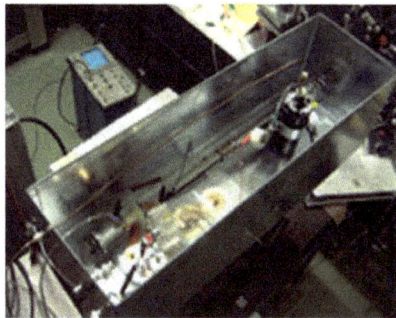

(b)

Figure 12. Longitudinally excited short-pulse CO₂ laser with capacitor-transfer circuit (Uno et al, 2009). (a) Schematic diagram. (b) Photograph.

after several µs from the start of discharge, the laser oscillates by a gain Q switch and a spike laser pulse is formed. At that time, the main discharge finishes mostly. Low current after the main discharge and the energy transfer of the upper-level N_2 with the long lifetime causes the laser pulse tail of several tens µs. The pulse tail can be eliminated by eliminating the after current or using of pure CO_2 gas to eliminate the long N_2 tail. Fig. 13 and 14 shows the laser pulse in the discharge formation time of 32.8 µs and 0.97 µs, respectively, at the same discharge tube with the inner diameter of 13 mm and the length of 45 cm and the same gas pressure of 2.9 kPa (CO_2: N_2: He= 1: 1: 2) (Uno et al, 2009). Fig. 13 shows the laser pulse with the spike pulse width of 103.3 ns, the pulse tail length of 61.9 µs, and the output energy of 50 mJ by the discharge formation time of 32.8 µs. The pulse tail length was defined from the end of the spike pulse to the end of the pulse tail. Fig. 14 shows the laser pulse with the spike pulse width of 96.3 ns, the pulse tail length of 17.2 µs, and the output energy of 30 mJ by the discharge formation time of 0.97 µs. These results showed the fast discharge caused the decrease of the pulse tail.

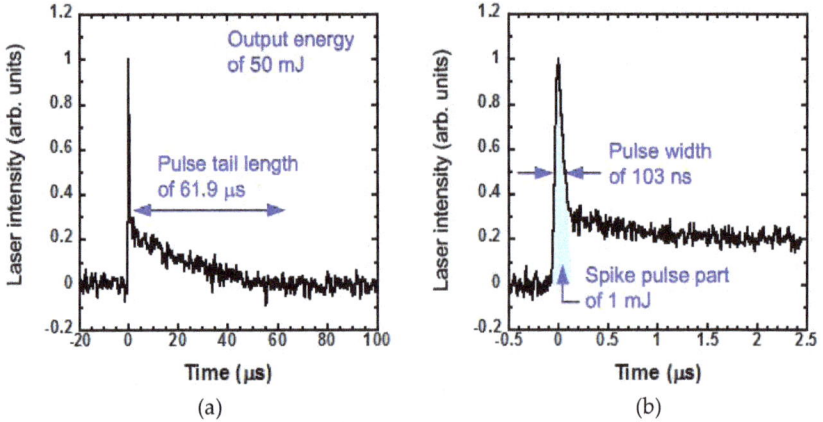

Figure 13. Laser pulse waveforms at slow discharge (discharge formation time of 32.8 µs and gas pressure of 2.9 kPa (CO_2: N_2: He= 1: 1: 2)) (Uno et al, 2009). Intensity of vertical axis is normalized by peak strength. (a) Overall waveform. (b) Magnified time scale view of spike pulse.

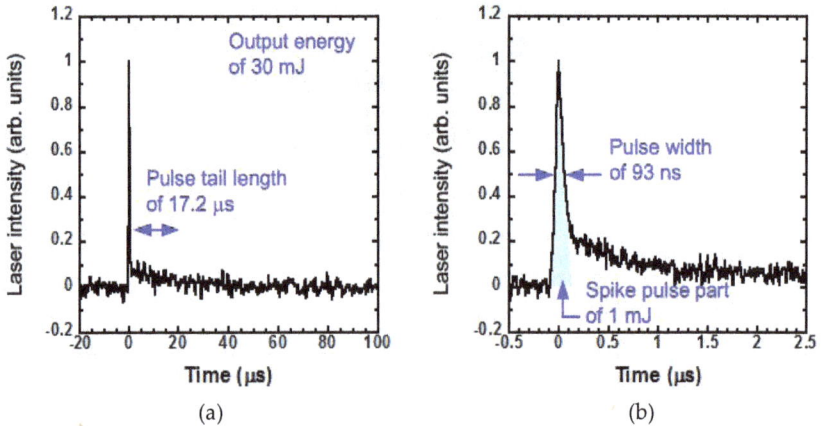

Figure 14. Laser pulse waveforms at fast discharge (discharge formation time of 0.97 µs and gas pressure of 2.9 kPa (CO_2: N_2: He= 1: 1: 2)) (Uno et al, 2009). Intensity of vertical axis is normalized by peak strength. (a) Overall waveform. (b) Magnified time scale view of spike pulse.

The capacitor-transfer circuit like Fig. 12 developed for the transversely excited excimer laser (Miyazaki et al, 1986). The capacitor-transfer circuit has a buffer capacitance in parallel with the laser tube. The buffer capacitance functions to increase the applied voltage due to the storage capacitance. However, because the transverse excitation scheme has a narrow electrode gap and a small discharge impedance, a buffer capacitance is required to sustain the discharge. On the other hand, the longitudinal excitation scheme has a wide electrode gap and a large discharge impedance and does not require a buffer capacitance to sustain the discharge. Therefore, a circuit without a buffer capacitance, called a direct-drive circuit like Fig. 15, has been developed (Uno et al, 2012a). Fig. 15 shows a longitudinally excited CO$_2$ laser with a direct-drive circuit (Uno et al, 2012a). A negative several hundreds V pulse was generated by the power supply with a silicon-controlled rectifier and was fed to a

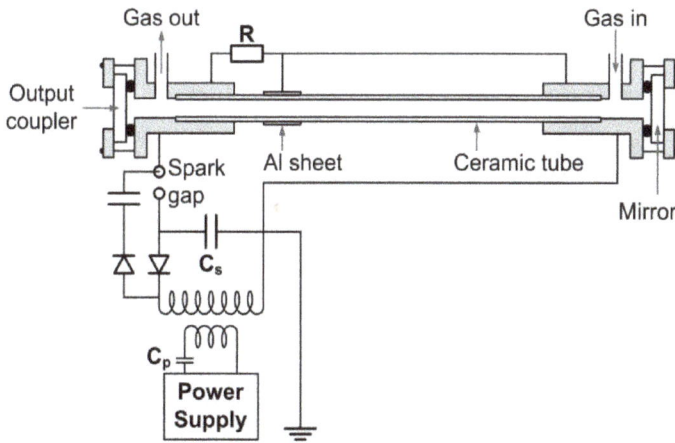

(a)

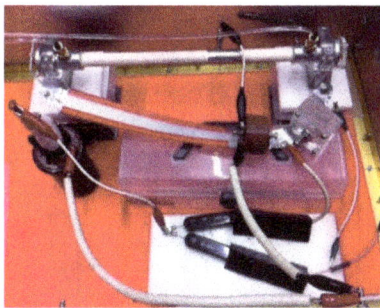

(b)

Figure 15. Longitudinally excited short-pulse CO$_2$ laser with direct-drive circuit (Uno et al, 2012a). (a) Schematic diagram. (b) Photograph.

transformer, which had a primary capacitance. First, a negative voltage pulse was generated and was used to charge a storage capacitor through a rectifier. Then, a positive voltage pulse generated by the overshoot of the transformer was applied to the trigger electrode of the spark gap through a rectifier and a small capacitor. The storage capacitor was charged, the spark gap was switched, and the high voltage was applied to the laser tube. When the voltage reached the breakdown threshold, a rapid discharge took place in the laser tube. This system produces a short pulse CO_2 laser with a spike pulse.

For example, in our study (Uno et al, 2012a), the laser tube was the almost same as the above longitudinally excited CO_2 laser with the capacitor-transfer circuit. A different point was a partially preionization. Part of the discharge tube was covered with an Al sheet for a high discharge starting voltage. In the excitation circuit, the primary capacitance C_p was 10.2 μF and the storage capacitance was 700 pF.

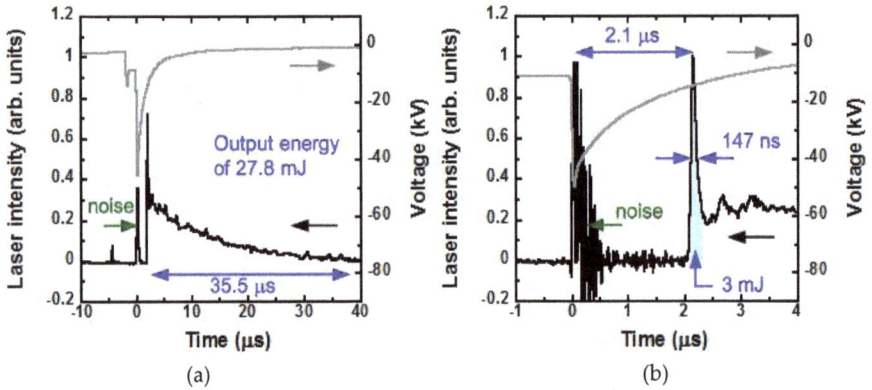

Figure 16. Discharge voltage and CO_2 laser pulse waveforms at 3.4 kPa (mixed gas, CO_2 : N_2 : He = 1 : 1 : 2) (Uno et al, 2012a). Black and gray lines represent laser pulse and discharge voltage, respectively. (a) Overall waveform. (b) Magnified time scale view of spike pulse.

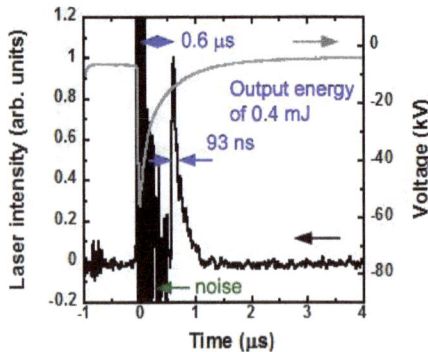

Figure 17. Discharge voltage and CO_2 laser pulse waveforms at 2.2 kPa (pure CO_2) (Uno et al, 2012a). Black and gray lines represent laser pulse and discharge voltage, respectively.

Fig. 16 shows the discharge voltage and laser pulse waveforms at the mixed gas (CO$_2$: N$_2$: He= 1: 1: 2) pressure of 3.4 kPa. The discharge starting voltage was –60.2 kV, and the fall time of the discharge was 4.4 μs. The laser oscillation began 2.1 μs after the start of discharge. The laser pulse had a sharp spike, like that from TEA and Q-switched CO$_2$ lasers. The spike pulse width was 147 ns (FWHM), and the pulse tail length was 35.5 μs. The laser output energy was 27.8 mJ, and the energy of the spike pulse part was estimated to be 3.1 mJ. Fig. 17 shows the discharge voltage and laser pulse waveform at the pure CO$_2$ gas pressure of 2.2 kPa. The discharge starting voltage was –60.8 kV, and the fall time of the discharge was 1.9 μs. Laser oscillation began 0.6 μs after the start of discharge. The laser output energy was 0.4 mJ. The laser pulse contained a spike pulse only, and the width was 93 ns (FWHM). Pulse-tail-free oscillation was realized by the use of pure CO$_2$ gas.

Figure 18 shows the simplest short-pulse longitudinally excited CO$_2$ laser (Uno et al, 2012b). This system has a laser tube, a capacitor, a resister, and a pulse power supply with a low-voltage silicon-controlled rectifier, and a set-up transformer only. The excitation circuit does not have a high-voltage switch; instead, the laser tube plays the role of the switch. Therefore, this system is almost same as the above long-pulse longitudinally excited CO$_2$ laser. However, the step-up transformer has the fast rise time of about 3 μs. The fast transformer produces the fast discharge which causes the short laser pulse.

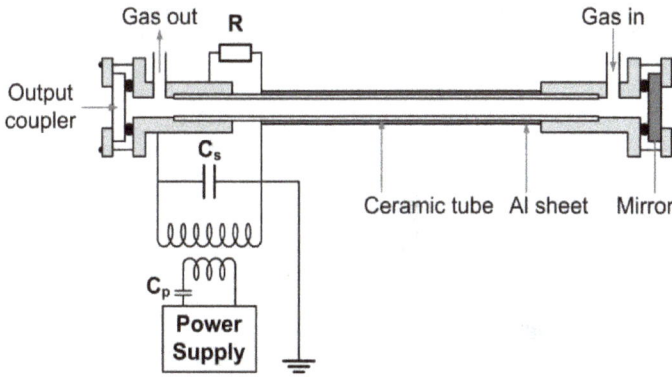

(a)

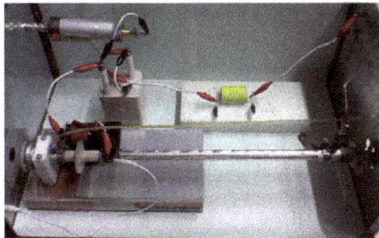

(b)

Figure 18. Short-pulse longitudinally excited CO$_2$ laser with switchless circuit (Uno et al, 2012b). (a) Schematic diagram. (b) Photograph.

For example, in our study (Uno et al, 2012b), the laser tube was the same as the above longitudinally excited CO_2 laser with the capacitor-transfer circuit (Uno et al, 2009). In the excitation circuit, the primary capacitance C_P was 4.7 µF, the storage capacitance C_s was 700 pF, and the resistor R was 10 MΩ.

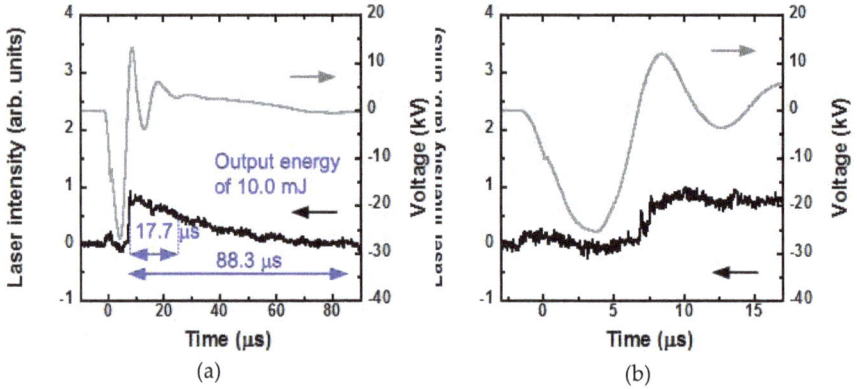

Figure 19. Discharge voltage and CO_2 laser pulse waveforms at 3.8 kPa (CO_2: N_2: He= 1: 1: 2) (Uno et al, 2012b). Black and gray lines represent laser pulse and discharge voltage. (a) Overall waveform. (b) Magnified time scale view of the start of discharge.

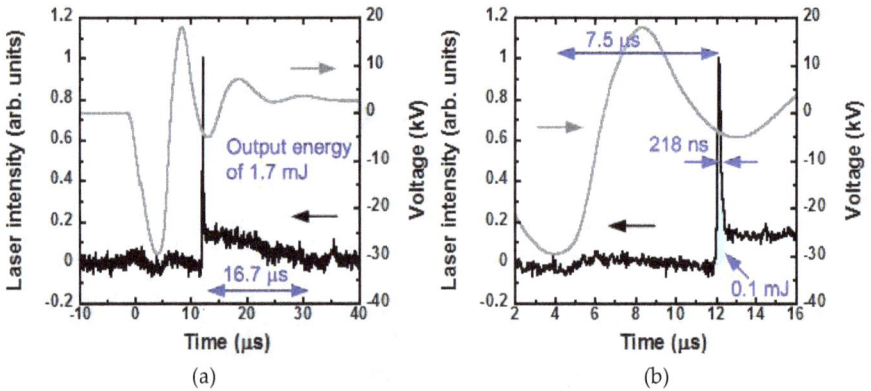

Figure 20. Discharge voltage and CO_2 laser pulse waveforms at 9.0 kPa (CO_2: N_2: He= 1: 1: 2) (Uno et al, 2012b). Black and gray lines represent laser pulse and discharge voltage, respectively. (a) Overall waveform. (b) Magnified time scale view of the start of discharge.

The long pulse oscillation takes place at gas pressure less than 3.8 kPa (CO_2: N_2: He= 1: 1: 2) and the short pulse oscillation takes place at gas pressure more than 4.2 kPa. Fig. 19 shows the discharge voltage and laser pulse waveforms at 3.8 kPa. The discharge voltage reached -25.8 kV at a raise time of 3.4 µs. In the low pressure, the breakdown voltage is low. The discharge starts when the applied voltage reaches the breakdown voltage. The fall time of the main discharge was 2.8 μs. Laser oscillation took place 3.5 μs after the start of discharge. At this time, the discharge continues. The laser pulse did not have a spike pulse and is a conventional long pulse oscillation. The laser output energy was 10.0 mJ. The laser pulse width was the full width at half maximum (FWHM) of 17.7 μs and the pulse length of 88.3 μs. Fig. 20 shows the discharge voltage and laser pulse waveforms at 9.0 kPa. The discharge voltage reached -29 kV at a raise time of 3.1 μs. In the high pressure, the breakdown voltage is high and the applied voltage reaches maximum. The fast discharge is influenced by the impedance matching of the excitation circuit and the discharge space. The fall time of main discharge was 2.2 μs. The laser oscillation took place 7.5 μs after the start of discharge. The laser pulse had a spike pulse width of 218 ns and a pulse tail length of 16.7 μs. The laser output energy was 1.7 mJ. The energy of the spike pulse part was estimated to be 0.1 mJ. It may be necessary to improve the output energy for various supplications.

5. Application

5.1. CW and long pulse

5.1.1. Medical applications

80 wt% of soft biological tissue is water. Evaporation of water is a factor important for soft tissue excision (Awazu, 2008). In laser irradiation, evaporation of water takes place at the critical temperature of 101 K (Auerhammer et al, 1999). The evaporation of water causes the soft tissue excision around the laser irradiation part. The tensile strength of soft tissue is about 10 MPa on skin and several MPa on corneas, blood vessel, and muscle (Duck, 1990). The laser excision of soft tissue is carried out by a high mechanical power which causes a phase transition because 1 atm is about 0.1 MPa.

A CO$_2$ laser is effectively absorbed in water. In 9.2 – 11.4 μm of the CO$_2$ laser wavelength band, the absorption coefficient is 10^3 cm^{-1} (Zarrabi & Gross, 2011). Therefore, in the CO$_2$ laser, the penetration depth to soft biological tissue is short and about 50 μm. The CO$_2$ laser evaporates only surface of soft tissue. Heat influence of deeper tissue is small and the damage to normal tissue is minimally suppressed. Edema and sharp pain after an operation are little because the heat influence is restrictive. In the CO$_2$ laser irradiation, shrink, dehydration, and carbonization of surface tissue take place. They cause water evaporation and vaporization of inner tissue. In this process, a protein denaturation layer of residual tissue is certainly generated. The denaturation layer functions as a solidification layer for an arrest of hemorrhage.

At present, a CO$_2$ laser of about 3 W a pulsed CW of (10 – 1000 ms and a super pulse of 0.1 – 1 ms) is used for the skin treatment such as eliminating a mole, a wart, macule (Gotkin et al, 2009) and wrinkle the aurinasal treatment such as hay fever and allergic rhinitis (Takeno et al, 2009), the eye treatment such as blepharochalasis and droopy eyelid (Rokhsar et al, 2008), and the dental treatment such as section and evaporation of soft tissue, hemostasis, and stomatitis (Zand et al, 2009). The advantages of CO$_2$ laser surgery are alleviation of bleeding, pain and edema, and shortening of operation and recovery time.

5.1.2. *Industrial applications*

When laser is irradiated to a material surface, a part is reflected by the reflectance of the material to the laser wavelength at the material surface, and the remainder is absorbed at the material surface. Then, the temperature of the material surface increases. A part of the heat is radiated by the heat radiation rate of the material, and the remainder conducts to inside by the heat conductivity of the material and makes the internal temperature increase. The surface temperature is fixed by the balance between the thermal energy lost by heat conduction and radiation and the thermal energy given by the laser. If the surface temperature does not reach the melting point of the material, a surface treatment takes place. If the surface temperature reaches the melting point, the melting takes place. When the heat input stops by movement or a stop of the laser beam, the material is cooled. Solidification causes welding. Eliminating the molten material (e.g. gas blowing) causes cutting. When the surface temperature exceeds the boiling point of the material by increasing the thermal energy given by the laser beam, a hole called keyhole begins to be formed. The laser reaches the inside of the material through the keyhole. The penetration causes deep melting process called keyhole welding.

The laser cutting is a specialty application of CO_2 lasers. In the laser cutting, a beam control is important. In space, the beam with TEM_{00} mode or the beam with low order mode of axial symmetry like TEM_{00} mode is required. In time, the control of the beam power (W/cm^2) is required. In the laser cutting of metals like steel, a fine cutting without heat influence, strain, and dross is required. It is very important to optimally control heat input by control of the peak power, the repetition rate, the duty ratio, and so on. These factors must be optimally controlled according to the kind of material, the thickness, and the cutting speed. A metal with a few mm of thickness can be cut by the CO_2 laser with the peak power of 100 – 200 W, the repetition rate of about 300 Hz, the cutting speed of about 30 cm/min, and the duty ratio of about 30%. In non-metallic materials such as woods, cloth, paper, ceramics, glass, gum, plastics, polyimide, epoxy, polycarbonate, vinyl and so on, the absorption of CO_2 laser light is very large. Most of these non-metallic materials can be processed (cutting, hole making, marking, etc.) by a low output CW-CO_2 laser emitted less than 300 W. The laser processing is used for the cutting of automobile instrument panels and clothes, the hole making of the material (e.g. ceramics), which the mechanical processing is difficult for, and so on.

5.2. Short pulse

5.2.1. *Applications by commercial TEA-CO$_2$ lasers*

An ablation process by irradiation of a short-pulse CO_2 laser is used for hole-making a printed board, marking to a resin board, and cleaning a metal part. In the hole-making of a printed board, improving density of trace is the purpose. Making a via-hole with a diameter of 50 mm or less requires several tens kW of the peak power, a few µs of the pulse width and a few kHz of repetition rate. Thus, TEA-CO_2 lasers are used for the-hole making. The laser marking stamps the model number and name of a product, the date of manufacture, a lot number, etc. In a TEA-CO_2 laser, image transfer with mask is used because the TEA-CO_2

laser produces high output power. CO_2 lasers are used for marking to resin boards, such as glass epoxy. In the laser cleaning, TEA-CO_2 lasers are used for dirt and coat removal on metal material. Dirt and a coat are removed by laser ablation because they tend to absorb the CO_2 laser light than metal. This process does not give the damage by heat to a surface of metal, because the laser light is reflected when the metal material appears. The laser cleaning does not use a medical fluid and blast material, and is a dry process.

5.2.2. Future applications

At present, drilling of hard tissue (enamel and dentine) uses an Er:YAG laser with the wavelength of 2.94 μm that is well correspond with the absorption wavelength of water. In 2.94 μm of the Er:YAG laser wavelength, the absorption coefficient is 10^4 cm^{-1} (Zarrabi & Gross, 2011). The laser light is focused on water and micro-explosion of water drills hard tissue. CO_2 laser can also drill hard tissue because the CO_2 laser (9.2 – 11.4 μm) is well absorbed in water. However, a long-pulsed CO_2 laser which is used in the present dental clinic causes carbonization of hard tissue. Fig. 21 (a) and (b) shows the human tooth surface irradiated by a long-pulsed longitudinally excited CO_2 laser with the pulse width of 30 μs and the output energy of 80 mJ. The dentin which contains heat-sensitive protein was carbonized by thermal influence because of the long-pulsed CO_2 laser. On the other hand, Fig. 21 (c) and (d) shows the human tooth surface irradiated by a short-pulsed longitudinally excited CO_2 laser with the spike pulse width of about 100 ns, the pulse tail length of about 60 μs, and the output energy of 80 mJ (Uno et al, 2009). The short-pulsed CO_2 laser resulted in almost no carbonization not only for the enamel but also for the dentine.

Additionally, CO_2 lasers are also absorbed in the human tooth (Heya et al, 2003). The dentine has a weak acid resistance and becomes a saprodontia easily. The dentine is reformed to a hydroxyapatite that has a strong acid resistance by the laser with the wavelength of 8.8 – 10.6 μm (Heya et al, 2003). CO_2 lasers can change the dentine to the hydroxyapatite because the wavelength corresponds to the wavelength of the CO_2 laser. Moreover, the dentin covers dental pulp and is porous. Exposure of the dentine causes dentinal hypersensitivity. The irradiation of the CO_2 laser melts and blocked the dentine surface. Thus, by the CO_2 laser, the medical treatment of dentinal hypersensitivity is also possible. The longitudinally excited CO_2 laser is expected as a complex dental care machine because the laser can produce the long pulse for the soft tissue treatment and the short pulse for the hard tissue treatment with a low-cost and portable device.

Glass marking can be effectively performed by a low-power CO_2 laser because glass absorbs infrared light well. Additionally, the glass marking is energy saving because a CO_2 laser has high electro-optical conversion efficiency. However, a long-pulsed CO_2 laser produces cracks on the surface of glass by heat influence. Fig. 22 (a) shows the glass surface irradiated by a long-pulsed longitudinally excited CO_2 laser with the pulse width of about 30 μs and the output energy of about 45 mJ. The processing part looks white by cracking. On the other hand, a short-pulsed CO_2 laser produces crackless marking. Fig. 22 (b) shows the glass

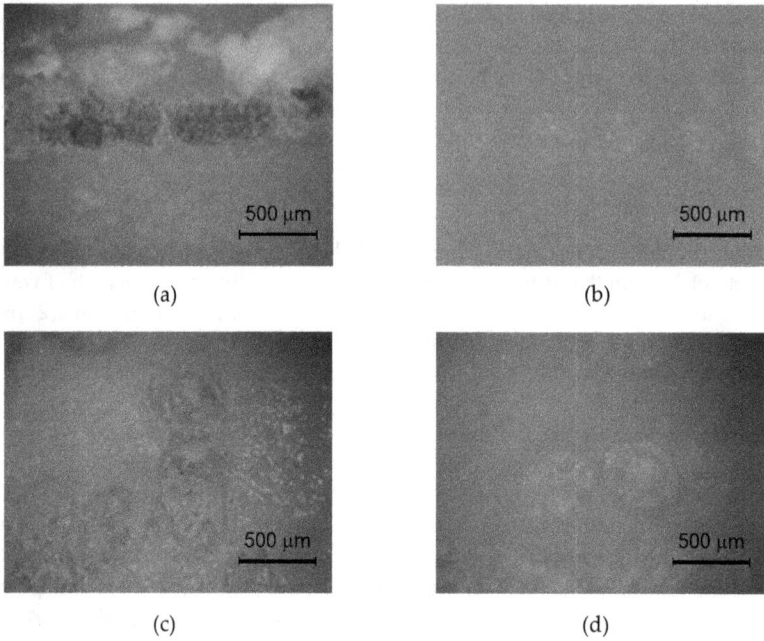

(a)   (b)

(c)   (d)

Figure 21. Tooth surface irradiated longitudinally excited CO_2 laser (Uno et al, 2009). (a) Dentine surface by long-pulsed CO_2 laser irradiation. (b) Enamel surface by long-pulsed CO_2 laser irradiation. (c) Dentine surface by short- pulsed CO_2 laser irradiation. (d) Enamel surface by short- pulsed CO_2 laser irradiation.

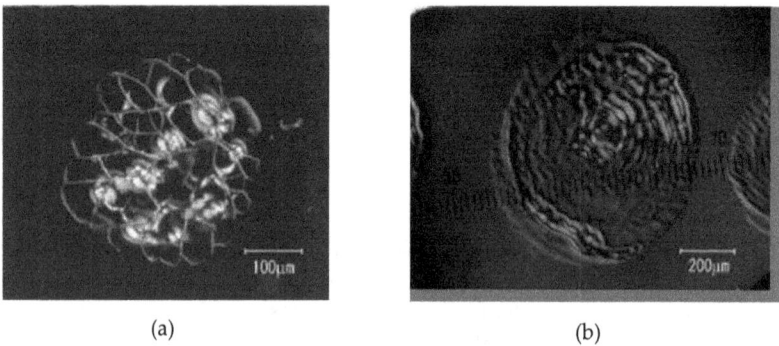

(a)   (b)

Figure 22. Glass surface irradiated longitudinally excited CO_2 laser (Uno et al, 2009). (a) Long-pulse CO_2 laser irradiation. (b) Short-pulse CO_2 laser irradiation.

surface irradiated by a short-pulsed longitudinally excited CO_2 laser with the spike pulse width of about 100 ns, the pulse tail length of about 60 μs and the output energy of about 45 mJ (Uno et al, 2009). The processing part is not visible with the naked eye because the part does not have a crack. In fact, the short-pulsed CO_2 laser produces stealth marking. However, the irradiation fluence is limited by the heat influence of the pulse tail. Therefore, the development of a short-pulsed CO_2 laser without pulse tail is desired. The short-pulsed longitudinally excited CO_2 laser with a low-cost and portable device will produce a device for product identification and traceability by marking for every process in a factory, and management of a sample by marking in a research institution.

6. Conclusion

In this chapter, the longitudinally excited CO_2 laser has been described. The excitation system and mechanism of the CO_2 laser, the feature of longitudinal excitation system and the excitation discharge, and the application of the CO2 laser have been explained. Especially, in the longitudinally excited CO_2 laser pumped by pulse discharge, new excitation circuits have been introduced and the relation between the discharge and the laser pulse has been explained. The longitudinally excited CO_2 laser can produce CW oscillation, long pulse oscillation, or short pulse oscillation with a compact, simple, and low-cost device. The development of a maintenance-free and warm-up-free device can be expected by the simple structure. Therefore, a user-friendly CO_2 laser which anyone can use easily always anywhere will contribute to various fields.

Author details

Kazuyuki Uno
University of Yamanashi, Japan

7. References

Auerhammer, J. M., Walker, R., Meer, A. F. G. & Jean, B. (1999). Dynamic Behavior of Photoablation Products of Corneal Tissue in the Mid-IR: A Study with FELIX. *Applied Physics B*, Vol. 68, (1999), pp. 119–119.

Awazu, K. (2008). *Infrared Laser for Biomedical Engineering*. in Japanese, Osaka University Press, Osaka, ISBN: 978-4-87259-160-6.

Battou, K., Ameur, K. A. & Ziane, O. (2008). Q-switch of a continuously pumped CO_2 laser with a scanning coupled-cavity Michelson mirror. *Optics Communications*, Vol. 281, (2008), pp. 5234-5238.

Bethel, J. W., Baker, H. J. & Hall, D. R. (1998). A new scalable annular CO_2 laser with high specific output power. *Optics Communications*, Vol. 125, (1998), pp. 352–358.

Biswas, A. K., Bhagata, M. S., Rana, L. B., Verma, A. & Kukreja, L. M. (2010). Indigenous development of a 2 kW RF-excited fast axial flow CO_2 laser. *PRAMANA*, Vol. 75, (2010), pp. 907-913.

Burkhard, P., Gerber, T. & Luthy, W. (1981). XeF excimer laser pumped in a longitudinal low-pressure discharge. *Applied Physics Letters*, Vol. 39, (1981), pp. 19–20.

Chung, H.-J., Lee, D.-H., Hong, J.-H., Joung, J.-H., Sung, Y.-M., Park S.-J. & Kim, H.-J. (2002). A simple pulsed CO_2 laser with long milliseconds pulse duration. *Review of Scientific Instruments*, Vol. 73, (2002), pp. 484-485.

Cleeschinsky, D., Dammasch, D., Eichler, H. J. & Hamisch, J. (1981). XeF-LASER WITH LONGITUDINAL DISCHARGE EXCITATION. *Optics Communications*, Vol. 39, (1981), pp. 79–82.

Deshmukh, S. V. & Rajagopalan, C. (2003). High-power multibeam CO_2 laser for industrial applications. *Optics & Laser Technology*, Vol. 35, (2003), pp. 517-521.

Duck, F. A. (1990). *Physical Properties of Tissue*. Academic Press, London,ISBN: 978-0122228001.

Eichler, H. J., Hamisch, J., Nagel, B. & Schmid, W. (1985). KrF laser with longitudinal discharge excitation. *Applied Physics Letters*, Vol. 46, (1985), pp. 911–913.

El-Osealy, M. A. M., Jitsuno, T., Nakamura, K. & Horiguchi, S. (2002a). Gain characteristics of longitudinally excited F_2 lasers. *Optics Communications*, Vol. 205, (2002), pp. 377–384.

El-Osealy, M. A. M., Jitsuno, T., Nakamura, K., Uchida, Y. & Goto, T. (2002b). Oscillation and gain characteristics of longitudinally excited VUV F_2 laser at 40 Torr total pressure. *Optics Communications*, Vol. 207, (2002), pp. 255–259.

El-Osealy, M. A., Ido, T., Nakamura, K., Jitsuno, T. & Horiguchi, S. (2001). Oscillation and gain characteristics of high power co-axially excited N_2 gas lasers. *Optics Communications*, Vol. 194, (2001), pp. 191–199.

Furuhashi, H., Hiramatsu, M. & Goto, T. (1987). Longitudinal discharge XeCl excimer laser with automatic UV preionization. *Applied Physics Letters*, Vol. 50, (1987), pp. 883–885.

Gotkin, R. H., Sarnoff, D. S., Cannarozzo, G., Sadick, N. S. & Armenakas, M. A. (2009). Ablative Skin Resurfacing With a Novel Microablative CO_2 Laser. *Journal of Drugs in Dermatology*, Vol. 8, (2009), pp. 138–144.

He, Z., Zhang, Y., Zhang, H., Zhang, Q., Liao, J., Zhou, Y., Liu, S. & Luo, X. (2010). Study of Optimal Cavity Parameter in Optically Pumped D_2O Gas Terahertz Laser. *Journal of Infrared, Millimeter, and Terahertz Waves*, Vol. 31, (2010), pp. 551-558.

Heya, M., Sano, S., Takagi, N., Fukami, Y. & Awazu, K. (2003). Wavelength and Average Power Density Dependency of the Surface Modification of Root Dentin Using an MIR-FEL. *Lasers in Surgery and Medicine*, Vol. 32, (2003), pp. 349–358.

Hill, A. E. (1968). MULTIJOULE PULSES FROM CO_2 LASERS. *Applied Physics Letters*, Vol. 12, (1968), pp. 324-327.

Hocker, L. O. & Phi, T. B. (1976). Pressure dependence of the atomic fluorine laser transition intensities. *Applied Physics Letters*, Vol. 29, (1976), pp. 493–494.

Javan, A., Bennett, Jr. W. R. & Herriott, D. R. (1961). Population Inversion and Continuous Optical Maser Oscillation in a Gas Discharge Containing a He-Ne Mixture. *Physical Review Letters*, Vol. 6, (1961), pp. 106–110.

Komatsu, K., Matsui, E., Kannari, F., & Obara, M. (1991). Low Pressure Atomic Xenon Laser Excited by Self-Sustained Longitudinal Discharge. *The Review of Laser Engineering*, in Japanese, Vol. 19, (1991), pp. 490–495.

Lee, D.-H., Chung H.-J. & Kim, H.-J. (2000). Comparison of dc and ac excitation of a sealed CO$_2$ laser. *Review of Scientific Instruments*, Vol. 71, (2000), pp. 577-578.

Loy, M. M. T. & Roland, P. A. (1977). Simple longitudinally pulsed CO$_2$ laser and its application in single-mode operation of TEA lasers. *Review of Scientific Instruments*, Vol. 48, (1977), pp. 554-556.

Miyazaki, K., Hasama, T., Yamada, K., Fukatsu, T., Eura, T. & Sato, T. (1986). Efficiency of a capacitor-transfer-type discharge excimer laser with automatic preionization. *Journal of Applied Physics*, Vol. 60, (1986), pp. 2721-2728.

Nagai, H. (2000). *Laser Process Technology*, in Japanese, Optoronics Co. Ltd., Tokyo, ISBN: 978-4-902312-36-2.

Newman, L. A. (1978). XeF* and KrF* waveguide lasers excited by a capacitively coupled discharge. *Applied Physics Letters*, Vol. 33, (1978), pp. 501–503.

Patel, C. K. N. (1964). SELECTIVE EXCITATION THROUGH VIBRATINAL ENERGY TRANSFER AND OPTICL MASER ACTION IN N$_2$-CO$_2$. *Physical Review Letters*, Vol. 13, (1964), pp. 617-619.

Photonics.com. http://www.photonics.com/LinearCharts/Default.aspx?ChartID=1

Rokhsar, C. K., Ciocon, D. H., Detweiler, S. & Fitzpatrick, R. E. (2008). The Short Pulse Carbon Dioxide Laser Versus the Colorado Needle Tip with Electrocautery for Upper and Lower Eyelid Blepharoplasty. *Lasers in Surgery Medicine*, Vol. 40, (2008), pp. 159–164.

Rosa, J., J. Eichler, H. & Herweg, H. (1986). ArF laser excited in a capacitively coupled discharge tube. *Journal of Applied Physics*, Vol. 54, (1986), pp. 1598–1599.

Soukieh, M., Ghani, B. A. & Hammadi, M. (1999). Mathematical modeling TE CO$_2$ laser with SF$_6$ as a saturable absorber. *Optics & Laser Technology*, Vol. 31, (1999), pp. 601-611.

Takeno, S., Hirakawa, K., Ishino, T. & Goh, K. (2009). Surgical treatment of the inferior turbinate for allergic rhinitis: clinical evaluation and therapeutic mechanisms of the different techniques. *Clinical & Experimental Allergy Reviews*, Vol. 9, (2009), pp. 18–23.

Terai, K., Murata, T. & Tamagawa, T. (1993). Characteristics of RF Excited CO$_2$ Lasers. *The Review of Laser Engineering*, in Japanese, Vol. 21, (1993), pp. 475–484.

Tian, Z., Sun, Z. & Qu, S. (2005). Tunable pulse-width, electro-optically cavity-dumped, rf-excited Z-fold waveguide CO$_2$ laser. *Review of Scientific Instruments*, Vol. 76, (2005), 083110.

Tripathi, A. K., Gupta, N. M., Chatterjee, U. K. & Bhawalkar, D. D. (1994). Development of a sealed-off cw CO$_2$ laser using a supported gold catalyst. *Review of Scientific Instruments*, Vol. 65, (1994), pp. 3853-3855.

Ueno, Y., Soumagne, G., Sumitani, A., Endo, A. & Higashiguchi, T. (2007). Enhancement of extreme ultraviolet emission from a CO$_2$ laser-produced Sn plasma using a cavity target. *Applied Physics Letters*, Vol. 91, (2007), 231501.

Uno, K., Akitsu, T. & Jitsuno, T. (2012a). Longitudinally excited CO$_2$ laser with short laser pulse using direct-drive circuit. *Journal of Engineering and Technology*, Vol. 2, (2012) pp. 101-106.

Uno, K., Jitsuno, T. & Akitsu, T. (2012b). Simple short-pulse CO$_2$ laser excited by longitudinal discharge without high-voltage switch. *Journal of Infrared, Millimeter, and Terahertz Waves*, Vol. 33, (2012), pp. 485-490.

Uno, K., Nakamura, K., Goto, T. & Jitsuno, T. (2006). Longitudinally Excited N_2 Laser Pumped by Lamplike Discharge. *Japanese Journal of Applied Physics*, Vol. 45, (2006), pp. 1651-1653.

Uno, K., Nakamura, K., Goto, T. & Jitsuno, T. (2008). Longitudinally excited N_2 lasers without high-voltage switches. *Review of Scientific Instruments*, Vol. 79, (2008), 063107.

Uno, K., Nakamura, K., Goto, T. & Jitsuno, T. (2008). Red-F* Laser and VUV-F_2 Emission Pumped at Low Pressure by Longitudinal, Lamp-Like Discharge. *Plasma and Fusion Research,* Vol. 3, (2008), 037.

Uno, K., Nakamura, K., Goto, T. & Jitsuno, T. (2009). Longitudinally Excited CO_2 Laser with Short Laser Pulse like TEA CO_2 Laser. *Journal of Infrared, Millimeter, and Terahertz Waves*, Vol. 30, (2009), pp. 1123–1130.

Xie, J., Guo, R., Li, D., Zhang, C., Yang, G. & Geng, Y. (2010). Theoreteical calculation and experimental study of acousto-optically Q-switched CO_2 laser. *Optics Express*, Vol. 18, (2010), 12371.

Yamanaka, C., Nakai, S., Matoba, M., Fujita, H., Kawamura, Y., Daido, H., Inoue, M., Fukuyama, F. & Terai K. (1981). The LEKKO VIII CO_2 gas laser system. *IEEE Journal of Quantum Electronics*, Vol. 17, (1981), pp. 1678-1688.

Zand, N., Ataie-Fashtami, L., Djavid, G. E., Fateh, M., Alinaghizadeh, M.-R., Fatemi, S.-M. & Arbabi-Kalati, F. (2009). Relieving pain in minor aphthous stomatitis by a single session of non-thermal carbon dioxide laser irradiation. *Lasers in Medical Science*, Vol. 24, (2009), pp. 515–520.

Zarrabi, A. & Gross, A. J. (2011). The Evolution of Lasers in Urology : Lasers: A Short History and Simplified Physics. *Therapeutic Advances in Urology*, Vol. 3, (2011), pp. 81–89.

Zhou, Z., Zeng, Y. & Qiu, M. (1983). XeCl excimer laser excited by longitudinal discharge. *Applied Physics Letters*, Vol. 43, (1983), pp. 347–349.

Cutting and Shooting

Chapter 5

Ultrashort Laser Pulses Machining

Ricardo Elgul Samad, Leandro Matiolli Machado,
Nilson Dias Vieira Junior and Wagner de Rossi

Additional information is available at the end of the chapter

1. Introduction

Over the past several years, the necessity for micromachining technologies has been growing from scientific research to industry. Segments from medical appliances, microelectronics and the automotive world demand a great variety of applications such as micromotors, microfluidic circuits, MEMS (Micro Electrical Mechanical System), medical devices, electronic tooling, particle filters, micromolds and microvalves, among others [1]. This massive growth in the micromachining segment demands a variety of new micromachining methods.

Techniques to machine the surface of materials are continually being improved. Recently, mechanical micromachining with accurate positioning systems and very small drills were used to trim out materials, covering the range from some millimeters down to a few tens of nanometers [2]. A similar technique that also applies direct micromachining and requires a highly accurate positioning system but no contact, is Electron-Discharge Micromachining (EDM), in which material is removed by an electrode guided along the desired path, very close to the surface of a conducting material submerged in a dielectric fluid; a spark is established between the electrode and the material being processed, removing matter by melting and evaporation [3, 4], with micrometric resolution. Similarly to EDM, the Focused Ion Beam (FIB) technique scans an ion beam, with spot sizes ranging from 10 to 500 nm, over a specimen, etching the material [5]. Additionally to etching, this technique can be used to build new structures by localized deposition using the ion beam-induced decomposition of an organo-metallic gas [6]. Examples of the use of a FIB system are the milling of angled cuts into the suspension ligaments of MEMS accelerometers, and the milling of large depth-to-width aspect ratios tracks to make photonic band gap lattices in $As_{40}S_{60}$ [5, 7]. A disadvantage of this technique is that it does not allow batch production due to the small amount of material removed during machining.

Photolithography, a technique that is available for integrated circuits fabrication since the 1970's, has been improved, and nowadays is used to create structures ranging from micrometers, as in the case of MEMS devices, down to tens of nanometers for integrated circuits [8]. This technique selectively removes material from thin films (nitride/oxide /polysilicon) deposited on silicon substrates, and also from the bulk of the substrate [9]. The process consists in using UV light and optical projection to transfer a pattern from a mask to a light sensitive material (photoresist) film pre-deposited on the substrate surface, selectively curing it. The non-cured photoresist is then removed by a developer, and the substrate exposed portions are etched either by a liquid (wet etching) or by a plasma (dry etching). The repetition of many cycles of photoresist deposition and etch can create complex structures, in a process known as *Surface Micromachining*. Different patterns of polycrystalline silicon can be selectively etched and are used as sacrificial layers to create suspended parts attached to the substrate, building 3D structures on the substrate surface [10]. This process enables the creation of structures such as accelerometers [11] and neural probe arrays [12]. When the etching removes a large quantity of material from the bulk of the substrate, the process is known as *Bulk Micromachining* [9], and is used to create devices such as piezoresistive sensors [13]. Photolithography allows batch production, but it is considered the most indirect method, since material is not removed by a tool.

Although largely used, the described techniques have some disadvantages, such as tool wearing (mechanical micromachining and EDM), high complexity (photolithography) and necessity of processing in vacuum (FIB). Besides, when machining different materials, the etching rates can strongly depend on the material and the tool, and the processing parameters, or even the tool, have to be changed. Additionally, the technique can be restricted to a specific material class, such as metals in the EDM case. Other limitations include the small etching speed in the techniques that require precise positioning (mechanical machining, EDM, FIB), preventing batch processing, and the high complexity of the photolithography, which does not allow fast prototyping. In view of those limitations, the ultrashort pulse laser emerges as a valuable tool for micromachining.

Just a few years after the laser invention, it was already being used as a tool to ablate [14] and machine [15] a wide variety of materials, including metals [16], dielectrics [17], semiconductors [18], composites [19] and biological tissues [20]. At that time, the machining was based on thermal processes arising from the material heating by the laser [21], and the ablation occurred as a consequence of melting and vaporization resulting from phase transitions due to the heat deposited into the material being processed [22, 23].

When machining with laser pulses, the laser beam is focused on the material being processed, the light is absorbed and the material undergoes a physical change around the absorption region. The dimensions of the structures that can be machined depend, on a first approximation, on the focused beam diameter. A diffraction limited TEM_{00} Gaussian beam [24] can be focused to diameters close to the laser wavelength, ranging from a hundred nanometers (Excimer UV lasers) to tens of microns (CO_2 lasers). Nevertheless, even with the smallest attainable laser spot on the material surface, the interaction region follows the heat diffusion volume guided by the process dynamics. For long pulses, which last more than the typical period of lattice vibration (around tens of picoseconds), the dynamics of the laser-matter interaction depends on the laser parameters (wavelength, pulse energy, repetition rate) and on the physical characteristics (absorption, heat capacity, thermal conductivity) of the material being machined. In this long pulse regime, the machining process is a consequence of heating (that results in melting, evaporation, sublimation), depending on the energy absorbed per unit of time and volume, and how it flows inside the material. When the laser radiation is absorbed, the material is heated and thermodynamic processes are responsible for heat diffusion and phase transformations. As the pulse duration shortens, higher intensities are faster achieved, the spatial heat propagation is decreased and phase transformations that lead to material removal occur more efficiently; in this regime, the ablation threshold decreases with the square root of the pulse duration, demonstrating that the ablation is a thermal process [25]. The excess heat that is not used to jettison material flows into the surroundings increasing its temperature, and eventually creating a Heat Affected Zone (HAZ), in which phase transitions modify the material properties, usually in a detrimental way. As a consequence, smaller structures require shorter pulses to be etched, which also produce a reduced HAZ. Nevertheless, although microsecond to sub-nanosecond laser pulses generates intensities high enough to machine metals, semiconductors and even some transparent dielectrics, the heat conduction expands the affected region far beyond the focused beam diameter, creates a large HAZ, and can also generate structures several times bigger than the wavelength used. Collateral effects from heat, like burr, debris and molten material, spread the interaction area limiting the machining precision. Additionally, albeit the high intensities attainable in this temporal regime allow the etching of transparent materials such as sapphire and silica, large amounts of energy are necessary to reach the ablation threshold, resulting in thermal stresses and damages such as chipping and micro-cracking [26, 27]. To handle these problems, different wavelengths are needed for efficient absorption by different materials. For dielectrics, lasers with wavelengths in the UV range are used for better absorption; some glasses have greater absorption in the far infrared [28], hence CO_2 lasers are often employed.

Femtosecond (10^{-15} s) pulses come up as a new possibility for micromachining with some advantages over longer pulses. The machining performed by these ultrashort pulses relies on their very brief duration, which is shorter than the thermal vibration period of the lattice. This small duration minimizes the energy transfer to the material, resulting in an almost nonthermal etching that minimizes the HAZ and preserves the properties of the surrounding material. Also, the ultrashort pulses duration creates very elevated intensities that promote a highly nonlinear interaction with matter, resulting in extremely localized ablations that can etch the material with nanometric precision, allowing the fabrication of very precise, minute structures. Additionally, ultrashort pulses promote a non-selective ablation regarding the material class (dielectric, metallic, etc.), allowing the machining of all kinds of materials with the same laser. The ability of ultrashort pulses to machine any material, even transparent ones, without changing the laser wavelength and the recent availability of relatively cheap, stable, high average power systems that increase the processing speed, are making these pulses very versatile tools to process materials for many high technology applications and devices. Also, in this temporal scale heat diffusion is minimized, and, with the adequate laser parameters, heat effects can be completely eliminated. Furthermore, is worth mentioning that, although it is not strictly a micromachining technique, the creation of micro and nanostructures is possible by the nonlinear interaction of the ultrashort pulses with a liquid solution. In this technique an ultrashort pulses beam is focused, through an optical a microscope optics, into the liquid solution containing a photosensitive polymer [29] or a photon induced reduction composite material [30]; the multiphotonic absorption that happens at the beam focal region promotes the highly localized solidification of the solution, allowing the manufacture of micro and nanometric structures.

The disadvantages in micromachining with femtosecond pulses comes from the high degree of precision demanded, with consequent necessity of tight focusing that imposes very accurate and expensive 3-axes positioning systems. Also, the low average powers that prevented high batch production are being overcome by modern systems with dozens of watts, allowing increased throughputs.

The invention, in the second half of the 1980's, of the Chirped Pulse Amplification Technique [31, 32], the KLM Ti:Sapphire laser [33], and the diode pumped solid state lasers [34] disseminated the use of microjoule-millijoule (μJ-mJ) ultrashort pulses to many laboratories around the world. The widespread availability of these systems in the decades of 1990-2000, resulted in the emergence of many new applications, including the machining by ultrashort pulses. The most used ultrashort pulse lasers nowadays for machining are Ti:Sapphire and Yb:fiber. The Ti:Sapphire systems can deliver pulses as short as a few tens of femtoseconds, centered at 800 nm, with mJ level energies, and repetition rates up to a few kHz; the Ytterbium based lasers generate pulses centered around 1030 nm lasting hundreds of femtoseconds, and, although limited in energy to the μJ range, operate at tens of MHz repetition rates, increasing its machining speed. Recent developments in laser technology design offer high average power systems [35], which in conjunction with beam conforming devices [36] and high speed scanning systems enable the use of ultrashort pulses machining in batch production, without limiting its fast prototyping capabilities.

2. Ablation by ultrashort laser pulses

The typical characteristics of ultrashort pulses for ablation of solids are energies in the range from tens to hundreds of microjoules, pulse widths around 100 fs and focalization to 20 μm radius spot sizes, generating intensities in the range 10^{12}-10^{14} W/cm^2.

When an ultrashort pulse impinges on a solid, the fact that its duration is shorter than the lattice ions vibrational period means that most of the pulse energy will be transferred to the material electrons, heating them. The majority of the electronic thermal energy will be carried away with the ablated material, and the small remaining portion will be added to the energy directly coupled into the ions, heating the lattice, with the possibility of creating a small HAZ. The control of the irradiation conditions can reduce this HAZ to almost zero.

2.1. Ablation physical mechanisms

The underlying mechanism for ultrashort pulses ablation is almost the same for all material classes (metals, semiconductors, dielectrics, polymers, etc.): when the pulse impinges on the material surface, seed electrons are accelerated by the pulse electrical field into a quivering motion, and either generate free electrons in the conduction band by collisions in an exponential avalanche process [37, 38], or are ejected from the surface due to the acquired kinetic energy [39]. The ejection process has a high occurrence probability on the 100 nm surface layer for electrons that acquired a few eV of kinetic energy, and leaves a charge imbalance that can produce a Coulomb explosion [39-41] of the lattice ions after the pulse. Simultaneously to the surface electrons ejection, the avalanche process occurs in deeper layers in the material, increasing the free electrons density and temperature, while the lattice is kept at a lower temperature (two-temperatures model [42, 43]); if the pulse energy and intensity are high enough, the free electrons density reaches a critical value (around 10^{21} cm^{-3} [38, 44]), and then the electrons transfer its energy to the surrounding ions. This relaxation quickly heats the neighboring lattice above its vaporization temperature, creating an unstable phase that undergoes a violent adiabatic expansion (phase explosion) [45-47], which removes material from the surface, carrying most of the thermal energy with it. Also, spallation [47] and fragmentation [45] can occur, although those are not dominant effects. The energy that is not taken away flows to the lattice, heating it. The dominating mechanism will depend on the pulse characteristics (energy, duration) and material.

For metals, the seed electrons are the conduction band free electrons. For the other materials, the pulse leading edge excites electrons from the valence band either by multiphoton ionization [25, 48] or tunneling induced by the laser field [49, 50]. Once the free electrons are present, a metallization occurs, and the electrons heating evolves deterministically in time [38, 50] in almost the same way in all materials [51].

Depending pulse energy, two ablation domains can be identified: the low and high fluence ones. In nonmetals, these regimes are defined by the dominant ablation mechanism: in the low fluence regime the Coulomb explosion predominates [41], while the phase explosion prevails for the high fluence ablation. In metals, Coulomb explosion does not occur because, at the intensities used in machining, the surface charge accumulation is effectively quenched [40] by the electronic mobility, suppressing the positive ions explosion. The electric field needed to promote positive charge accumulation in metals (~10^{10} V/m) demands intensities around 10^{18} W/cm^2 [52], pushing the Coulomb explosion threshold way above the phase explosion one. In metals, the low and high fluence regimes occur when the HAZ is shorter or longer than the optical absorption depth (Beer's Law [53]), respectively. The regimes threshold can be predicted using the two temperatures model [42, 43] that describes the electronic heating, diffusion and lattice heating.

For all material classes, the main difference between the two ablation regimes is that in the high fluence there will be fusion in the interaction region, and the remaining material can cool down to an amorphous or polycrystalline phase, whose physical characteristics (mechanical, optical, etc.) can differ from the starting material ones [54]. Also, the HAZ will be bigger in the high fluence regime.

Since, by the mechanisms explained, the ultrashort pulses ablation process depends primarily on the electrons answer to the laser field excitation and not on the thermodynamic properties that arise as a consequence of the atomic lattice, the ablation threshold does not show a square root dependence on the pulse duration [25], as in thermal ablation processes; additionally, the ablation by ultrashort pulses has a nonselective character, and the only parameter that has to be known to etch a material is its ablation threshold fluence, F_{th}. As a general rule, since in nonmetals a portion of the pulse energy is used to create free electrons, these material usually present higher ablation thresholds than metals for the same laser conditions. Also, a single material can present two values for the ablation threshold, one for low and other for high fluence, as can be seen in Figure 1, and they have to be known and taken into account when machining the material.

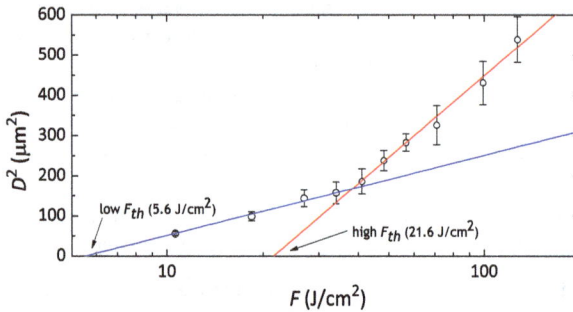

Figure 1. Diameter squared of the region ablated by ultrashort pulses on the surface of Sapphire as a function of the pulse fluence, for single-shot 100 fs pulses [55]. The ablation threshold fluence is given by the fluence at which the ablated diameter is zero, obtained from the fit by equation (1) [56]. The low and high fluence regimes can be clearly identified by two different slopes on the data points.

2.2. Ablation threshold determination

The determination of a given material ablation threshold fluence by ultrashort pulses is usually done using the "zero damage" method, introduced by Liu in 1982 [56]. This method consists in using a TEM₀₀ Gaussian beam to ablate the material at various positions in the sample surface, using different pulse energies. Then the diameters of the ablated craters, D, are measured and their square values are plotted as a function of the pulse fluence, F, as show in Figure 1. The data is then fitted by:

$$D^2 = 2w^2 \ln\left(\frac{F}{F_{th}}\right), \tag{1}$$

where w is the beam spot size at the sample surface and F_{th} is the ablation threshold fluence, at which the ablated crater diameter is zero. To execute this measurement, the beam propagation parameters have to be known to calculate its spot size and fluence at the sample surface, and many measurements must be done. Experimentally, this method is demanding because it requires the precise knowledge and stabilization of the beam parameters, a good sample positioning system to maintain the beam spot size constant at the sample surface for all the measurements, and since it can last a long time, up to hours, it is prone to be affected by laser instabilities. Additionally, frequently the use of an electron or atomic force microscope is needed to determine the crater diameter for fluences close to the threshold, at which the damage size is close to zero and is difficult to be measured.

A few years ago we introduced the Diagonal Scan (D-Scan) technique [57, 58], an alternative and simple method to measure the ablation threshold for ultrashort pulses. The method consists in moving a sample longitudinally and transversely (z and y directions in Figure 2a) across the beamwaist of a focused beam, from a position before the beamwaist, where there is no ablation, to a position after the ablation stops. In this way, a symmetrical profile with two lobes, as the one shown in Figure 2b, will be etched on the sample surface. If the etched profile does not present two lobes, the measurement has to be repeated with a higher pulse energy or with a tighter focusing lens [57].

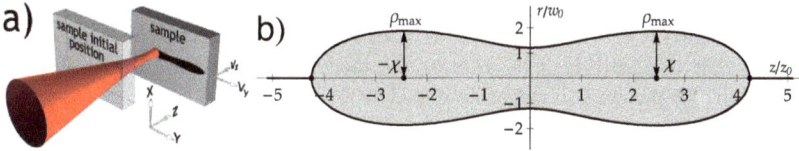

Figure 2. a) Scheme of the D-Scan method. b) profile etched on the sample surface by the diagonal movement across the beam waist position.

It can be shown [57] that the sample ablation threshold fluence is given by:

$$F_{th} = \frac{1}{e\pi} \frac{E_0}{\rho_{max}^2} \cong 0.117 \frac{E_0}{\rho_{max}^2}, \tag{2}$$

where E_0 is the pulse energy and ρ_{max} is the half maximum transversal dimension of the profile etched.

The determination of the ablation threshold involves knowing the pulse energy and the measurement of the profile maximum transversal dimension (typically a few tens of microns), which can be done in an optical microscope, and the use of equation (2). Compared to the "zero damage" method, the D-Scan is easier to be experimentally performed once it demands only one scan that can be done in a few seconds and a geometrical measurement, replacing a series of measurements that can last a long time, the knowledge of the laser spot size at the sample, and a fitting.

2.3. Ablation by many pulses and incubation effects determination

In a given material, the ultrashort pulses ablation threshold can depend on the presence of defects, dopants, impurities, excitons, etc. [59, 60], which either create intermediate levels in the bandgap or modify the local electronic density. As a consequence, seed and free electrons are created more easily than in the ideal material, and the ablation requires less energy to occur, lowering the ablation threshold value.

The defects can either be intrinsic to the material, or externally originated, such as color centers created by ultrashort pulses [61]. In this case, when etching solids with superimposing pulses, modifications are induced in the portions that are irradiated but not ablated, lowering these regions F_{th} for subsequent pulses due to defects accretion. The following pulses experience a decreasing F_{th} until the defects density saturates and the ablation threshold reaches a constant value. These cumulative phenomena are known as incubation effects [62-64], and the ablation threshold fluence modification caused by them must be taken into account when machining a material.

A few models were proposed by different authors to describe the incubation mechanisms. One of those is the probabilistic defect accumulation model [62] that, although widely used for different material classes [65-67], does not predict the ablation threshold saturation; other is the exponential defect accumulation model, in which the lowering of the ablation threshold increases the defect creation probability for the next pulse, until the defects saturation is reached and a constant value of F_{th} for the superposition of many pulses is established [59, 60, 63]. In this exponential model, the ablation threshold for the superposition of N pulses, $F_{th,N}$, can be described by [63]:

$$F_{th,N} = (F_{th,1} - F_{th,\infty})e^{-k(N-1)} + F_{th,\infty},\qquad(3)$$

where $F_{th,1}$ is the single shot ablation threshold fluence and $F_{th,\infty}$ is the saturated ablation threshold, below which there is no defect accumulation. If $F_{th,1} > F_{th,\infty}$, incubation effects are present and the ablation threshold decreases as more pulses hit the material; if $F_{th,1} < F_{th,\infty}$ laser conditioning occurs, and the material becomes progressively more difficult to ablate as the pulse superposition grows. In equation (3), k is the incubation parameter, which characterizes the strength of defects accumulation, and is equal or greater than zero. If $k=0$ the ablation threshold is constant and do not depend on the pulses superposition. As k grows, fewer pulses are needed to reach the incubation effects saturation.

To quantify the incubation effects for a given material, its ablation threshold is measured for various pulse superpositions, including the single shot case, and then equation (3) is fitted to the data. Usually this is carried out using the "zero damage" method, which produces a graphic like the one show in Figure 1 to determine F_{th} for each superposition, demanding a lot of experimental time and data analysis.

It is possible to use the D-Scan Method to determine the ablation threshold for the superposition of an arbitrary number of pulses, $F_{th,N}$, and for that the pulse superposition N that etches the profile maximum transversal dimension, $2\rho_{max}$, has to be known. To calculate this superposition, we hypothesize that it is given by the sum of the intensities generated at ρ_{max} by all the pulses that hit the sample during a D-Scan, normalized by the intensity generated by the pulse centered at $\circ$ (Figure 2b). Under these assumptions, and also considering that the longitudinal and transversal translation speeds (v_z and v_y in Figure 2a) are chosen to etch an elongated profile, it can be shown [68] that the pulses superposition N is given by:

$$N = \vartheta_3\left(0, e^{-\left(\frac{v_y}{f\,\rho_{max}}\right)^2}\right),$$ (4)

where f is the pulses repetition rate and ϑ_3 is the Jacobi Theta Function of the third kind [69]. It can also be shown [68] that, when the transversal speed is low or the pulses repetition rate is high, situation in which a large value of N is expected, equation (4) can be simplified to:

$$N = \sqrt{\pi}\,\frac{f\,\rho_{max}}{v_y} \approx 1.8\frac{f\,\rho_{max}}{v_y}, \qquad \text{for} \quad \frac{v_y}{f\,\rho_{max}} \approx 0.$$ (5)

Using the previous results, the determination of the ablation threshold for an arbitrary pulses superposition requires performing a D-Scan, then using the experimental values of ρ_{max}, v_y and f in equations (2) and (4) to provide $F_{th,N}$ and N. This is a fast experimental procedure since the D-scan can be done in a few seconds for high values of N (and in a few minutes for small ones), and its repetition for varying values of v_y and f quickly provides the ablation threshold for various superpositions, leading to a prompt determination of the incubation parameter from equation (3).

3. Ultrashort pulses machining main applications

The following sections describe ultrashort pulses machining applications, starting from low energy and fluence ones that produce surface modifications, going to high fluence usages that create 3D structures.

3.1. Low and high fluence regimes

Irradiation of metallic or semiconductor surfaces at fluences near the threshold can results in the formation of the so called "ripples", or "Laser Induced Periodic Structures" (LIPSS), which are regularly aligned, long and periodic structures. Their formation is not yet completely understood and is subject of intense study [70]. Typical periodicity is smaller than the laser wavelength, and for enough number of pulses it can evolve to trenches as deep as 1 µm, meaning an aspect ratio of 10 or more. The spacing and orientation of the ripples seems to depend mainly on the beam properties, as energy, number of shots, polarization and angle of incidence, and then, in principle can be controlled to some extent.

Further surface modifications, however, take place as the number of pulses or their energy increase, and the ripple pattern is broken by the formation of cone-like structures, which present a typical feature size and spacing at the order of few micrometers or even smaller. The shape and size of these structures can be further changed with the increase of fluence and number of superimposed pulses, and the use of an acid atmosphere has also been applied to increase the cones height and sharpness. Figure 3 shows examples of LIPSS and cone like structures produced by low energy ultrashort pulses from a Ti:Sapphire laser.

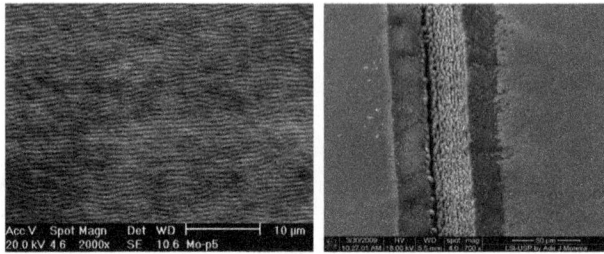

Figure 3. a) Laser Induced Periodic Structures (LIPSS) formed on molybdenum surface. b) Track of cone like structures (white central track) formed on silicon surface.

Up to this point, a minimum amount of material has been ejected from the surface and its topography has not been significantly recessed. To produce substantial depressions it is necessary to further increase the number of shots or the pulse energy. The shape of the created depression resembles the beam intensity profile, and for Gaussian beams the minimum diameter obtained by focusing can be as small as the laser wavelength. If the material and laser conditions allow, heating effects can be neglected, and holes and structures can be directly machined with this size. This is the low fluence ablation regime.

Processing in the low fluence domain can avoid heat accumulation, but the extraction of material is not efficient and only few hundreds of nanometers are etched by each laser pulse. An enormous increase in ejection efficiency is obtained when the fluence is raised to higher values and machining is performed at the high fluence regime. In this regime, heat is accumulated in the affected zone, and several associated phenomena occur. The most important one is the phase explosion, in which the local temperature raises so rapidly that there is not enough time for the material to undergo phase transition. The temperature goes well above the vapor value while the material remains in liquid phase. At this point, a severe explosion happens taking material away from the heated pool.

Machining with high fluence indeed offers high material removal efficiency, but the accumulated heat results in almost the same drawbacks as the ones verified with longer laser pulses that produce thermal ablation. Hence, like in conventional laser processing, this regime leads to debris formation, burr and molten material resolidification on the processed area sidewalls. A HAZ can also appear on the vicinity of the processed volume, causing phase transitions in metals and index of refraction changes in crystals and dielectrics. Due to the very low pulses energy, these effects can extend to a very small region of the produced structure and are irrelevant in many practical cases.

Heating effects also increase the ablated crater size, and the minimum diameter obtained can be far bigger than the focused spot size. On the other hand, low fluence processing can result in affected areas no longer restricted by the diffraction limit. The deterministic character of the laser-material interaction means that ablation will only take place in the portion of the beam cross section in which the intensity is above threshold. Of course this is only possible with a precise control of the laser energy and focus position from shot to shot, and a variation of more than 2% can prevent this possibility.

As already discussed, the complex nature of the ultrashort pulses and matter interaction is responsible for the onset of many different phenomena that depend on the particular characteristics of the laser and the material. The choice of these parameters enables the control of the affected region to produce a rather large amount of diverse structures. Among these possibilities, machining with femtosecond laser pulses can be roughly divided in three great areas: topographic surface modifications, selective ablation of surfaces, and direct writing of microstructures in surfaces and 3D structures in the bulk of transparent materials.

Topographic surface modifications occur when small amounts of material are extracted, resulting in very shallow recesses on the surface, nevertheless affecting its physical and chemical properties. In selective ablation of surfaces, very thin layers of one type of material can be completely etched without affecting the layer bellow it. Direct writing of surface microstructures results in the creation of pre-designed contours and profiles, produced by the relative movement of the laser beam and sample. 3D structures in the bulk of transparent materials are formed when the laser beam passes through the surface of a transparent material and is focused into it, creating complex 3D structures.

3.2. Topographic surface modifications

Nano and micro cone-like structures have been used to trap light and enhance the absorption of reflective surfaces. In particular, this was done by Mei et al. [71] on the surface of silicon with the purpose of using it as an improved sun light absorber. In their experiment, very thin and flexible pieces of silicon were irradiated by femtosecond pulses followed by HF etching. This acid etching enhances the micro spikes sharpness produced by the laser and increases the structure number of reflections, resulting in enlarged light absorption. The irradiation was done in the presence of sulfur hexafluoride (SF_6), which further improved light absorption by creating impurity bands due to sulfur atoms doping during laser irradiation.

Halbwax et al. [72] produced slightly different geometries of such micro structures on n-type silicon doped phosphorus. This was done by varying processing parameters in vacuum. Even using high fluence irradiation, spikes and other structures were formed with size and spacing in the few micrometers range. After boron implantation, these structured surfaces presented an increase of more than 30% in light absorption induced photocurrents, showing its potential for use in photovoltaic cells.

Nano and micro structures have also been used to change physical and chemical properties of titanium and ceramics surfaces in order to improve their biocompatibility for medical implants [73]. This is possible by creating very small structures on their surfaces, enabling, to some extent, the control of cellular growth and adhesion, inhibiting or promoting proliferation of certain types of cells. In this respect, Vorobyev et al. [74], produced several kinds of sub-micrometric structures on pure titanium surfaces by femtosecond laser irradiation. By the use of low and high laser fluences, and varying the number of overlapped laser shots, they produced diverse structures like nanostructured LIPSS, sphere-like nanoprotrusions and columnar microstructures, whose featured size ranged from few dozens of nanometers to hundreds of micrometers.

The same kind of texturing has also been studied to improve the tribological properties of several materials. This has been specially done on thin films of hard materials such as TiN and diamond-like carbon [75], in which nanostructured LIPPS changed the micro and macro frictional properties and hardness of these materials. Another consequence of the formation of micro arrays as cone-like or pillars, is their interaction with water, keenly increasing the contact angle of the liquid with the surface. By controlling the aspect ratio of these arrays, super-hydrophobic surfaces have been produced [76, 77], creating self-cleaning surfaces and which are also believed to take an important role in cell control processes.

As the presence of micro and nano structures on metal surfaces changes its interaction with light, some research has been done to control the surface reflectivity, and hence the perception of its color [78, 79]. The control of shape, size and spacing of gratings on surfaces is a well-known way to control the reflectivity of light; therefore, ripples formed by nano-structured LIPSS obtained by femtosecond laser irradiation are a natural approach to accomplish such a task. These characteristics can be controlled by handling some process parameters, as fluence, polarization direction, angle of incidence and scanning speed. The final shape comprises self-produced micro/nano gratings maintaining an almost fixed pitch and direction, covering a large area. Besides using LIPSS, color in metals has also been done by directly machining small grooves and arrays of micro-holes in their surfaces. Consequently, structuring metal surfaces in scales from nanometers up to hundreds of micrometers allow the control of the optical properties, from UV to terahertz.

3.3. Selective ablation of surfaces

The ability of ultrashort laser pulses to ablate very thin layers of material can be used to selectively remove coatings overlaid on a substrate. Layers as thin as a few hundreds of nanometers can be removed by each laser shot, and the process can be repeated up to the point where all the unwanted material is removed and the substrate is stricken by the laser beam. This process can be a precise and useful tool in many applications [80-84], as cleaning, restoration of art objects, decontamination and decoating of cutting tools.

The differences in physical properties between coating and substrate determine the feasibility, precision and complexity of the process. If the material to be removed absorbs the laser wavelength and presents a lower ablation threshold than the substrate, then the process is relatively simple and there is no need for feedback control. The pulse fluence is adjusted to be above that of the coating and below that of the substrate; an excess of pulses on the substrate surface will not affect it. On the other hand, if the coating damage threshold is close or lower than the substrate one, then a feedback must control the ablation process to prevent damage to the substrate. This can be accomplished by measuring, in real time, the spectrum of the emitted plasma (LIBS [81]) and observing the evolution of characteristics emissions of the coating and the substrate.

3.4. Direct writing microstructures in surfaces

Regardless of the topographic structures formed on a surface hit by femtosecond pulses, they can be used in a bigger scale to process complex structures previously designed, just like in a milling machine. Tiny holes and channels with high aspect ratios can be produced in almost any kind of material, with very high dimensional resolution, no heat effects, and beautiful cosmetic appearance.

Due to the relative low average power of the femtosecond laser systems, large processed areas or high ablated volumes still present a very low productivity, and are only considered when other methods fail. This happens when tiny structures must be produced on sensitive or transparent materials, when heating effects and dross must be avoided, and when sub-micron lateral size and resolution must be attained.

Although some processes use femtosecond laser assistance to produce sub-micron holes [85], these have also been directly drilled on thin films by precisely controlling the fluence just above the threshold [86]. Gaussian beams tightly focused produce a steep spatial intensity (fluence) distribution which precisely determines the diameter within which the fluence is above the threshold of the material. So the intended diameter of a hole can be achieved by using proper laser fluence and few hundreds nanometers holes can be directly drilled in thin films [87, 88].

Alumina is a very sensitive material that frequently presents cracks, delaminations and striations on the laser processed surface. However, some authors [89, 90] have managed to produce high aspect ratio and low tapered holes in alumina wafers suitable for micro-vias used in micro-electronic circuits. In trepanation mode, high fluences can be used in conjunction with other processing parameters in a way that minimizes heat accumulation and avoid recast molten material, cracks and delamination. Drilling in percussion mode produces smaller diameters, but these are dependent on the laser fluence. Unlike with longer pulses, however, the hole sizes almost do not depend on the number of overlapped pulses, which results in an easier and more precise production. Although percussion drilled holes present more tapered profiles, high aspect ratios can be obtained in relatively thick materials; typical diameters of a few dozens of micrometers are easily done on hundreds of micrometers thick alumina plates.

Structures with complex shapes in almost any kind of material can also be done by cutting and sculpting cavities with femtosecond lasers. The exact knowledge of the phenomena occurring during the laser material interaction allows the production of previously designed shapes in the same way as in a milling machining, with very high precision, but in a much smaller scale. Lateral and vertical accuracies depend on the fluence used and on the aspect ratio figure, but sub-micron or a few micrometers resolution is easily attained. Plates with thicknesses in excess of 0.5 mm have been cut to produce microparts and functional elements used, for instance, in MEMS and other microsystems. Carving into bulk or deposited films can also be done to produce trenches, microchannels and molds with applications in microelectronics, microfluidics and MEMS.

Just like in a milling process, machining with ultrashort laser pulses has been done in complete CAD/CAM systems where not only the toolpath has to be controlled, but also the energy and the number of overlapped pulses. Unlike a real milling tool, the size of a damage produced by a fixed laser focused spot increases as the number of shots accumulates (up to a certain point) on the same spot, and the same occurs by the increase in the pulse energy. Hence, the process control software must be fed with the fundamental parameters governing the interaction of the material with ultrashort pulses.

Although scanning mirrors can be used, the most common femtosecond processing systems use stable and precise 2D or 3D translation stages, with a still beam focused by a short focal length lens. Some cases demand vacuum processing, but generally the machining is performed in regular atmosphere with no shrouding gases. The pulse energy is controlled by a combination of a polarization rotator and an analyzer, and the pulses overlapping is calculated by combining the scanning speed with the laser repetition rate and the focused beam diameter.

As isolated components or as parts of a mechanical microsystem, tiny pieces have been machined in a variety of materials and shapes to an increasing number of new applications, where sensors and actuators for MEMS and microchannels for microfluidics are among the most important examples. Silicon, glasses and several kinds of ceramics, polymers and metals have been used to produce structures for these and other applications.

Silicon is one of the most largely studied materials for ultrafast laser machining, and silicon wafers have been cut and etched by laser to produce micro-molds to be used in a large number of micro electromechanical systems and electronic devices [91, 92].

Many kinds of polymers have also been machined by femtosecond lasers, mainly to produce small channels for rapid prototyping of microfluidic devices, and optical gratings. Suriano et al. [93] determined process protocols for laser machining of some polymers, showing the feasibility to obtain good shapes from some kinds of thermoplastic polymeric materials. Despite good geometrical and cosmetic results, some degradation, probably due the use of high fluences, was observed; physicochemical modifications in the processed area caused some darkening which was explained in terms of oxidation, dehydrogenation or thermal depolymerization.

Microstructured optical fibers were transversally machined to provide access to its core with minimum interference in their optical and guiding properties [94]. Micrometric transversal holes machined through polymeric cladding and glass core enable filling the fiber with fluids to take advantage of their optical properties for uses in sensors and in flexible microfluidic devices. Figure 4 shows an example of a square hole machined on the surface of a hollow fiber to allow gas filling to modify its optical properties.

In the work of Alemohammad et al. [95] a series of micro-grooves were transversally and accurately machined on the cladding of a fiber Bragg grating. This structure increases the transmission sensitivity of the fiber to the index of refraction of the external medium, making these engineered fibers a powerful tool for measurements of temperature and liquids concentrations. This is a good example of how controlled such machining can be, since it was precisely done (22 μm wide × 32 μm deep grooves with 50 μm pitch) on a thin polymer curved surface without damaging the core beneath it.

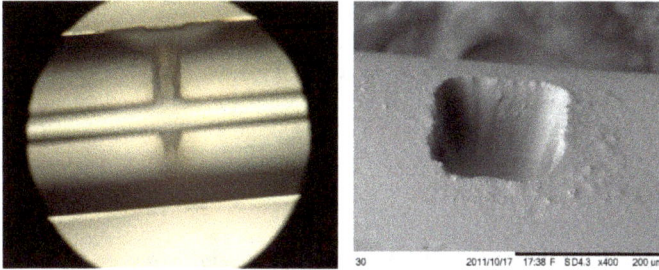

Figure 4. Hole bored in a hollow fiber with ultrashort pulses, using trepanation. a) fiber lateral view optical micrograph, and b) hole entrance view electron micrograph.

Another example of accuracy in laser processing of fibers is the machining of their core end faces [96]. Arrays of sub-micrometric structures boost the signal of a surface enhanced Raman scattering fiber probe, increasing its sensitivity for the detection of many biological and chemical substances. Minuscule cantilevers directly machined in one end of the fiber provided a very sensitive way of movement detection [97].

3.5. Micromachining in metals

Machining metals with ultrashort laser pulses is not as advantageous as machining dielectrics. The presence of free electrons, besides inhibiting Coulomb ablation, widens the heat affected region due to their high mobility. In this sense, the metal thermal properties play a fundamental role in the process characteristics and taking these into account, reproducible, small and accurate structures can still be produced [98].

The control of the pulse duration and wavelength can further improve efficiency and accuracy when machining some metals [99-101]. In a general way, as the pulse shortens (below 1 ps), the finishing improves and the ablation rate increases. Low fluence regime gives the best quality results, but at the same time implies a low ablation rate. Hence, work in this regime is only advantageous with systems offering high repetitions rates at moderate energies [102].

Another approach to increase efficiency is to work in the high fluence regime, where thermal processes, including phase explosion, fully control ablation. In this case, unlike pico or nanosecond pulses, almost all molten material is blown away in the form of nanometric particles, leaving a smooth and clean surface with almost no debris or dross and a negligible HAZ. Indeed, heat can be accumulated also from pulse to pulse as function of the laser repetition rate and thermal properties of the metal. Additionally, for sufficiently high repetition rates, the plume from the previous pulse can also provide shielding that scatters and absorbs pulse energy. Due to this, when processing with high repetition rates, from hundreds of kilohertz to some megahertz, efficiency and accuracy may vary according to the metal and process characteristics.

There are many examples of machining with femtosecond pulses to produce small devices in metals; some of the most common are microcantilevers [103], coronary stents [104] and pieces of NiTi shape memory alloy [105] used in MEMS. Percussion hole drilling in metals [106, 107] is also another application that is quickly growing. Thicknesses up to 0.5 mm and accuracy of few micrometers are reported in many different metals and applications, most of them without shrouding gas or vacuum, using fundamental wavelengths (800 nm or 1030 nm) and pulsewidths from 100 fs to 800 fs. The parts produced frequently do not need re-work or etching for cleaning and better finishing.

3.6. 3D structures in the bulk of transparent materials

The ablation threshold of the bulk of a transparent medium is considerable higher than of its surface (about twice). Even so, damages forming complex 3D structures can be produced in materials transparent to the wavelength used, by tightly focusing the laser beam below their surfaces. The requirement is to ionize and produce plasma in a volume inside the material, without affecting the surface above. To accomplish this, high numerical apertures (N.A.>0.5) objectives must be used to focus the beam under the surface, in a condition where the fluence is below the threshold on the surface, and at the same time overcomes the volume ablation threshold at the focus. In general, for glasses, nonlinear absorptions are induced for intensities in the range of $1\text{-}5\times10^{13}$ W/cm^2, which can be easily reached by any combination between the main processing parameters (pulse energy and duration, and beam diameter at focus).

For a laser beam with good fundamental spatial mode, the diameter of the focused spot can be close to its wavelength (800 nm for Ti:Sapphire laser, 1 μm for Yb:fiber). Practical temporal pulsewidths range from 30fs to 200fs for Ti:Sapphire and from 200 fs to 400 fs for Yb:fiber. Hence, less than 10 nJ are enough to create damages in the bulk of glasses when using Ti:Sapphire lasers, and this figure increases to 30–160 nJ when using Yb:fiber lasers. Under these conditions, induced absorption occurs in a volume of less than 1μm^3, and a single femtosecond pulse is enough to create damage in the bulk of any transparent material [108]. Within this volume, super-hot dense plasma is formed, creating very strong shock waves that destroy any atomic arrangement, resulting in modified structures or even voids.

The use of UV fs pulses enables the creation of even smaller defects into transparent materials, since its wavelength is shorter than near IR. Furthermore, the higher UV photon energy also increases the probability of multiphoton absorption, decreasing the breakdown intensity threshold. Using this approach, Dubov et al. [109] reported the production of 250 nm periodic nanostructures using 30 nJ, 300 fs laser pulses.

For intensities below the threshold, only nondestructive effects can occur, such as changes in the index of refraction of the material and color centers creation [61]. The presence of these defects produces intermediate levels in the band gap of the dielectric, resulting in incubation effects that usually lower the material damage threshold.

Although not a strict micromachining, permanent modification in the optical index of refraction have been extensively used to create light guiding structures in the bulk of transparent media [110, 111]. The relative movement of the laser beam focus inside the material produces a pathway with cross section and index of refraction contrast suitable to guide light as waveguides embedded in the material bulk. The focus movement can be done in the same direction as the beam propagation, or perpendicular to it. In the first case, the modified region presents a cross section with circular geometry, and in the other this geometry is elliptical. In spite of this disadvantage, the transversal geometry is preferred once the sample and waveguide lengths are not limited by the depth of focus as occurs in the longitudinal method.

The refractive index change depends on the laser parameters and material composition, and is induced by many different mechanisms not yet fully understood. The change can be either positive or negative, depending on the material properties, and is caused mainly by thermal effects, defects creation and material densification. High repetition rates (a few MHz) are generally used, causing heat accumulation with consequent melting and fast cooling in the focal volume scanned by the laser. These rapid phase transitions result in local structural modifications, giving rise to changes in the optical index of refraction that can be over 1%.

Many passive and active guiding structures have been incorporated in devices mainly used in communications and biochemical applications. Couplers, splitters and delay lines are some of the more typical, but more sophisticated structures, as interferometers and diffraction gratings, have been also produced.

Focusing the laser beam in the bulk of a transparent material, just above its damage threshold, causes local structural changes that affect its physical and chemical properties. In the case of sapphire and glasses, their chemical resistance to HF acid and KOH are considerable lowered, reaching a ratio of $1:10^4$ for sapphire and $1:10^2$ for glasses. This phenomenon has been used to produce hollow channels inside these materials when irradiating with femtosecond laser pulses. After laser irradiation, the material with complex modified 3D structures inside, is etched with a solution of HF acid (or KOH) for up to 48 h to remove the modified material, and results in a hollow structure.

Another approach to make 3D hollow structures inside transparent materials is by focusing the laser beam on the rear surface of a glass sample that is in contact with water [112]. Movements of the focused spot in the vertical direction create a microchannel entering the material's bulk while the water removes the debris formed during irradiation. The process is then continued and complex structures can be produced in all directions.

4. Case studies

The following sections present results obtained in our laboratories regarding the determination of ultrashort pulses ablation threshold and incubation parameters in various materials, and the micromachining in the low and high fluence regimes.

4.1. Silicon ablation threshold measurement

The "zero damage" method was used to determine the ablation threshold for crystalline silicon. For this, 30 μJ maximum energy, 100 fs (FWHM) pulses, centered at 800 nm, at 1 kHz maximum repetition rate, generated by a Ti:Sapphire CPA system (Odin, from Quantronix) were focused on the sample surface by a 38 mm lens. The sample was fixed to a computer controlled translation stage. Many measurements were performed to obtain $D^2 \times F_{th}$ data for different pulses superpositions N, as shown in Figure 5a for $N=1$. In this graph the low and high fluence ablation regimes are clearly seen, and the fits by equation (1) provide the values 0.32 J/cm² and 8.7 J/cm² for the low and high fluence ablation thresholds, respectively. Crystalline silicon shows a curve $D^2 \times F_{th}$ with the same characteristics seen in dielectrics, as exemplified by the sapphire results shown in Figure 1, with an abrupt slope change when thermal effects start to dominate and the material removal rate sharply increases.

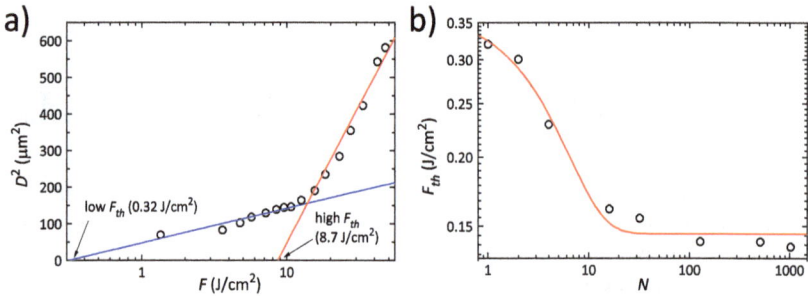

Figure 5. a) Square diameter $D^2 \times F_{th}$ for single pulse ablation ($N=1$) of silicon; b) Silicon damage threshold as fuction of the pulses superposition.

Figure 5b shows the low fluence ablation threshold measured for different pulses superpositions N, together with a fit by equation (3), which provides $F_{th,1}=0.32$ J/cm², $F_{th,\infty}=0.15$ J/cm² and $k=0.22$. It can be observed that the defects accumulation is strong in this material, and saturation of the ablation threshold occurs for less than 50 pulses.

4.2. Molybdenum ablation threshold measurement

The "zero damage" method was also used to determine the ablation threshold of metallic molybdenum, repeating the experiment performed with crystalline silicon described in section 4.1. The diameter of the ablated region as a function of the pulse fluence, for the superposition of 4 pulses on the surface of polished molybdenum, is shown in Figure 6a, together with a fit by equation (1), providing $F_{th}=0.019$ J/cm². This graph shows that the fluence increase does not exhibit an abrupt slope change from the low to the high fluence regime, instead a smooth transition is observed, demonstrating the onset of new processes as the fluence grows.

The ablation threshold dependence on the pulse superposition data, together with a fit by equation (3) is shown on Figure 6b, and, as is the case with dielectrics and semiconductors

(Figure 1 and Figure 5b), it can be seen that the damage threshold decreases as the number of overlapped pulses grows. The values returned by the fit are $F_{th,1}$=0.051 J/cm^2, $F_{th,\infty}$=0.0136 J/cm^2 and k=0.84, evidencing that saturation is quickly reached for less than 10 pulses. Nevertheless, more data points around this value would provide a better result.

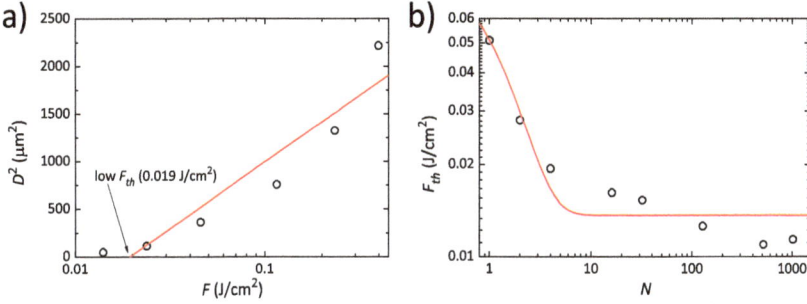

Figure 6. a) Square diameter dependence on the ablation threshold fluence for N=4; b) ablation threshold fluence $F_{th,N}$ as a fuction of the pulses number.

Comparing the single shot fluence ablation thresholds obtained by us for sapphire (5.6 J/cm^2, Figure 1), silicon (0.32 J/cm^2, Figure 5a) and molybdenum (0.051 J/cm^2, Figure 6b), it can be seen that, as the easiness to generate free and seed electrons increase (bandgap decreases), the ablation threshold value drops, as expected by the reasons discussed in section 2.1.

The behavior seen in Figure 6 implies that machining metals with ultrashort pulses is more challenging than machining dielectrics, demanding a better knowledge and control of the process parameters, and sometimes requiring a rework for finishing. The absence of ablation by Coulomb explosion means that always there will be some heat flow to the surrounding area and the conditions of machining must be adjusted to minimize its effects.

4.3. Validation of the D-scan method for ablation threshold and incubation parameter determination

To demonstrate that the D-Scan is a valid method to determine the ablation threshold for an arbitrary pulse superposition, we measured the ablation threshold of BK7 samples by the "zero damage" and D-Scan methods, and compared the results. A Ti:Sapphire CPA system (Femtopower Compact Pro CE-Phase HP/HR, from Femtolasers) was used, continuously generating 100 fs (FWHM) pulses centered at 785 nm with 40 nm of bandwidth (FWHM), at a maximum repetition rate of 4 kHz. The pulses were focused by a 38 mm lens for both methods, and all measurements were performed in air. For the "zero damage" method, the F_{th} was measured for single pulses and for the superposition of 2, 4, 16, 32, 120, 510 and 1020 pulses, and the results, calculated using equation (1), are show as red circles in Figure 7 [68].

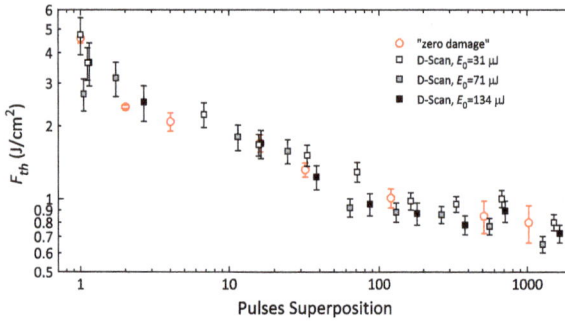

Figure 7. BK7 Ablation threshold measured by the traditional (red circles) and by the D-Scan (squares) method [68]. The good data agreement shows the methods equivalency to determine ablation thresholds in the ultrashort pulses regime.

For the D-Scan measurements, the sample was irradiated by 31, 71 and 134 μJ pulses, at various combinations of repetition rates (50, 100, 500, 1000, 2000 and 4000 Hz) and transversal displacement speeds (6, 12, 25, 50 and 100 mm/min) resulting in superpositions ranging from single pulse to almost 2000 pulses. The Results are show as squares in Figure 7 [68]. This graph shows that the ablation thresholds obtained by both methods agree for superpositions spanning more than 3 orders of magnitude. This implies that the superpositions measured by both methods, although conceptually different (in the "zero damage" the superposition is the number of pulses that completely overlap at the same spot, while for the D-Scan it is given by the sum of the intensity of many pulses spatially apart), are the same, indicating that incubation effects are a linear sum of the intensities that hit a spot. This intensity linear sum effect is also corroborated by the fractionary superpositions that can be observed in Figure 7 D-Scan results for the superposition of less than 10 pulses, which are consistent with the values obtained for integer values of N.

4.4. D-scan technique determination of the incubation parameter of optical materials

The D-Scan method was used in our laboratories to quickly determine the incubation parameter for two common optical materials, sapphire and suprasil glass. The same Ti:Sapphire CPA system described in section 4.3 was used, continuously generating 95 μJ, 25 fs (FWHM) pulses. The beam was focused by a 75 mm lens, and the samples were fixed to a 3-axes computer controlled translation stage that executed the diagonal motion. The measurements were done using various combinations of the pulses repetition rate (100, 500, 1000 and 4000 Hz) and transversal displacement speeds (10, 5, 2.5, 1.25, 1, 0.5, 0.25, 0.2, 0.1, 0.05, 0.04, 0.03, 0.02 and 0.01 mm/s), and it took less than 1 hour to etch more than 35 profiles in each sample. The ρ_{max} value for each profile was measured in micrographs taken in and optical microscope, and the ablation threshold and superposition values for each profile were then calculated using equations (2) and (4). The results are exhibited in Figure 8, and show that in about 1 hour of laboratory time, it is possible to perform measurements that produce data spanning more than four orders of magnitude of N, revealing the dynamics of the defects accumulation until saturation is reached. One of the strengths of the D-Scan technique, which is fast data acquisition, becomes evident when comparing the graphs exhibited in Figure 8 with the ones shown in Figure 5b and Figure 6b: lots of data points are obtained (in a short time), allowing a detailed analysis of the incubation effects, and deciding which model is better to describe them. This kind of analysis is very arduous to be performed with data gathered with the "zero damage" method, due to its complexity that demands a lot of laboratory and data analysis time.

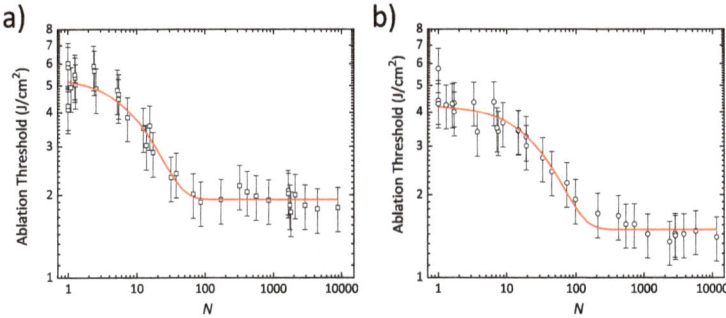

Figure 8. Ablation threshold as a function of the pulse superposition measured using the D-Scan technique for a) suprasil and b) sapphire samples. In each graph, equation (3) is fitted to the data.

To determine the samples ablation thresholds and incubation parameter, equation (3) was fitted to each sample data (red curves on Figure 8), and the parameters obtained are shown in Table 1. The parameters in this table highlight what can be seen in Figure 8: laser created defects have a faster accumulation rate in suprasil (higher k), in which saturation of the ablation threshold is reached for less than 100 pulses, in contrast to sapphire where approximately 500 pulses are needed to saturate F_{th}. Nevertheless, saturation fluence is lower in sapphire, indicating that defects electrons are easier excited in this material.

Sample	$F_{th,1}$ (J/cm^2)	$F_{th,\infty}$ (J/cm^2)	k
Suprasil	5.1±0.1	1.92±0.05	0.064±0.008
Sapphire	4.2±0.1	1.49±0.04	0.022±0.003

Table 1. Fit parameters for each sample

4.5. Direct micromachining of microchannels in BK7

In direct machining of cavities, multipulse ablation is often needed to remove the necessary amount of material, and incubation effects occur. These effects must be taken into account when establishing the machining parameters to ensure adequate control of the ablation process. Within this context, this section presents a study on BK7 etching, in which this approach was used and produced microchannels with shapes, dimensions, finishing and HAZ complying with the requirements for microfluidic devices [113].

To evaluate the efficacy of the method, two different strategies were used, and in both approaches, several parallel lines were swept by the focused laser spot to create a channel with the desired width. For each scanning, a different lateral offset distance d (Figure 9) was used.

The first strategy, named hard machining, consists in working in the high fluence regime, with a high material removal rate, and low longitudinal displacement speed (large longitudinal pulse superposition N). The second approach, dubbed smooth machining, uses a high displacement speed (small N) and operates in the low fluence regime, removing small amounts of material per pulse. This condition sweeps the surface with a smaller number of shots when compared to the hard machining, making it necessary to repeat the process many times (many sweeping layers) to reach the desired depth [113].

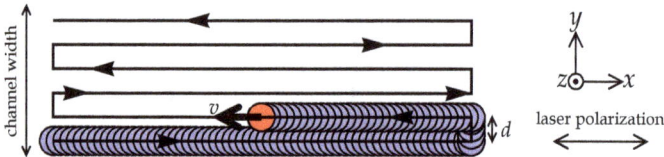

Figure 9. Diagram of the beam path for one etched layer. The beam polarization and scanning direction are parallel to the channel longitudinal axis.

For the hard machining, microchannels were etched with 1 kHz repetition rate, 33 µJ maximum energy pulses, constant scanning speed at 1.6 mm/min chosen to produce a pulses superposition $N=500$, and varying offsets d for each different channel. These irradiation conditions lead to asymmetric, shallow and irregular channels, as can be seen on the transversal section OCT images shown in Figure 10. In this image, the ablated regions are clearly seen for each track, together with very deep and tapered modified regions below the ablated surface. These regions, which cannot be seen by optical microscopy, are not related to catastrophic damage, and can only be observed by OCT due to changes in the index of refraction caused by nonlinear effects resulting from the pulses high intensity [114, 115].

The subsurface modified regions are a consequence of incubation effects. As the pulses overlap increases, F_{th} is lowered, and a greater fraction of the pulse energy is used to cause more severe effects on the processed material. The large amount of ablated material also produces a skewed wall for each individual track. As consequence, the sloped walls of the machined track reflect the subsequent pulses to the adjacent track that follows it, making the ablated area erratic and unpredictable [113].

In the smooth ablation strategy, the lateral displacement d was changed in each layer to always maintain a fixed overlap of 25% of the created damage diameter, since the size of the ablated area changes due to incubation processes. In each layer, the total width of the machined track takes into account the many sweeps to keep the lateral overlap constant, spreading the amount of pre-damaged material smoothly and homogeneously. When this is not considered, the deposited energy is more quickly absorbed in places where accumulation of pre-damages is greater, resulting in a non-homogeneous ablation.

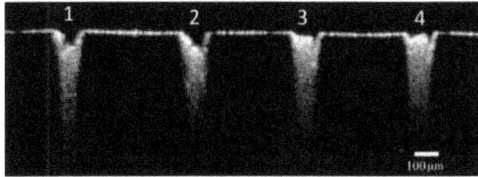

Figure 10. Cross section OCT image of the channels etched by "hard machining". The lateral displacements are d=10, 20, 30 and 40 μm for channels 1 to 4, respectively.

The width D of the ablated track during each sweep depends on the incubation resulting from the *previous* layers etched by N_{n-1} pulses superimposed, and the longitudinal overlap of the *actual* scanning, N_{actual}, where n is the layer number. Within these considerations, the total superposition to calculate D is $N_n=N_{n-1}+N_{actual}$. Thus, D is a function of the beam radius on the machined surface, w, the peak fluence of each pulse, F, and the ablation threshold $F_{th,N}$ for total incubation contribution N, and can be obtained using equation (1) (substituting $F_{th}=F_{th,N}$).

Finally, the value of the longitudinal superposition of the actual scanning, N_{actual}, can be determined by employing equation (4) using the scanning speed v, the repetition rate f and considering that the $\rho_{max}=w$ (beam radius on the sample surface).

A clear improvement in the channels shape occurred when the smooth machining ablation was used, as shown in the OCT image on Figure 11. The channels cross section shows a flat bottom and straight walls, almost perpendicular to the surface. Refractive index modifications below the channels are absent, evidencing that the material properties are preserved in the immediate vicinity of the machined regions.

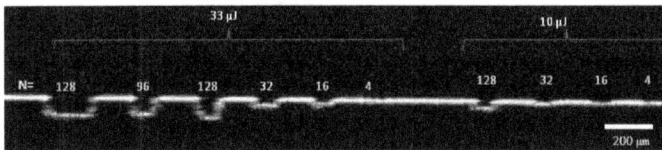

Figure 11. Cross section OCT image of the channels etched by the smoot machining strategy. The total number of overlapped pulses, N, is indicated above each channel. Pulse energies of 10 μJ and 33 μJ were used.

5. Conclusions

In this chapter we presented an overview of micro and nano machining with ultrashort laser pulses. A historical context was provided, the basic physical mechanisms governing the ablation of solids by femtosecond pulses were described, some relevant real-world applications were outlined and experimental results obtained by us were presented. The text references many important and cornerstone works, and the reader who wants to expand his knowledge and have a deeper understanding in this field should start by reading these books, reviews and papers.

The most important parameters to machine a material with ultrashort pulses are its ablation threshold fluence and incubation parameter. Two techniques to determine these parameters were described, the traditional "zero damage" and the D-Scan methods. Experimental results obtained by both techniques were presented, showing that the methods are equivalent, and both can be used to determine these parameters. Nevertheless, we recommend the use of the D-Scan technique because it is experimentally simpler, provides quick results, and the material modifications caused by it are closer to the final machining results once it is executed in a moving sample.

Author details

Ricardo Elgul Samad, Leandro Matiolli Machado,
Nilson Dias Vieira Junior and Wagner de Rossi
Instituto de Pesquisas Energéticas e Nucleares – IPEN-CNEN/SP, Brazil

6. References

[1] Gad-el-Hak M (2006) MEMS : introduction and fundamentals : the MEMS handbook. Boca Raton: CRC Press/Taylor & Francis. p.

[2] Dornfeld D, Min S, Takeuchi Y (2006) Recent Advances in Mechanical Micromachining. CIRP Ann.-Manuf. Techn. 55: 745-768.

[3] Pradhan B, Masanta M, Sarkar B, Bhattacharyya B (2009) Investigation of electro-discharge micro-machining of titanium super alloy. Int. J. Adv. Manuf. Tech. 41: 1094-1106.

[4] Kaminski P C, Capuano M N (2003) Micro hole machining by conventional penetration electrical discharge machine. Int. J. Mach. Tool Manu. 43: 1143-1149.

[5] Langford R M, Dale G, Hopkins P J, Ewen P J S, Petford-Long A K (2002) Focused ion beam micromachining of three-dimensional structures and three-dimensional reconstruction to assess their shape. J. Micromech. Microeng. 12: 111-114.

[6] Malek C K, Hartley F T, Neogi J (2003) Fast prototyping of high-aspect ratio, high-resolution X-ray masks by gas-assisted focused ion beam. Microsyst. Technol. 9: 409-412.

[7] Dale G, Langford R M, Ewen P J S, Reeves C M (2000) Fabrication of photonic band gap structures in $As_{40}S_{60}$ by focused ion beam milling. J. Non-Cryst. Sol. 266–269, Part 2: 913-918.

[8] Chou S Y, Krauss P R, Renstrom P J (1996) Imprint lithography with 25-nanometer resolution. Science 272: 85-87.

[9] Kovacs G T A, Maluf N I, Petersen K E (1998) Bulk micromachining of silicon. Proceedings of the IEEE 86: 1536-1551.

[10] Bustillo J M, Howe R T, Muller R S (1998) Surface micromachining for microelectromechanical systems. Proceedings of the IEEE 86: 1552-1574.

[11] Boser B E, Howe R T (1996) Surface micromachined accelerometers. IEEE J. Solid-St. Circ. 31: 366-375.

[12] Takeuchi S, Suzuki T, Mabuchi K, Fujita H (2004) 3D flexible multichannel neural probe array. J. Micromech. Microeng. 14: 104.

[13] Fuller L F, Sudirgo S (2003) Bulk micromachined pressure sensor In: Proceedings of the 15th Biennial University/Government/Industry Microelectronics Symposium (IEEE, 2003), 317-320

[14] Honig R E, Woolston J R (1963) Laser-Induced Emission of Electrons, Ions, and Neutral Atoms from Solid Surfaces. Appl. Phys. Lett. 2: 138-139.

[15] Boot H A H, Clunie D M, Thorn R S A (1966) Micromachining with a pulsed gas laser. Electron. Lett. 2: 1.

[16] Anisimov S I (1968) Vaporization of Metal Absorbing Laser Radiation. Sov. Phys. JETP-USSR 27: 182-183.

[17] Kocher E, Tschudi L, Steffen J, Herziger G (1972) Dynamics of laser processing in transparent media IEEE J. Quantum Elec. QE 8: 120-&.

[18] Bourg H, Frederick R W (1975) Laser Machining of Silicon. J. Electrochem. Soc. 122: C260-C260.

[19] Longfell J (1971) High Speed Drilling in Alumina Substrates with a CO_2 Laser. Am Ceram Soc Bull 50: 251-&.

[20] McGuff P E, Deterling R A, Gottlieb L S, Bushnell D, Fahimi H D (1964) Surgical Applications of Laser. Ann. Surg. 160: 765-&.

[21] Cohen M I, Unger B A, Milkosky J F (1968) Laser Machining of Thin Films and Integrated Circuits. AT&T Tech. J. 47: 385-405.

[22] Steen W M (2003) Laser material processing. New York: Springer. 408 p.

[23] Ion J C (2005) Laser processing of engineering materials : principles, procedure and industrial application. Oxford: Elsevier/Butterworth-Heinemann. 556 p.

[24] Kogelnik H, Li T (1966) Laser Beams and Resonators. Appl. Opt. 5: 1550-1567.

[25] Stuart B, Feit M, Rubenchik A, Shore B, Perry M (1995) Laser-Induced Damage in Dielectrics with Nanosecond to Subpicosecond Pulses. Phys. Rev. Lett. 74: 2248-2251.

[26] Lan B, Hong M H, Ye K D, Wang Z B, Chong T C (2003) Laser microfabrication of glass substrates by pocket scanning. In: Miyamoto I, Ostendorf A, Sugioka K, Helvajian H,

editors. Fourth International Symposium on Laser Precision Microfabrication. pp. 133-136.

[27] Ben-Yakar A, Byer R L (2002) Femtosecond laser machining of fluidic microchannels for miniaturized bioanalytical systems. In: Sugioka K, Gower M C, Haglund R F, Pique A, Dubowski J J, Hoving W, editors. Photon Processing in Microelectronics and Photonics. pp. 212-217.

[28] Lane D W (1990) The optical properties and laser irradiation of some common glasses. J. Phys. D Appl. Phys. 23: 1727.

[29] Correa D S, Cardoso M R, Tribuzi V, Misoguti L, Mendonca C R (2012) Femtosecond Laser in Polymeric Materials: Microfabrication of Doped Structures and Micromachining. IEEE J. Sel. Top. Quantum Elec. 18: 176-186.

[30] Vora K, Kang S, Shukla S, Mazur E (2012) Fabrication of disconnected three-dimensional silver nanostructures in a polymer matrix. Appl. Phys. Lett. 100: 063120.

[31] Strickland D, Mourou G (1985) Compression of amplified chirped optical pulses. Opt. Commun. 56: 219-221.

[32] Diels J-C, Rudolph W (2006) Ultrashort laser pulse phenomena : fundamentals, techniques, and applications on a femtosecond time scale. Burlington, MA: Academic Press. 652 p.

[33] Sibbett W, Lagatsky A A, Brown C T A (2012) The development and application of femtosecond laser systems. Opt. Expr. 20: 6989-7001.

[34] Keller U (2010) Ultrafast solid-state laser oscillators: a success story for the last 20 years with no end in sight. Appl. Phys. B-Las. Opt. 100: 15-28.

[35] Russbueldt P, Mans T, Weitenberg J, Hoffmann H D, Poprawe R (2010) Compact diode-pumped 1.1 kW Yb:YAG Innoslab femtosecond amplifier. Opt. Lett. 35: 4169 -4171.

[36] Umhofer U, Jäger E, Bischoff C (2011) Refractive and diffractive laser beam shaping optics. Laser Technik Journal 8: 24-27.

[37] Bloembergen N (1974) Laser-induced electric breakdown in solids. IEEE J. Quantum Elec. QE10: 375-386.

[38] Du D, Liu X, Korn G, Squier J, Mourou G (1994) Laser-induced breakdown by impact ionization in SiO_2 with pulse widths from 7 ns to 150 fs. Appl. Phys. Lett. 64: 3071-3073.

[39] Reif J, Costache F (2006) Femtosecond laser interaction with solid surfaces: explosive ablation and self-assembly of ordered nanostructures. In: Rempe G, Scully M O, editors. Advances in Atomic Molecular, and Optical Physics, V. 53. San Diego: Elsevier Academic Press Inc. pp. 227-251.

[40] Stoian R, Rosenfeld A, Ashkenasi D, Hertel I V, Bulgakova N M, Campbell E E B (2002) Surface charging and impulsive ion ejection during ultrashort pulsed laser ablation. Phys. Rev. Lett. 88: 097603.

[41] Bulgakova N M, Stoian R, Rosenfeld A, Hertel I V, Marine W, Campbell E E B (2005) A general continuum approach to describe fast electronic transport in pulsed laser irradiated materials: The problem of Coulomb explosion. Appl. Phys. A-Mat. Sci. Proc. 81: 345-356.

[42] Kanavin A P, Smetanin I V, Isakov V A, Afanasiev Y V, Chichkov B N, Wellegehausen B, Nolte S, Momma C, Tunnermann A (1998) Heat transport in metals irradiated by ultrashort laser pulses. Phys. Rev. B 57: 14698-14703.

[43] Singh N (2004) Relaxation between electrons and surface phonons of a homogeneously photoexcited metal film. Pramana-J. Phys. 63: 1083-1087.

[44] Stuart B C, Feit M D, Herman S, Rubenchik A M, Shore B W, Perry M D (1996) Nanosecond-to-femtosecond laser-induced breakdown in dielectrics. Phys. Rev. B 53: 1749-1761.

[45] Lorazo P, Lewis L J, Meunier M (2003) Short-pulse laser ablation of solids: From phase explosion to fragmentation. Phys. Rev. Lett. 91: 225502.

[46] Perez D, Lewis L J (2003) Molecular-dynamics study of ablation of solids under femtosecond laser pulses. Phys. Rev. B 67: 184102.

[47] Zhigilei L V (2003) Dynamics of the plume formation and parameters of the ejected clusters in short-pulse laser ablation. Appl. Phys. A-Mat. Sci. Proc. 76: 339-350.

[48] Kautek W, Kruger J, Lenzner M, Sartania S, Spielmann C, Krausz F (1996) Laser ablation of dielectrics with pulse durations between 20 fs and 3 ps. Appl. Phys. Lett. 69: 3146-3148.

[49] Keldysh L V (1965) Ionization in the field of a strong electromagnetic wave. Sov. Phys. JETP-USSR 20: 1307-1314.

[50] Joglekar A P, Liu H, Spooner G J, Meyhofer E, Mourou G, Hunt A J (2003) A study of the deterministic character of optical damage by femtosecond laser pulses and applications to nanomachining. Appl. Phys. B-Las. Opt. 77: 25-30.

[51] Gamaly E G, Rode A V, Luther-Davies B, Tikhonchuk V T (2002) Ablation of solids by femtosecond lasers: ablation mechanism and ablation thresholds for metals and dielectrics. Phys. Plasmas 9: 949-957.

[52] Borghesi M, Romagnani L, Schiavi A, Campbell D H, Haines M G, Willi O, Mackinnon A J, Galimberti M, Gizzi L, Clarke R J, Hawkes S (2003) Measurement of highly transient electrical charging following high-intensity laser–solid interaction. Appl. Phys. Lett. 82: 1529.

[53] Born M, Wolf E (1999) Principles of optics : electromagnetic theory of propagation, interference and diffraction of light. New York: Cambridge University Press. 952 p.

[54] Hirayama Y, Obara M (2002) Heat effects of metals ablated with femtosecond laser pulses. Appl. Surf. Sci. 197: 741-745.

[55] Machado L M (to be published in 2012) Micromachining of optical components with femtosecond laser. Ph.D. Thesis (Universidade de São Paulo, São Paulo).

[56] Liu J M (1982) Simple technique for measurements of pulsed Gaussian-beam spot sizes. Opt. Lett. 7: 196-198.

[57] Samad R E, Vieira N D (2006) Geometrical method for determining the surface damage threshold for femtosecond laser pulses. Las. Phys. 16: 336-339.

[58] Samad R E, Baldochi S L, Vieira Jr N D (2008) Diagonal scan measurement of Cr:LiSAF 20 ps ablation threshold. Appl. Opt. 47: 920-924.

[59] Mero M, Clapp B, Jasapara J C, Rudolph W, Ristau D, Starke K, Kruger J, Martin S, Kautek W (2005) On the damage behavior of dielectric films when illuminated with multiple femtosecond laser pulses. Opt. Eng. 44: 051107.

[60] Costache F, Eckert S, Reif J (2008) Near-damage threshold femtosecond laser irradiation of dielectric surfaces: desorbed ion kinetics and defect dynamics. Appl. Phys. A-Mat. Sci. Proc. 92: 897-902.

[61] Courrol L C, Samad R E, Gomes L, Ranieri I M, Baldochi S L, de Freitas A Z, Vieira N D (2004) Color center production by femtosecond pulse laser irradiation in LiF crystals. Opt. Expr. 12: 288-293.

[62] Jee Y, Becker M F, Walser R M (1988) Laser-induced damage on single-crystal metal surfaces. J. Opt. Soc. Am. B 5: 648-659.

[63] Ashkenasi D, Lorenz M, Stoian R, Rosenfeld A (1999) Surface damage threshold and structuring of dielectrics using femtosecond laser pulses: the role of incubation. Appl. Surf. Sci. 150: 101-106.

[64] Martin S, Hertwig A, Lenzner M, Kruger J, Kautek W (2003) Spot-size dependence of the ablation threshold in dielectrics for femtosecond laser pulses. Appl. Phys. A-Mat. Sci. Proc. 77: 883-884.

[65] Lim Y C, Boukany P E, Farson D F, Lee L J (2011) Direct-write femtosecond laser ablation and DNA combing and imprinting for fabrication of a micro/nanofluidic device on an ethylene glycol dimethacrylate polymer. J. Micromech. Microeng. 21: 015012.

[66] Gomez D, Goenaga I (2006) On the incubation effect on two thermoplastics when irradiated with ultrashort laser pulses: Broadening effects when machining microchannels. Appl. Surf. Sci. 253: 2230-2236.

[67] Choi H W, Farson D F, Bovatsek J, Arai A, Ashkenasi D (2007) Direct-write patterning of indium-tin-oxide film by high pulse repetition frequency femtosecond laser ablation. Appl. Opt. 46: 5792-5799.

[68] Machado L M, Samad R E, de Rossi W, Vieira Junior N D (2012) D-Scan measurement of ablation threshold incubation effects for ultrashort laser pulses. Opt. Expr. 20: 4114-4123.

[69] Wolfram Research Inc. (retrieved 2012) Jacobi theta function J_3, http://functions.wolfram.com/EllipticFunctions/EllipticTheta3/06/01/03/.

[70] Reif J, Costache F, Varlamova O, Jia G, Ratzke M (2009) Self-organized regular surface patterning by pulsed laser ablation. Phys. Status Solidi C 6: 681-686.

[71] Mei H, Wang C, Yao J, Chang Y-C, Cheng J, Zhu Y, Yin S, Luo C (2011) Development of novel flexible black silicon. Opt. Commun. 284: 1072-1075.

[72] Halbwax M, Sarnet T, Delaporte P, Sentis A, Etienne H, Torregrosa F, Vervisch V, Perichaud I, Martinuzzi S (2008) Micro and nano-structuration of silicon by femtosecond laser: Application to silicon photovoltaic cells fabrication. Thin Solid Films 516: 6791-6795.

[73] Schlie S, Fadeeva E, Koroleva A, Ovsianikov A, Koch J, Ngezahayo A, Chichkov B N (2011) Laser-based nanoengineering of surface topographies for biomedical applications. Photonic. Nanostruct. 9: 159-162.

[74] Vorobyev A Y, Guo C (2007) Femtosecond laser structuring of titanium implants. Appl. Surf. Sci. 253: 7272-7280.

[75] Yasumaru N, Miyazaki K, Kiuchi J, Sentoku E (2011) Frictional properties of diamond-like carbon glassy carbon and nitrides with femtosecond-laser-induced nanostructure. Diam, Relat. Mater. 20: 542-545.

[76] Tang M, Hong M H, Choo Y S (2008) Hydrophobic Surface Fabrication by Laser Micropatterning. 779 p.

[77] Bizi-Bandoki P, Benayoun S, Valette S, Beaugiraud B, Audouard E (2011) Modifications of roughness and wettability properties of metals induced by femtosecond laser treatment. Appl. Surf. Sci. 257: 5213-5218.

[78] Vorobyev A Y, Guoa C (2008) Colorizing metals with femtosecond laser pulses. Appl. Phys. Lett. 92: 041914.

[79] Ahsan M S, Ahmed F, Kim Y G, Lee M S, Jun M B G (2011) Colorizing stainless steel surface by femtosecond laser induced micro/nano-structures. Appl. Surf. Sci. 257: 7771-7777.

[80] Mateo M P, Ctvrtnickova T, Fernandez E, Ramos J A, Yanez A, Nicolas G (2009) Laser cleaning of varnishes and contaminants on brass. Appl. Surf. Sci. 255: 5579-5583.

[81] Pouli P, Melessanaki K, Giakoumaki A, Argyropoulos V, Anglos D (2005) Measuring the thickness of protective coatings on historic metal objects using nanosecond and femtosecond laser induced breakdown spectroscopy depth profiling. Spectrochim. Acta B 60: 1163-1171.

[82] Dumitru G, Luscher B, Krack M, Bruneau S, Hermann J, Gerbig Y (2005) Laser processing of hardmetals: Physical basics and applications. Int. J. Refract. Met. H. 23: 278-286.

[83] Urech L, Lippert T, Wokaun A, Martin S, Madebach H, Kruger J (2006) Removal of doped poly(methylmetacrylate) from tungsten and titanium substrates by femto- and nanosecond laser cleaning. Appl. Surf. Sci. 252: 4754-4758.

[84] Hermann J, Benfarah M, Coustillier G, Bruneau S, Axente E, Guillemoles J F, Sentis M, Alloncle P, Itina T (2006) Selective ablation of thin films with short and ultrashort laser pulses. Appl. Surf. Sci. 252: 4814-4818.

[85] Atanasov P A, Takada H, Nedyalkov N N, Obara M (2007) Nanohole processing on silicon substrate by femtosecond laser pulse with localized surface plasmon polariton. Appl. Surf. Sci. 253: 8304-8308.

[86] Pronko P P, Dutta S K, Squier J, Rudd J V, Du D, Mourou G (1995) Machining of sub-micron holes using a femtosecond laser at 800 nm. Opt. Commun. 114: 106-110.

[87] Venkatakrishnan K, Tan B, Sivakumar N R (2002) Sub-micron ablation of metallic thin film by femtosecond pulse laser. Opt. Laser Technol. 34: 575-578.

[88] Lim C S, Hong M H, Lin Y, Chen G X, Kumar A S, Rahman M, Tan L S, Fuh J Y, Lim G C (2007) Sub-micron surface patterning by laser irradiation through microlens arrays. J. Mat. Process. Tech. 192: 328-333.

[89] Wang X C, Zheng H Y, Chu P L, Tan J L, Teh K M, Liu T, Ang B C Y, Tay G H (2010) Femtosecond laser drilling of alumina ceramic substrates. Appl. Phys. A-Mat. Sci. Proc. 101: 271-278.

[90] Li C, Lee S, Nikumb S (2009) Femtosecond Laser Drilling of Alumina Wafers. J. Electron. Mater. 38: 2006-2012.

[91] Wang Y, Dai N, Li Y, Wang X, Lu P (2007) Ablation and cutting of silicon wafer and micro-mold fabrication using femtosecond laser pulses. J. Laser Appl. 19: 240-244.

[92] Lee S, Yang D, Nikumb S (2008) Femtosecond laser micromilling of Si wafers. Appl. Surf. Sci. 254: 2996-3005.

[93] Suriano R, Kuznetsov A, Eaton S M, Kiyan R, Cerullo G, Osellame R, Chichkov B N, Levi M, Turri S (2011) Femtosecond laser ablation of polymeric substrates for the fabrication of microfluidic channels. Appl. Surf. Sci. 257: 6243-6250.

[94] van Brakel A, Grivas C, Petrovich M N, Richardson D J (2007) Micro-channels machined in microstructured optical fibers by femtosecond laser. Opt. Expr. 15: 8731-8736.

[95] Alemohammad H, Toyserkani E, Pinkerton A J (2008) Femtosecond laser micromachining of fibre Bragg gratings for simultaneous measurement of temperature and concentration of liquids. J. Phys. D Appl. Phys. 41: 185101.

[96] Lan X, Han Y, Wei T, Zhang Y, Jiang L, Tsai H-L, Xiao H (2009) Surface-enhanced Raman-scattering fiber probe fabricated by femtosecond laser. Opt. Lett. 34: 2285-2287.

[97] Said A A, Dugan M, de Man S, Iannuzzi D (2008) Carving fiber-top cantilevers with femtosecond laser micromachining. J. Micromech. Microeng. 18: 035005.

[98] Nolte S, Momma C, Jacobs H, Tunnermann A, Chichkov B N, Wellegehausen B, Welling H (1997) Ablation of metals by ultrashort laser pulses. J. Opt. Soc. Am. B 14: 2716-2722.

[99] Le Harzic R, Breitling D, Weikert M, Sommer S, Fohl C, Valette S, Donnet C, Audouard E, Dausinger F (2005) Pulse width and energy influence on laser micromachining of metals in a range of 100 fs to 5 ps. Appl. Surf. Sci. 249: 322-331.

[100] Ancona A, Doering S, Jauregui C, Roeser F, Limpert J, Nolte S, Tuennermann A (2009) Femtosecond and picosecond laser drilling of metals at high repetition rates and average powers. Opt. Lett. 34: 3304-3306.

[101] Doering S, Ancona A, Haedrich S, Limpert J, Nolte S, Tuennermann A (2010) Microdrilling of metals using femtosecond laser pulses and high average powers at 515 nm and 1030 nm. Appl. Phys. A-Mat. Sci. Proc. 100: 53-56.

[102] Schille J, Ebert R, Loeschner U, Scully P, Goddard N, Exner H (2010) High repetition rate femto second laser processing of metals. In: Heisterkamp A, Neev J, Nolte S, trebino R P, editors. Frontiers in Ultrafast Optics: Biomedical, Scientific, and Industrial Applications X.

[103] Zhang Q, Guo X P, Dai N L, Lu P X (2009) Corrosion and Fatigue Testing of Microsized 304 Stainless Steel Beams Fabricated by Femtosecond Laser. J. Mater. Sci. Technol. 25: 187-193.

[104] Momma C, Knop U, Nolte S (1999) Laser Cutting of Slotted Tube Coronary Stents – State-of-the-Art and Future Developments. Prog. Biom. Res. 4: 45-51.

[105] Li C D, Nikumb S, Wong F (2006) An optimal process of femtosecond laser cutting of NiTi shape memory alloy for fabrication of miniature devices. Opt. and Las. in Eng. 44: 1078-1087.

[106] Weck A, Crawford T H R, Wilkinson D S, Haugen H K, Preston J S (2008) Laser drilling of high aspect ratio holes in copper with femtosecond, picosecond and nanosecond pulses. Appl. Phys. A-Mat. Sci. Proc. 90: 537-543.

[107] Campbell B R, Palmer J A, Semak V V (2007) Peculiarity of metal drilling with a commercial femtosecond laser. Appl. Surf. Sci. 253: 6334-6338.

[108] Gamaly E G, Juodkazis S, Nishimura K, Misawa H, Luther-Davies B (2006) Laser-matter interaction in the bulk of a transparent solid: Confined microexplosion and void formation. Phys. Rev. B 73: 214101.

[109] Dubov M, Bennion I, Nikogosyan D N, Bolger P, Zayats A V (2008) Point-by-point inscription of 250 nm period structure in bulk fused silica by tightly focused femtosecond UV pulses. J. Opt. A: Pure Appl. Opt. 10: 025305.

[110] Ams M, Dekker P, Marshall G D, Withford M J (2009) Monolithic 100 mW Yb waveguide laser fabricated using the femtosecond-laser direct-write technique. Opt. Lett. 34: 247-249.

[111] Dharmadhikari J A, Dharmadhikari A K, Bhatnagar A, Mallik A, Singh P C, Dhaman R K, Chalapathi K, Mathur D (2011) Writing low-loss waveguides in borosilicate (BK7) glass with a low-repetition-rate femtosecond laser. Opt. Commun. 284: 630-634.

[112] Li C, Chen T, Si J, Chen F, Shi X, Hou X (2009) Fabrication of three-dimensional microchannels inside silicon using a femtosecond laser. J. Micromech. Microeng. 19:

[113] Machado L M, Samad R E, de Freitas A Z, Vieira Jr N D, de Rossi W (2011) Microchannels Direct Machining using the Femtosecond Smooth Ablation Method. Phys. Procedia 12: 67-75.

[114] Taylor R, Hnatovsky C, Simova E, Rayner D, Mehandale M, Bhardwaj V, Corkum P (2003) Ultra-high resolution index of refraction profiles of femtosecond laser modified silica structures. Opt. Expr. 11: 775-781.

[115] Saliminia A, Nguyen N T, Chin S L, Vallee R (2004) The influence of self-focusing and filamentation on refractive index modifications in fused silica using intense femtosecond pulses. Opt. Commun. 241: 529-538.

Chapter 6

Interaction of Femtosecond Laser Pulses with Solids: Electron/Phonon/Plasmon Dynamics

Roman V. Dyukin, George A. Martsinovskiy, Olga N. Sergaeva,
Galina D. Shandybina, Vera V. Svirina and Eugeny B. Yakovlev

Additional information is available at the end of the chapter

1. Introduction

Femtosecond lasers bring new opportunities in a variety of technological applications [1] in micro- and nanotechnologies, including electronics, mechanics, medicine and biology. Technologies, based on femtosecond effects, are used, for example, to make light absorbers for solar energy devices [2], for direct fabrication of integrated optical components [3], enhancing performance of photo-electronic devices [4], friction reduction and improvement of mechanical wear resistance [5], surface conditioning of medical implants [6], etc. Further development of the above technologies requires deeper understanding of the physical processes occurring under the ultrashort laser pulse action on different materials.

Changes in the material optical properties under the action of intense radiation represent the key feature of the interaction of laser radiation with condensed media. Dynamics of optical properties of solids under the action of femtosecond laser pulse determines a number of physical effects which are of the great interest for both fundamental science and new applications. In particular, the feedbacks, which are being formed in this case, fundamentally change the properties of condensed matter [7].

During the action of femtosecond pulse on solids the electronic subsystem undergoes intensive photoexcitation while the lattice stays cold. The processes of excitation of the electrons and release of the absorbed energy are spaced in time. High intensity of the laser radiation results in modification of the state of the electron subsystem thus significantly changing the optical properties of the medium [8]. Studies of the femtosecond pulses effects on semiconductors and insulators [9, 10] showed that the concentration of nonequilibrium carriers generated by laser radiation is so high that the surface layer acquires properties of the metals during the pulse.

It was observed experimentally that semiconductors can be disintegrated during the femtosecond laser pulse action [11]. In [11] it was proposed a mechanism of destruction based on the crystal lattice destruction by the electric field resulting from the violation of quasi-neutrality in the irradiated area due to the external electron emission - Coulomb explosion. Conditions for Coulomb explosion occurrence in metals were not found.

The currently used experimental approaches such as femtosecond pump-probe technology [12, 13] and mass-spectroscopy (for example, [14]) provide measurement of integral characteristics, but have limited capability for retrieval of dynamics of the processes. The limitations of the experiment approach are being compensated by extensive use of mathematical modeling, where the fast non-linear processes are simulated in a wide range of the initial data.

Two-temperature model, which is traditionally used to describe the ultrashort laser pulse interaction with matter, has proved its validity in various conditions. The phenomenological two-temperature model of parabolic type was proposed in the 50's of last century by M.I. Kaganov, I.M. Lifshitz, L.V. Tanatarov [15]. It has been used by S.I. Anisimov to describe transient phenomena in a nonequilibrium electron gas and lattice under the submicrosecond laser action [16]. The model represents the primary approach to mathematical description of the nonequilibrium heating of the condensed medium by the action of short- and ultrashort-pulse laser radiation. According to the model it is assumed that the energy absorbed by free electrons increases their temperature, then the interaction of the heated electrons with the lattice results in increasing lattice temperature. Heat transfer takes place through the heat conduction mechanisms.

When analyzing the effect of femtosecond laser pulses on matter one has to consider the following: applicability of two-temperature model to description of the electron temperature, which is determined by the electron equilibrium and applicability of the notion of temperature; and taking into account multi-quantum effects in description of electron emission.

The applicability of a two-temperature model for description of action of femtosecond laser pulse on metals. Femtosecond pulse action on the metal can be generally described as follows. Absorption of photons by free electrons in metals results in increase in the electron kinetic energy and the energy distribution becomes nonequilibrium. This well-known feature determines the behavior of metals in a wide spectral range. One can use the diffusion approximation for obtaining qualitative characterization of photoexcitation of solids during the action of femtosecond pulse. With this approach the distribution of free electrons is described by the integral concentration $n(z, t)$, which varies in time and space (along the axis z, directed into the material) due to photo- and thermionic emission from the surface layers and electron diffusion. It is assumed that the thermalization of the electron gas occurs so fast that the notion of electron temperature can be immediately applied. "Hot" electrons contribute to the photo- and thermal emission: they withdraw a part of the energy stored in the electronic subsystem, thus reducing its temperature and, eventually, the temperature of solid as a whole. At the same time, change in the electrons concentration in the surface region results in a change in the optical characteristics of the material.

It is known that the light is absorbed by the conduction electrons in metals. Depending on the concentration of conduction electrons n and the wavelength of the incident light λ the electrons can be considered as free provided $n \gg 1/\lambda^3$, and the free electron model cannot be applied if $n \ll 1/\lambda^3$. In typical conditions in metals, where $n \approx 10^{22}$ cm^{-3}, and for the light wavelength of $\lambda \approx 1\,\mu m$ the free electron model works well. It is also well known that the electron gas in metals is degenerated at practically all temperature range, and the distribution function of the electron gas just slightly differs from the distribution function at the absolute zero.

The Fermi energy ε_F for metals is very high. For example, for copper ε_F=7.1 eV, for silver ε_F=5.5 eV. For this reason heat effects engage the electrons, whose energy lies in a narrow energy range $\approx 2k_B T$ (k_B - Boltzmann constant) near the Fermi level.

The concentration of electrons, which absorb the incident radiation in metal, can be estimated by using the relation $n = h\nu N/\varepsilon_F$, where ν – the frequency of the incident light. For example, n/N=25% of the conduction electrons are affected in copper under $h\nu$=1.7 eV.

The electron transfer part of excessive energy, which they receive due to light absorption, though their collision to other electrons, ions and lattice defects (dislocations, grain boundaries, etc.). Heating of the metal is determined by efficiency of the collisions, which depends on the particles that exchange with energy. Typically the following relation takes place: $\nu_{ee} > \nu_{ei} > \nu_{ep}$, where ν_{ee} – the frequency of electron - electron collisions, ν_{ei} – the frequency of collisions between electrons and phonons, ν_{ep} – the frequency of electrons collisions with impurities and defects in the metal. The electron-phonon mechanism is quite feasible, since the Fermi energy of electrons in metals is high and is essentially represented by the energy of translational motion of free electrons. The Fermi velocity is $\nu_F = (2\varepsilon_F/m_e)^{1/2} = 1{,}5 \cdot 10^8$ cm/s is also high compared to the speed of sound in metals, that is typically: $\nu_F \sim 10^5$ cm/s.

A single collision may not be enough for the electron, which absorbed a photon, to release the excessive energy, i.e. energy relaxation is a multi-stage process of a diffusion character. For this reason, the energy redistribution occurs not only at the skin layer ($\delta_s \sim 10^{-6}$ cm), but in the deeper layer of $l_d = (D/\nu_{ee})^{1/2} = 10^{-5}$ cm, where D is the electron diffusion coefficient.

Thus there is a heating of the metal. The heat, which is released in the layer l_D, is further transferred depthward into the material through heat conduction. The characteristic time of the absorbed energy transfer in metal is

$$\tau_{ei} = 1/\nu_{ei} = 10^{-12}\text{-}10^{-11} \text{ s.}$$

The electron gas and the lattice of the metal are two weakly interacting subsystems. Under the conditions:

$$t \gg \tau_{ee},\qquad(1)$$

(τ_{ee} – time to establish an equilibrium energy distribution in the electron gas) and

$$\nu_{ii} \ll \nu_{ei},\qquad(2)$$

(v_{ii} – frequency of ion-ion collisions) the electron gas and the lattice can be described separately with electron T_e and lattice T_i temperatures. Condition (1) infers rapid redistribution of the absorbed energy between the conduction electrons, and (2) means that the energy transferred to the lattice by electrons, rapidly redistributes between the ions.

Let us excessively consider the relaxation rates v_{ef}, v_{ee}, v_{ei}, v_{ii}. The relations between these relaxation rates substantially determine the processes in metals during the absorption of radiation.

The frequency of collisions between electrons and photons v_{ef} is proportional to the power density of laser radiation q absorbed by the metal. It can be estimated using the relation:

$$v_{ef} = \frac{\alpha q}{h v n},$$ (3)

where α is the absorption coefficient in metals: $\alpha \sim 10^5$ cm^{-1}.

The frequency of electron-electron collisions v_{ee} in the metal is mainly determined by the number of electrons in the Fermi smearing and is calculated using the relation:

$$v_{ee} = v_F \sigma_{ee} n_e \left(\frac{k_B T_e}{\varepsilon_F}\right)^2,$$ (4)

where v_F – the electron velocity on the Fermi surface $\sim 10^8$ cm/s; σ_{ee} – electron - electron interaction cross-section, $\sigma_{ee} \approx 5 \cdot 10^{-16}$ cm^2; $k_B T_e$ – Fermi smearing region. Then, at $T_e \sim 10^3$ K, the value of $v_{ee} \approx 10^{14}$ s^{-1}, and the time to establish an equilibrium distribution of the electron gas $\tau_{ee} \sim 1/v_{ee} \sim 10^{-14}$ s. The rate of energy transfer from the electron gas to the lattice and the lattice temperature is determined by the heat source and heat transfer coefficient of the electrons with the lattice β_{ei}.

The energy, which the lattice obtains from the electron gas per unit volume per unit time is $\sim \beta_{ei}(T_e-T_i)$, $\beta_{ei} \sim 10^{10}$ W/(cm^{3}K).

The frequency of electron-ion relaxation can be expressed in terms of heat transfer coefficient

$$v_{ei} = \frac{\beta_{ei}}{\rho_i c_i},$$ (5)

where $\rho_i c_i$ – the volumetric heat capacity of the lattice [10^{-7} W·s/(cm^{3}K)]. Substituting numerical values one obtains $v_{ei} \sim 10^{11}$ s^{-1} and $\tau_{ei} \sim 10^{-11}$ s.

Comparing the expressions (3) and (4), one can show that (1) is always satisfied with the flux densities $q_0 \le 10^9$ W/cm^2 and the electron gas in a metal is described by temperature T_e. Similarly, the condition (2) is also satisfied, therefore temperature T_i can be introduced to describe the thermal state of the lattice.

Thus, the electron relaxation time is estimated as $\tau_r \sim 1/v_{ee}$, where v_{ee} – the frequency of electron-electron collisions. According to the estimates given above, its magnitude is about

10^{-14}–10^{-13} s, which is comparable with the pulse duration (tens of femtoseconds). However, further studies [16-19] have shown that the electron-electron relaxation time can be reduced up to 10^{-16} s at the electron gas temperatures of ~ 100 000 K, achieved by the action of ultrashort laser pulse. In other words, two-temperature model could be applied to analyze the effects of femtosecond laser pulse.

Multiphoton absorption. Although the theory of multiphoton absorption is pretty well developed [17], there is a certain difficulty regarding definition of multi-photon absorption cross-sections of real media, when this theory is used for the analysis of multiphoton absorption during femtosecond laser action.

In terms of quantum mechanics multiphoton process can be represented as a series of successive transitions of an electron to the virtual states [18]. Only the initial and final states are real in this case. The energy conservation law is valid with an accuracy of a natural width of the energy level only for the initial and final states. For virtual states the energy conservation law takes place with the accuracy, that is determined by the energy-time uncertainty relation $\delta E \cdot \delta t \geq \hbar$. At each virtual state the quantum system lives for the time:

$$\delta t \geq \hbar / \delta E. \tag{6}$$

Absorption of another photon makes the system transit to the next state.

It means that the quantum system can potentially absorb a photon of any energy, however the lifetime of the quantum system, absorbing a photon, would differ. If the quantum system absorbs a photon with energy $h\nu = \Delta E_{mn}$, the system transits into a real state, where the lifetime δt is determined by the probability of spontaneous decay of this state. If the quantum system absorbs photon with energy $h\nu \neq \Delta E_{mn}$, the system transits into a virtual state, where the lifetime is determined by the energy-time uncertainty relation: $\delta t \geq \hbar / \delta E$, $\delta E = |h\nu - \Delta E_{mn}|$.

With this approach one can use the following method for estimation of multiphoton absorption cross sections for various medium. One-photon absorption cross section σ_1 for metals can be determined by the known absorption coefficient α and the concentration of free electrons n: $\sigma_1 = \alpha/n$. Estimating the lifetime of an electron on a virtual level with the photon energy of about 1 eV $\tau_0 \sim 10^{-16}$ s and assuming the absorption cross section of excited electrons in all virtual levels to be equal to σ_1, one obtains two-photon absorption cross section to be $\sigma_2 = \sigma_1^2 n \tau_0$. Similar consideration provides $\sigma_3 = \sigma_1^3 n \tau_0^2$. Consequently for $m+1$-photon absorption cross section is $\sigma_{m+1} = \sigma_1^{m+1} n \tau_0^m$. For example, for $\alpha = 10^5$cm^{-1} and $n = 10^{22}$cm^{-3} one can obtain $\sigma_1 = 10^{-17}$cm^2, $\sigma_2 = 10^{-28}$cm s, $\sigma_3 = 10^{-61}$cm^{3}s^2.

2. Effect of electron emission on metal heating and destruction by femtosecond laser pulse

Let us analyze the influence of hot electron emission on heating and destruction of metals based on a two-temperature model and using as example the numerical simulation of Coulomb explosion.

In theoretical studies dealing with the different variants of two-temperature model, the most important aspect is to define the nonlinear optical and thermal properties in a wide temperature range, as well as quantitative characteristics of the electron-electron and electron-phonon interaction, controlling the temperature of the electron gas and the energy exchange between electrons and lattice, respectively.

In terms of the numerical analysis classical version of a two-temperature model is a system of interrelated differential equations of heat conduction, the accuracy of whose solution depends strongly on adaptation of the computational grid to the required numerical solution.

Let us consider the two-temperature model of metal heating by a femtosecond laser pulses on metals [19-22]. The model consists of a system of heat-conduction equations for the electrons and the phonons (lattice) subsystems, where thermophysical properties depend on electron temperature T_e and electron concentration n, and the equation describing the temporal evolution of the electron concentration.

$$c_e\left(T_e,n\right)\frac{\partial T_e}{\partial t}-\frac{\partial}{\partial z}\left[\lambda_e\left(T_e,n\right)\frac{\partial T_e}{\partial z}\right]=-\beta_{ei}\left(T_e,n\right)\left(T_e-T_i\right)+q_v, \qquad (7)$$

$$c_i\frac{\partial T_i}{\partial t}-\frac{\partial}{\partial z}\left[\lambda_i\frac{\partial T_i}{\partial z}\right]=\beta_{ei}\left(T_e,n\right)\left(T_e-T_i\right), \qquad (8)$$

where λ_e, λ_i, c_e, c_i are the electron and lattice heat conductivity and heat capacity, β_{ei} – electron-ion energy transfer coefficient, q_v – the absorbed power density released in the electron subsystem.

Dependence of optical and thermo-physical properties of metal [23] on the electron temperature and concentration is taken into account for electron heat capacity: $c_e=\frac{\pi^2 k_b^2 n(z,t)T_e}{2\varepsilon_F}$, electron heat conductivity: $\lambda_e=v_e^2\,\tau_{ee}c_e/3$, electron-ion energy transfer coefficient: $\beta_{ei}=c_e/t_{ei}$, electron velocity: $v_e=\sqrt{3k_bT/m_e}$, the electron mean free path: $l_e=1/(n\sigma\sqrt{2})$. The absorption coefficient of metals depends only on the concentration of free electrons: $\alpha(n)=(\alpha/n_0)n$, n_0 is the initial concentration of free electrons.

The boundary conditions:

$$\lambda_e\frac{\partial T_e}{\partial z}\bigg|_{z=0}=-j_e,\ \lambda_i\frac{\partial T_i}{\partial z}\bigg|_{z=0}=0,\ T_e\big|_{z=\infty}=T_i\big|_{z=\infty}=T_n,$$

where j_e – is the heat flow carried away by the emitted electrons, T_n – the initial temperature. Evolution of the electron density distribution is described by the following diffusion equation:

$$\frac{\partial n(z,t)}{\partial t}=D\frac{\partial^2 n(z,t)}{\partial z^2}, \qquad (9)$$

with initial and boundary conditions:

$$D\frac{\partial n}{\partial z}\bigg|_{z=0} = -F_e, \ n\big|_{z=0} = n_0, \ n\big|_{z=\infty} = n_0,$$

where $n_0 = 10^{22}$ cm^{-3} is the initial concentration of electrons evenly distributed over the volume, F_e is the flow of the electrons resulted from thermionic emission and photoemission $F_e = F_t + F_{mpho}$.

The thermionic emission is determined by the law of Richardson:

$$F_t = -BT_e^2\big|_{z=0} \exp\left(-\frac{\varphi_e}{k_bT_e\big|_{z=0}}\right) \exp(-z/l_e)/q_e,$$

here B is the Richardson coefficient, φ_e is the work function, q_e is the electron charge.

External photoelectric effect implies that the energy of absorbed photons is used by the electrons to overcome the work function, i.e. the minimum energy required for the electron to escape from the surface. The work function in metals is several eV (for silver $\varphi_e = 4{,}28$ eV). Therefore, for $hv = 1.55$ eV one should expect the three-photon absorption. For calculating the photoemission let us assume that the free electrons, which are involved into the multiphoton process, reach the surface without energy loss and leave the metal. The emissive layer thickness is limited by the electron mean free path and also depends on the electron concentration and the emission coefficient. The flow of electrons (cm^{-2}/s) caused by photoemission for the m-photon absorption:

$$F_{mpho}\big|_{z=0} = -\int_0^\infty \sigma_m J^m \exp(-z/l_e)dz,$$

J is the absorbed photons flow, $J = q/hv$, σ_m is the multi-photon absorption cross section.

The emission of electrons leads to accumulation of positive charge on the metal surface and, therefore, to generation of the electric field. The electric field resulted from breaking of quasi-neutrality of the irradiated area can be calculated from the following equation [24]:

$$\frac{\partial E}{\partial z} = \frac{q_e}{\varepsilon\varepsilon_0}(n_i - n). \tag{10}$$

This electric field, which is induced by the charge separation, can reach extremely high magnitude and exceed the energy of atomic bonds resulting in Coulomb explosion. To determine the conditions for initiation of the Coulomb explosion the electric field (10) is compared to the threshold magnitude required for removal of an atom from the target. The estimation of the critical electric field [24]:

$$E_{th}\big|_{z=0} = \sqrt{\frac{2\Lambda n}{\varepsilon\varepsilon_0}},$$

where n_0- the concentration of atoms (cm^{-3}), $\Lambda = 2.951$ J/atom - the heat of sublimation, $\varepsilon = 4.9$ - relative permeability of silver, $\varepsilon_0 = 8.854 \cdot 10^{-14}$ F/cm - dielectric constant.

For the numerical solution of the heat conduction equation and the equation describing the electron density, which are non-stationary partial differential equations, the finite difference method was applied. To calculate the values of the temperature and the electron concentration explicit difference scheme was used. Though explicit scheme provides a relatively high speed of calculation, it has a serious disadvantage related to the need for satisfying the stability conditions, which impose limits on the amount of steps partitions with respect to coordinate and time.

Fig. 1 - 4 show the results for silver for the laser pulse shape $q=q_m\exp(-(t-t_m)^2/t_{m1}^2)$, $t_m=100$ fs, $t_{m1}=50$ fs. The calculation was performed until the beginning of the Coulomb explosion, when the electric field resulted from the charge separation exceeds the threshold required for the removal of atoms. For comparison, the simulation was performed without taking the emission into account, but preserving the dependence of material properties on temperature.

Fig. 1 shows that temperature of the electrons and the lattice increase, but does not reach its maximum during the pulse. The difference between the temperature, calculated taking into account the emission and without it at the beginning of Coulomb explosion $t=0.15$ fs is $\Delta T \approx 290$ K for temperature of the electrons (Fig. 1b) and $\Delta T \approx 60$ K for temperature of the lattice.

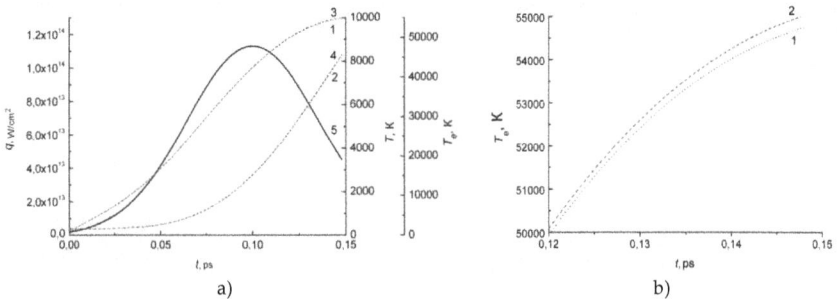

a)   b)

Figure 1. a – Transient behavior of electron and lattice temperature: (1) and (2) – with taking into account emission, (3) and (4) – disregarding emission. (5) – laser pulse shape $q=q_m\exp(-(t-t_m)^2/t_{m1}^2)$, $t_m=100$ fs, $t_{m1}=50$ fs, maximum power density $q_m=5\cdot10^{14}$ W/cm^2. b – an enlarged part of the temporal dependence of the electron temperature: (1) – with taking into account emission, (2) – disregarding emission

Fig. 2, 3 illustrate temporal behavior of the free electrons concentration, and the number of electrons emitted by photo-and thermionic emission for different pulse shapes. At the initial stage of pulse action photoemission dominates, but thermionic emission increases with the electron temperature rapidly, so thermionic emission begins to prevail over the photoemission. The maximum of the photoemission rate maximum corresponds to the maximum of laser power density. This indicates that the laser pulse shape significantly influences the dynamics of the processes.

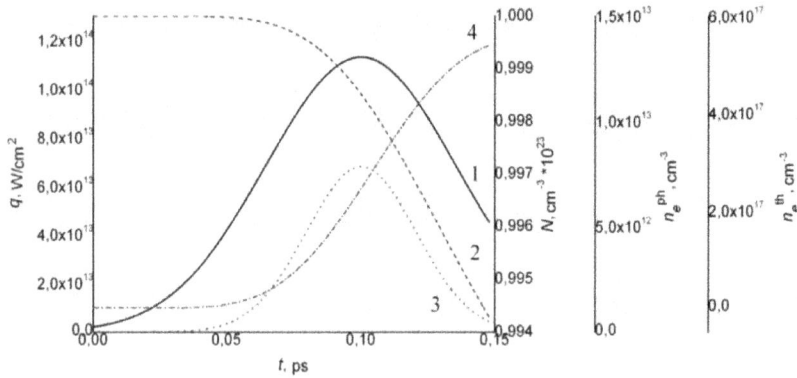

Figure 2. Laser pulse shape (1), the temporal evolution of the net surface electron density N (2) and the density of emitted electrons due to photoemission n_e^{ph} (3) and thermionic emission n_e^{th} (4). Laser pulse shape $q=q_m\exp(-(t-t_m)^2/t_{m1}^2)$, $t_m=100$ fs, $t_{m1}=50$ fs

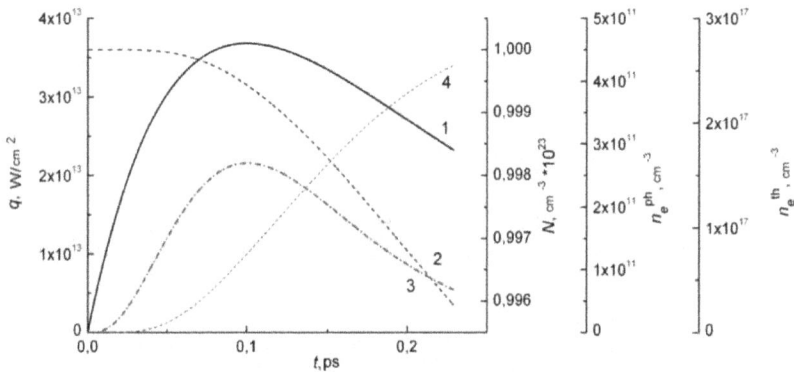

Figure 3. Laser pulse shape (1), the temporal evolution of the net surface electron density N (2) and the density of emitted electrons due to photoemission n_e^{ph} (3) and thermionic emission n_e^{th} (4). Laser pulse shape $q=q_m(t/t_m)\exp(-t/t_m)$, $t_m=100$ fs

The gradient of the electron density results in an electric field, which grows and reaches the Coulomb explosion threshold (Fig. 4).

Fig. 5 illustrates the nonlinear dependence of the moment when Coulomb explosion starts on the laser power density. It is seen that the dependence is nonlinear, and Coulomb explosion for a pulse duration ~ 100 fs can occur when $q > 10^{15}$ W/cm^2.

We can draw the following conclusions from the results of numerical simulation of the influence of electron emission on heating and destruction of metals irradiated by femtosecond laser pulse.

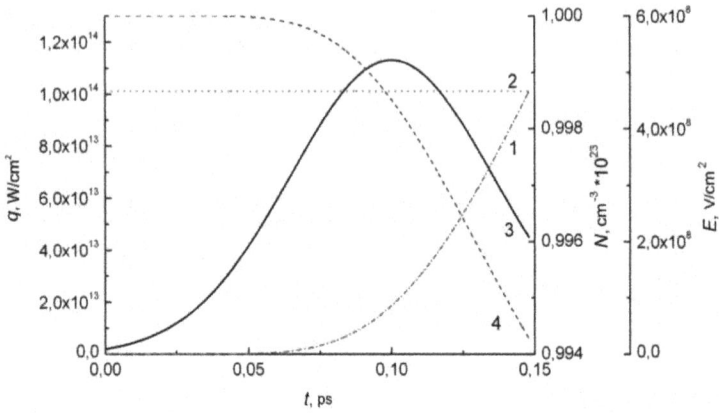

Figure 4. Temporal dependence of the electric field E (1), arising due to emission of electrons and the threshold field value E_{th} (2) corresponding to the beginning of Coulomb explosion. Laser pulse shape $q=q_m\exp(-(t-t_m)^2/t_{m1}^2)$, $t_m=100$ fs, $t_{m1}=50$ fs (3) and the temporal evolution of the net surface electron density (4)

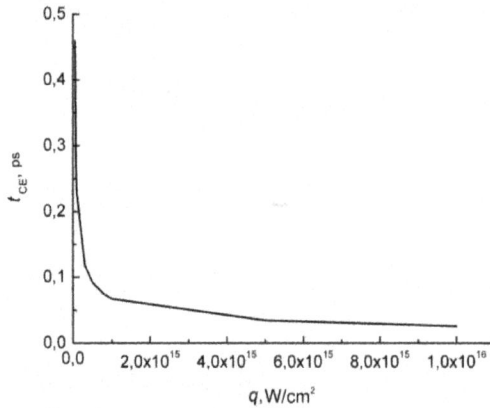

Figure 5. Time of Coulomb explosion onset as function of laser density for laser pulse shape $q=q_m\exp(-(t-t_m)^2/t_{m1}^2)$, $t_m=100$ fs, $t_{m1}=50$ fs

Various types of emission have different impacts on the concentration of emitted electrons (see Fig. 2-3): thermionic emission dominates over the photoemission and increases with the electron temperature increasing. The pulse shape significantly affects on the dynamics of the emission processes. However, according to the calculations the effect of emission processes on the electron gas temperature and the lattice temperature (Fig. 1a, b) is negligible. Also, numerical experiment showed that the occurrence of Coulomb explosion (caused by the emission processes during the pulse) in metals requires high-power incident radiation, which is impossible in the real exposure modes.

Emission processes have a significant impact on the processes of heating and destruction of the semiconductors, because the initial concentration of conductivity electrons in semiconductors can be below concentration of free electrons generated by the action of femtosecond laser radiation in contrast to metals. Let us consider the emission impact on the example of a femtosecond microstructuring of the silicon surfaces.

3. Effect of electron emission on changes in optical properties of semiconductors under the femtosecond laser pulse action

The formation of periodic surface structures (PSS) is a perfect evidence of an induced change in the surface optical properties under the femtosecond pulse action on semiconductor and wide-band dielectrics (Fig. 6). Being formed under different conditions PSS exhibit the same formation regularities: the structures orientation depends on the direction of polarization vector of the laser radiation; the structures period depends on the wavelength, incidence angle of the radiation and dielectric permeability of the medium. The observed regularities suggest that the mechanism of PPS formation is determined by electromagnetic field, which is the result of interference of the incident wave and the excited surface electromagnetic waves (SEW).

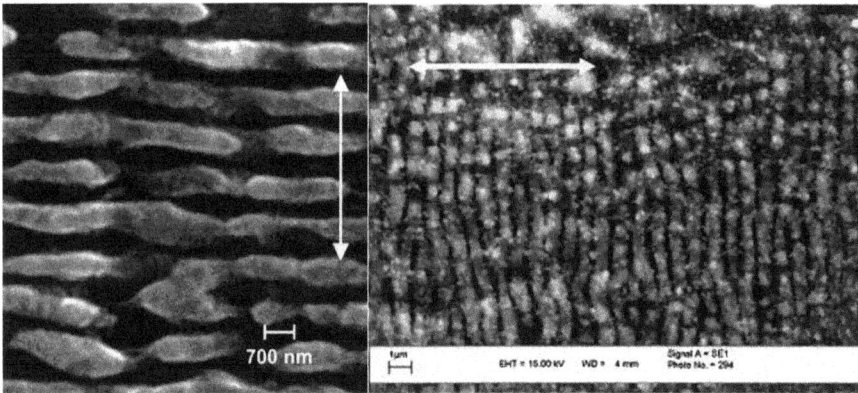

Figure 6. SEM image of the monocrystalline silicon surface irradiated by the laser pulse. Arrow indicates the direction of laser radiation polarization. Left: after irradiation by 1200 pulses, the energy density is ~ 1 J/cm². Right: after exposure by 300 pulses, the energy density is ~ 2 J/cm²

Microstructures shown in Fig. 6, were obtained on silicon irradiated by 80 fs laser pulses at a wavelength of 1.25 μm [25, 26]. Surface structures turned during rotation of the laser radiation polarization vector, and depending on the density of laser flux were oriented either perpendicular ($Q_0 \sim 1$ J/cm²) or parallel to the direction of polarization ($Q_0 \sim 2$ J/cm²).

It should be noted that low concentration of free electrons in semiconductor in the initial state does not provide the surface optical properties, which are necessary for excitation of the SEW. In case of longer laser pulses (longer than tens of picoseconds) experimentally observed excitation of SEW on semiconductors is related to properties of the melt formed on the surface due to laser heating. This explanation can't be used for femtosecond action because photoexcitation and thermal processes are separated in time and the surface does not melt during the laser pulse. The conditions for excitation of SEW during ultrashort laser pulse result from high concentration of nonequilibrium carriers, which are generated in the semiconductor by the light.

Dynamics of the optical properties change at the surface of semiconductors under femtosecond laser action is related to change of the non-equilibrium carriers plasma frequency. In order to analyze the behavior of the optical properties let us first consider the basic mechanisms of light absorption and recombination of the absorbed energy.

Total absorption coefficient in semiconductors can be considered as a sum of absorption coefficients associated with different mechanisms $\alpha = \sum_{i=1} \alpha_i$. First of all these are fundamental band-to-band absorption of light ($h\nu > E_g$, E_g is the band-gap) and intraband absorption, i.e. absorption by free carriers - electrons and holes (from now on we consider only the electrons for simplicity). Rate of relaxation of the crystal electron system from the excited state to the equilibrium state is determined by the recombination mechanisms with characteristic times $\sim 10^{-12}$-10^{-10}s.

During femtosecond laser pulse the light intensity can achieve very high level without destruction of the matter, therefore initiating multiphoton processes. In the wavelength range from IR to near UV the energy of a single photon is not sufficient for electron transition from the valence band to the conduction band ($h\nu < E_g$). Such transition takes place as a result of simultaneous absorption of several photons. In this case the rate (or probability) of multiphoton ionization is highly dependent on the laser power. The multiphoton ionization rate is proportional to σI^m, where I is the laser radiation intensity, σ_m - m-photon absorption cross-section. The required number of photons is determined by the lowest value of m, satisfying the relation $mh\nu > E_g$.

During femtosecond laser pulse only photoexcitation and fast electronic processes are observed, while recombination and lattice heating can be neglected, because the characteristic times of these processes are much higher than the pulse duration. This is one of the major differences between action of ultrashort and longer laser pulses.

Model. According to the model the dynamics of distribution of nonequilibrium electron concentration $n(z, t)$ during the femtosecond pulse is determined by generation of the nonequilibrium electrons due to two-photon absorption ($h\nu < E_g$) followed by their diffusion and participation in collision processes as described in the diffusion equation (11).

In this consideration nonequilibrium electrons are the electrons that transit from the valence band to the conduction band under photo-excitation, and then contribute to increasing electron gas temperature, emission, and finally recombination after the end of the pulse. Below by "electrons" we will understand "non-equilibrium electrons" and by "electron gas" we will understand "nonequilibrium electrons gas".

$$\frac{\partial n(z,t)}{\partial t} = \alpha_{2phi} J + D\frac{\partial^2 n(z,t)}{\partial z^2} - \frac{n(z,t)}{\tau_e},$$ (11)

where D is the diffusion coefficient of electrons in a solid, τ_e is the time of electron collisions.

Similarly to the above considerations for metals, the losses of electrons at the surface caused by external emission are taken into account in the model as follows:

$$-D\frac{\partial n(z,t)}{\partial z}\Big|_{z=0} = F_{2pho}\Big|_{z=0} + F_t\Big|_{z=0},$$ (12)

$$F_{2pho}\Big|_{z=0} = -\int_0^\infty \sigma_2 J^2 \exp(-z/l_e)dz.$$ (13)

Expression (13) describes two-photon emission. Single-photon absorption cross-section σ_1 is estimated assuming the maximum absorption coefficient, that corresponds to absorption by free carriers in metal $\sigma_1 \approx \alpha_{max}/n_{max} \sim 10^{-17}$ cm^2 (n_{max} is electron concentration in the metal). Photo-ionization of atoms is considered in terms of electron transition through virtual states. The electron life time (τ_0) at these virtual states is determined basing on the uncertainty relation between energy and time: $\tau_0 \sim 10^{-16}$ s. In this case two-photon cross-section is given by $\sigma_2 \approx \sigma_1^2 n(z, t)\tau_0 \sim 10^{-28}$ cm s. Expression (14) describes external thermo-emission (Richardson low).

$$F_t = -BT_e^2\Big|_{z=0}\exp\left(-\frac{\varphi_e}{k_b T_e\big|_{z=0}}\right)\exp(-z/l_e)/q_e$$ (14)

Temperature of electrons gas (T_e) is determined by heat conduction equations (15), where heat capacity of electron gas is a function of temperature and concentration.

$$\frac{\partial T_e}{\partial t} = a_e\frac{\partial^2 T_e}{\partial z^2} - \beta_{ei}(T_e - T_i)/c_e + \alpha_e J h\nu / c_e,$$

$$\frac{\partial T_i}{\partial t} = a_i\frac{\partial^2 T_i}{\partial z^2} - \beta_{ei}(T_e - T_i)/c_i,$$ (15)

$$c_e = \frac{\pi^2 k_b^2 n(z,t)T_e}{2\varepsilon_F},$$ (16)

where F_{2pho} is determined from (13) for two-photon photoeffect, F_t - from (14).

The Bouguer–Lambert differential law of Eq. (17) determines the intensity distribution $J(z, t)$ inside the solid (the z axis is directed depth ward). In this model absorptance of the material (A) is assumed to be constant.

$$\frac{\partial J(z,t)}{\partial z} = -(\alpha_{2phi} + \alpha_e + \alpha_{2pho})J(z,t),$$
$$J(0,t) = AJ_0(t); J_0(t) = q_0(t)/(h\nu), \tag{17}$$
$$q_0(t) = q_m(t/t_m)\exp(-t/t_m),$$

where $AJ_0(t)$ is the density of the absorbed photon flux, with a bell-shaped temporal distribution of intensity of the laser radiation, α_{2phi} is two-photon absorption coefficient of the inner photoeffect, α_{2pho} is two-photon absorption coefficient of the extrinsic photoeffect, and α_e is coefficient for absorption by free electrons, which in turn are defined as

$$\alpha_{2phi} = \sigma_1^2 n_{max} \tau_\nu J,$$
$$\alpha_{2pho} = \sigma_1^2 n(z,t)\tau_\nu J, \tag{18}$$
$$\alpha_e = \sigma_1 n(z,t).$$

Expressions (11-18) along with initial and boundary conditions at $t = 0$ and $z = \infty$

$$n\big|_{z=\infty} = n\big|_{t=\infty} = 0,$$
$$T_e\big|_{z=\infty} = T_i\big|_{z=\infty} = T_e\big|_{t=0} = T_i\big|_{t=0} = 0 \tag{19}$$

allow one to obtain spatial and temporal distribution of the electron concentration in semiconductor $n(z, t)$.

The first qualitative estimates were made for a simplified model, in which the emission mechanism is generalized and the emission flux in the expression (12) is taken into account using the emission factor μ without separation of photo- and thermal emission

$$-D\frac{\partial n(z,t)}{\partial z}\Big|_{z=0} = -\mu n.$$

Fig. 7 shows obtained variation of the electron concentration with depth during the laser pulse action on silicon. The following initial data were used for the calculation: $D = 80$ cm²/s, $\tau_e = 10^{-14}$s, $\mu = 2 \cdot 10^8$ cm/s, $Q_0 = 2$ J/cm², pulse duration is 80 fs, wavelength is 1.25 μm. The calculation results indicate that the semiconductor surface acquires properties of a metal during the laser pulse action. The maximum of electron density is located at some distance from the surface. It shifts from the surface into the bulk of material and its value increases during the pulse action. By the middle of the pulse the distribution of electron concentration stabilizes following the shape of the laser pulse.

The above estimates showed that dynamics of the optical properties of semiconductor under action of ultrashort pulse can be described within the same approach used for metals. According to Drude dispersion theory the dielectric permeability of photoexcited semiconductor can be determined by the plasma frequency of the electron gas (ω_p), incident radiation frequency (ω) and the frequency of electron collisions (γ), according to following expressions:

$$\varepsilon = \varepsilon' + i\varepsilon'',$$

$$\text{Re}\,\varepsilon = \varepsilon_n - \frac{\omega_p^2}{\omega^2 + \gamma^2}, \qquad (20)$$

$$\text{Im}\,\varepsilon = \frac{\omega_p^2 \gamma}{\omega\left(\omega^2 + \gamma^2\right)},$$

$$\omega_p = \sqrt{\frac{4\pi n(z,t) q_e^2}{m_e}}, \qquad (21)$$

where ε_n is the initial value of the semiconductor permittivity, and ε' and ε'' is the real and imaginary parts of the permeability.

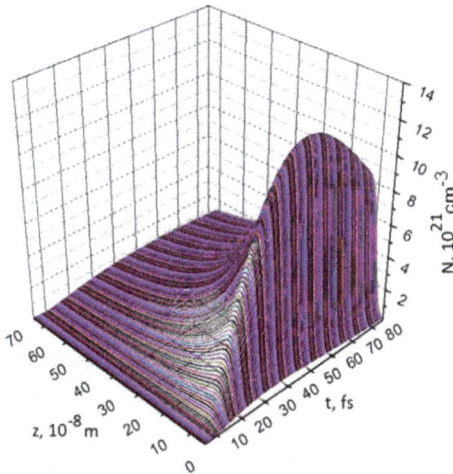

Figure 7. The dynamics of silicon photoexcitation

Let us consider dynamics of the dielectric permeability in the surface layer of semiconductor taking into account the change in the plasma frequency of non-equilibrium carriers by using a mathematical model (11-21). This will help us to identify the role of different emission processes in evolution of the optical properties during a femtosecond pulse.

Numerical simulation was performed for silicon ($\varepsilon_n = 12$, $\gamma = 10^{14}$ s^{-1}) with initial data given above. The calculation results are shown on Fig. 8-10.

When there is no emission surface the real part of dielectric permeability at the surface quickly decreases and becomes negative stabilizing by the end of the first quarter of the pulse (Fig. 10, curve 1). If external photo-emission is taken into account, the character of the dynamics of Reε does not change, but the magnitude at which the permeability stabilizes is increased (Fig. 8, curve 2). Thermionic emission strongly affects the dynamics of the permeability. Influence of the thermionic emission is small in the beginning of the pulse and value of Reε abruptly decreases. Few femtosecond later contribution of the thermionic emission grows and value of Reε, returned to its initial level after several damping oscillations (Fig. 8, curve 3). The observed inertia is typical for the thermionic emission mechanism, and the return of the dielectric permeability to the initial value means that all the "hot" electrons leave the surface as a result of thermal emission. If photo-emission is also taken into account escape of the electrons speeds up (Fig. 8, curve 4).

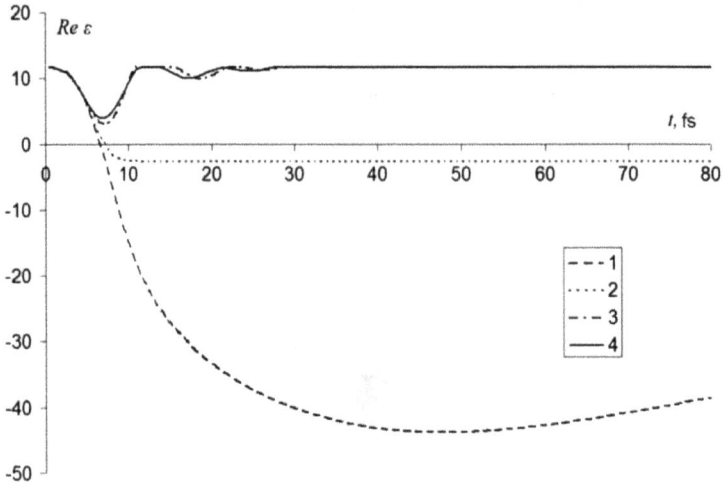

Figure 8. Dynamics of the real part of the permeability at the surface of silicon ($Q_0 = 2$ J/cm^2). 1 – no external emission, 2 – only photo-emission is taken into account, 3 – only thermionic emission is taken into account, 4 – combined action of photo-emission and thermal emission

Figure 9 shows dynamics of the real part of the permeability at the silicon surface for different values of the light flux density. If only two-photon photo-emission is taken into account, the surface acquires metal-like properties at the first femtoseconds of the pulse, and the permeability remains negative during the entire pulse. Increasing the radiation-flux density reduces the time to reach the steady-state level without changing the character of the dependence (Fig. 9a). When both photo-emission and thermionic emission are taken into account the picture changes (Fig. 9b). If $Q_0 \leq 1$ J/cm^2, Reε smoothly returns to its initial value (curve 1). When the energy density increases the permeability oscillates during transition to its initial value (curves 2 and 3). These oscillations result from dependence of heat capacity of the electron gas on electron concentration. The higher light flux induces higher temperature of the electron gas and contribution of thermionic emission increases. As a result the permeability returns to its initial value faster. At the same time increasing flux density raises the electron concentration and, accordingly, the heat capacity of the electron gas increases thus reducing its temperature and the contribution of the thermal emission. This results in oscillating of dielectric permeability.

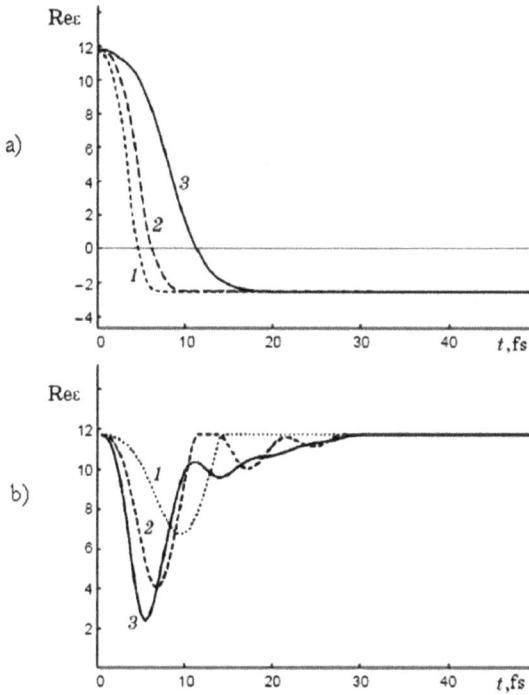

Figure 9. Dynamics of the real part of the permeability on the surface of silicon for various values of the light-flux energy density. (a) Photoemission only, (b) Combined action of photoemission and thermal emission. $1 - Q_0 = 1$ J/cm^2, $2 - Q_0 = 2$ J/cm^2, $3 - Q_0 = 3$ J/cm^2

Figure 10 shows the calculated distribution of permeability Re$\varepsilon(z)$, at the end of the pulse ($Q_0 = 2$ J/cm^2). If there is no emission, a very thin metal-like layer appears on the surface (curve 1). If external photo-emission is taken into account, a metal-like layer (several tens of nanometers thick, curve 2) is formed. The loss of the electrons due to thermionic emission qualitatively changes the permeability distribution depth wards (curve 3). A layer with smaller permeability than its initial value is formed at a distance of tens of nanometers from the surface. Combined action of photo-emission and thermionic emission shifts this layer deeper (curve 4).

Let us consider again the conditions necessary for the SEW excitation in order to compare the results of numerical simulation with the above experimental data on femtosecond silicon microstructuring associated with the SEW excitation.

It is known that the formation of PSS oriented perpendicular to the polarization of laser radiation usually results from excitation of surface plasmon-polaritons under the laser pulse action on the metal. Surface plasmon-polaritons are partially longitudinal electromagnetic waves of the TM-type propagating along the interface between two media with the wave electro-magnetic field being localized near the interface. Excitation of plasmon-polaritons is possible only if one of the media has positive dielectric permeability ($\varepsilon_1>0$), while the real part of dielectric permeability of the other is negative ($\text{Re}\varepsilon_2<0$). Also, the condition $|\text{Re}\varepsilon_2|>\varepsilon_1$ must be satisfied. A negative dielectric permeability in a metal is determined by a high concentration of free electrons. In the case of the relatively long laser pulse action (tens of picoseconds or more) on the semiconductor the appearance of metal-like optical response is usually associated with the properties of the melt formed on the surface due to laser heating.

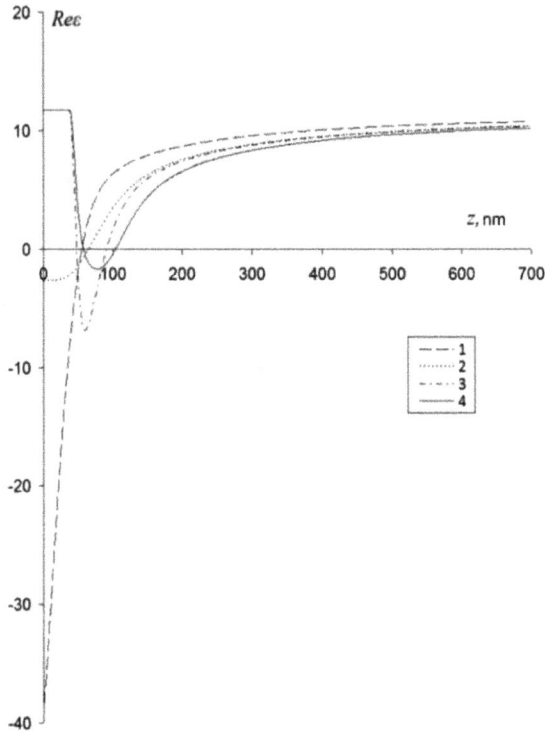

Figure 10. Spatial distribution (into the depthward of the semiconductor) of the real part of the permeability at the end of the laser pulse. 1 – no external emission of electrons, 2 – photo-emission only, 3 – thermionic emission only, 4 – combined action of photo-emission and thermal emission

Semiconductors typically have positive value of their dielectric permeability in the visible and IR range. Under the femtosecond laser pulse action the concentration of nonequilibrium carriers in semiconductors can become so high that dielectric permeability ε would change its sign and creates conditions for excitation of surface plasmon-polaritons. In this case, the formation of PSS perpendicular to the polarization vector is experimentally observed.

The formation of structures parallel to the polarization vector is associated with the excitation of surface waveguide modes (TE-polaritons). It is necessary to create an optically layered structure for excitation of a waveguide mode at the semiconductor surface (Fig. 11).

The refractive index of the waveguide layer in such optically layered structure (n_2) exceeds the refractive index values of adjacent layers (n_1, n_3), ($n_2 > n_1$, $n_2 > n_3$).

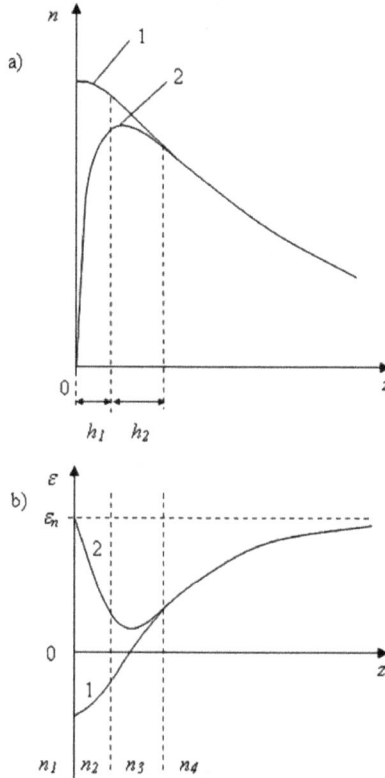

Figure 11. Schematic distribution of the electrons concentration in a semiconductor under the femtosecond laser pulse action (a) and the corresponding distribution of the dielectric permeability (b) for different emissions; n_1, n_2, n_3, n_4 - effective refractive indices of the layers as they are alternating along the coordinate z

In addition, it is necessary to provide a certain minimum thickness of the waveguide layer at a given frequency. If action of laser radiation results in formation of a layer with a refractive index n_3 and closer to surface a layer of thickness h_1 with a refractive index n_2 provided $n_2 > n_3 > n_1$, then [7]

$$h_{min} \approx \frac{\lambda}{2\pi\sqrt{n_2^2 - n_3^2}}\arccos\sqrt{\frac{n_2^2 - n_3^2}{n_2^2 - n_1^2}}. \tag{22}$$

In particular, for silicon at a wavelength of 1.25 µm the minimum thickness of the waveguide layer is $h_{min} \sim 70$ nm according to expression (22).

Let us consider the calculated spatial distribution of the silicon dielectric permeability under the femtosecond laser pulse action (see Fig. 10) in terms of the possible conversion of the incident light into surface plasmon-polaritons and waveguide modes. The surface polariton excitation requires transition of the semiconductor surface into a metal-like state, while formation of a dynamic optically-layered structure with a certain minimum thickness of the waveguide layer is necessary for the excitation and propagation of waveguide modes.

As follows from the numerical model (Fig. 10, curves 1-2) in case of relatively low emission a metal-like layer is formed at the surface. Within the layer thickness is about 50-60 nm dielectric permeability becomes negative. This provides the conditions for excitation of surface plasmon-polaritons (TM-type SEW), which is confirmed experimentally by formation of the microstructures perpendicularly to the polarization vector.

If emission rate is high as in case of thermo-emission (Fig. 10, curve 3) an optically layered structure is formed. Although the dielectric permeability does not change its sign, excitation of waveguide modes is possible.

Combination of both photo- and thermal-emission (Fig. 10, curve 4) results in formation of a dielectric layer of thickness ~ 60 nm and a metal-like layer of thickness about 40 nm. The refractive index of this dielectric layer is higher than both the refractive index of the air on one interface, and the refractive index of the metal-like layer on the other interface. Presence of such an optical structure enables excitation of waveguide mode, which results in the formation of periodic relief, which is parallel to the polarization vector on the incident radiation (see Fig. 6, right).

The above consideration showed that multiphoton emission and thermionic emission noticeably vary the optical properties of a semiconductor during a femtosecond pulse action. In particular, layer with different optical properties is formed at the surface and enables excitation of either surface polaritons or waveguide modes in the semiconductors. The considered model allows one to qualitatively and quantitatively interpret available experimental data. This approach allows one to use experimentally observed surface microstructures as relatively simple means of investigation of dynamics of the semiconductor surface properties under femtosecond action.

4. Conclusion

The results of numerical simulation have shown that influence of emission processes on the electron gas temperature and lattice temperature of metals is negligible. Therefore, the emission can be neglected when assessing the parameters of metals processing by femtosecond laser pulse, which simplifies the numerical calculations.

The Coulomb explosion occurrence in metals requires high-power incident radiation, which is impossible for the real exposure conditions.

However, in semiconductors both types of extrinsic emission noticeably change distribution of dielectric permeability near surface providing conditions for excitation of surface polaritons or waveguide modes depending on laser power magnitude.

The proposed method allowed to estimate the cross sections of multiphoton absorption in metals. For example, for metals absorption cross section: $\sigma_1 = 10^{-17} cm^2$ for one-photon absorption, $\sigma_2 = 10^{-28} cm$ s for two-photon absorption, $\sigma_3 = 10^{-61} cm^3 s^2$ for three-photon absorption.

Author details

Roman V. Dyukin, George A. Martsinovskiy, Olga N. Sergaeva, Galina D. Shandybina, Vera V. Svirina and Eugeny B. Yakovlev

Department of Laser-Assisted Technologies and Applied Ecology,
National Research University of Information Technologies, Mechanics and Optics,
Saint-Petersburg, Russia

Acknowledgement

This work was supported by grants of RFBR 09-02-00932-a, 09-02-01065-a and State Contract No P1134.

5. References

[1] Korte F., Nolte S., Chichkov B.N., Bauer T., Kamlage G., Wagner T., Fallnich C., Welling H. (1999) Far-Field and Near-Field Material Processing with Femtosecond Laser Pulses. Appl. Phys. A, 69 [Suppl.]: S7–S11.

[2] Vorobyev D.Y., Guo C.L., (2010) Metallic Light Absorbers Produced by Femtosecond Laser Pulses. Advances in Mechanical Engineering, 2010: 1-5.

[3] Chimmalgi A., Grigoropoulos C.P., Komvopoulos K. (2005) Surface Nanostructuring by Nano-Femtosecond Laser Assistant Force Microscopy. App. Phys. 97: 1043191 - 104319112.

[4] Myers R.A., Farrell R., Karger A.M., Carey J.E., Mazur E. (2006) Enhancing Near-Infrared Avalanche Photodiode Performance by Femtosecond Laser Microstructuring. Appl Opt. 45(35): 8825-31.

[5] Etsion I. (2005) State of the Art in Laser Surface Texturing. J. of Tribology, 127(1): 248-253.

[6] Oktem B., Kalaycioglu H., Erdoğan M., Yavaş S., Mukhopadhyay P., Tazebay U. H., Aykaç Y., Eken K., Ilday F. Ö. (2010) Surface Texturing of Dental Implant Surfaces with an Ultrafast Fiber Laser. in Conference on Lasers and Electro-Optics, OSA Technical Digest (CD) (Optical Society of America, 2010), JTuD15.

[7] Libenson M.N. (2007) [Laser-Induced Optical and Thermal Processes in Solids and Their Mutual Influence]. Saint-Petersburg: Nauka. 423 p. (in Russian)

[8] Libenson M.N. (2001) Non-Equilibrium Heating and Cooling of Metals Under Action of Super-Short Laser Pulse. Proc. SPIE. 4423: 1–7.

[9] Carey J.E., Crouch C.H., Mazur E. (2003) Femtosecond-Laser-Assisted Microstructuring of Silicon Surfaces. Optics&Photonics News. 14(2): 32-36.

[10] Shimosuma Y., Kazansky P.G., Qin J.R., Hirao K. (2003) Self-Organized Nanogratings in Glass Irradiated by Ultrashort Light Pulses. Phys. Rev. Lett. 91: 247405-247409.

[11] Kudryashov S.I., Emel'yanov V.I. (2001) Electron gas compression and Coulomb explosion in the surface layer of a conductor heated by femtosecond laser pulse. JETP Lett. 73(12): 666-670.

[12] Evans R., Badger A.D., Fallies F., Mahdieh M., Hall T.A., Audebert P., Geindre J.-P., Gauthier J. C., Mysyrowicz A., Grillon G., Antonetti A. (1996) Time- and Space-Resolved Optical Probing of Femtosecond-Laser-Driven Shock Waves in Aluminum. Phys. Rev. Lett. 77(16): 3359-3362.

[13] Del Fatti N., Arbouet A., Vall'ee F. (2006) Femtosecond Optical Investigation of Electron–Lattice Interactions in an Ensemble and a Single Metal Nanoparticle. Appl. Phys. B. 84: 175-181.

[14] Chen. L.M., Zhang J., Dong Q.L., Teng H., Liang T.J., Zhao L.Z., Wei Z.Y. (2001) Hot Electron Generation Via Vacuum Heating Process in Femtosecond Laser–Solid Interactions. Phys. of Plasmas. 8(6): 2925-2929.

[15] Kaganov M.I., Lifshitz I.M., Tanatarov L.V. (1956) [Relaxation between Electrons and Lattice]. J. Exp. Theor. Phys. 31(2): 232-237. (in Russian)

[16] Anisimov S.I., Kapeliovich B.L., Perelman T.L. (1974) [Electron Emission from Metal Surfaces Exposed to Ultrashort Laser Pulses]. J. Exp. Theor. Phys. 39(2): 375-377. (In Russian)

[17] Anisimov S.I., Benderskii V.A., Farkas G. (1977) Nonlinear Photoelectric Emission from Metals Induced by a Laser Radiation. Soviet Physics Uspekhi. 20(6): 467-488.

[18] Delone N.B. (1989) [Interaction of Laser Beam with Matter. The course of lectures]. Moscow: Nauka. 289 p. (In Russian)

[19] Kato S., Kawakami R., Mima K. (1991) Nonlinear Inverse Bremsstrahlung in Solid-Density Plasmas. Phys. Rev. A. 43(10): 5560-5567.

[20] Wang X.Y., Downer M.C. (1992) Femtosecond Time-Resolved Reflectivity of Hydrodynamically Expanding Metal Surfaces. Optics Letters. 17(20): 1450-1452.

[21] Rethfeld B., Kaiser A., Vicanek M., Simonc G. (2001) Nonequilibrium Electron and Phonon Dynamics in Solids Absorbing a Subpicosecond Laser Pulse. Proc. SPIE. 4423: 250-261.

[22] Rethfeld B., Kaiser A., Vicanek M., Simon G. (2002) Ultrafast Dynamics of Nonequilibrium Electrons in Metals under Femtosecond Laser Irradiation. Phys.l Rev. B. 65: 214303-214313.

[23] Ashcroft N.W., Mermin N.D. (1976) Solid State Physics. New York: Holt, Rinehart, and Winston. 826 p.

[24] Bulgakova N.M., Rosenfeld A., Ehrentraut L., Stoian R., Hertel I.V. (2007) Modeling of Electron Dynamics in Laser-Irradiated Solids: Progress Achieved Through a Continuum Approach and Future Prospects. Proc. SPIE. 6732: 673208-673223.

[25] Zabotnov S.V., Ostapenko L.A., Golovan L.A., Shandybina G.D., Timoshenko V.Yu., Kashkarov P.K. (2005) Third Optical Harmonic Generation of Silicon Surfaces Structured by Femtosecond Laser Pulses. Proc. SPIE. 6161: 0J1-0J5.

[26] Ostapenko I.A., Zabotnov S.V., Shandybina G.D., Golovan L.A., Chervyakov A.V., Ryabchikov Yu.V., Yakovlev V.V., Timoshenko V.Yu., Kashkarov P.K. (2006) Micro- and Nanostructuring of Crystalline Silicon Surface under Femtosecond Laser Pulses. Bulletin of RAS: Physics. 70: 1503–1506.

Chapter 7

Diagnostics of a Crater Growth and Plasma Jet Evolution on Laser Pulse Materials Processing

A. Yu. Ivanov and S. V. Vasiliev

Additional information is available at the end of the chapter

1. Introduction

The scheme of the experimental setup used in the study is presented in Fig. 1. The radiation of the GOR-100M ruby laser (1) (λ =0.694 μm) operating in the free oscillation regime (pulse duration $\tau \sim 1.2$ ms, Fig. 2) or rhodamine laser (λ =0.58 μm), pulse duration $\tau \sim 20$ μs) passed through the focusing system (2) and was directed through the hole in the electrode (3) onto the sample (4) that served as the second electrode and was mounted in air at a pressure of 10^5 Pa. The radiation spot diameter with sharp edges on the sample was varied in the course of the experiments from 1 to 2 mm.

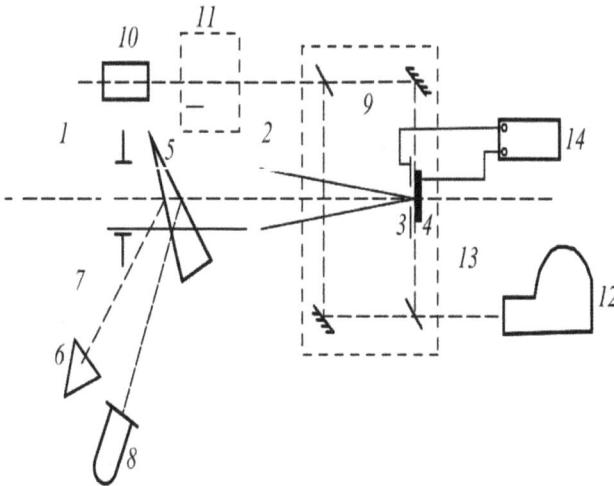

Figure 1. The experimental setup schematic diagram.

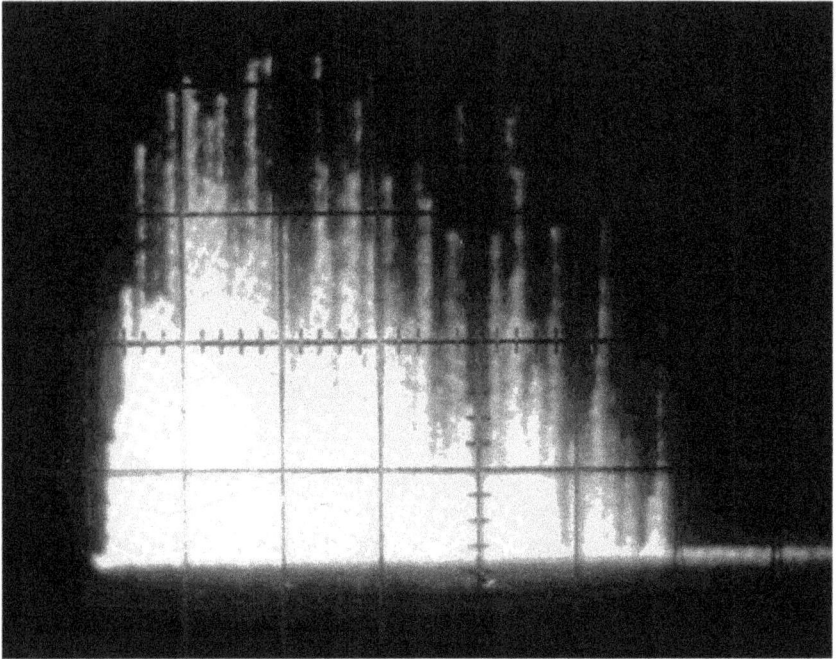

Figure 2. Oscillogram of the radiation pulse from the GOR-100M laser.
The scanning rate is 200 μs/div.

From the front face of the glass wedge *(5)* a part (4%) of laser radiation was directed into the IMO-2N energy meter (6), whose entrance window was located in the focal plane of the lens (7). The energy of the laser pulses varied from 5 to 60 J. The FEK-14 coaxial photodetector *(8)*, the signal from which was coupled to the S8-13 oscilloscope, was used to record the temporal shape of the laser pulse. The voltage was applied to the electrodes *(3, 4)* from the source *(14)*, built on the basis of the UN 9/27-13 voltage multiplier of the TVS-l10 unit. The source allowed the voltage variation within 25 kV and its stabilization in the course of the experiment.

To study the spatial and temporal evolution of the laser plasma plume in the course of laser radiation action on the sample, we used the method of high-speed holographic motion-picture recording. The interelectrode gap was placed in one of the arms of a Mach-Zehnder interferometer (9), which was illuminated with the radiation of the ruby laser *(10)* ($\lambda = 0.694$ μm) operating in the free oscillation regime. The pulse duration of the radiation amounted to ~ 400 μs. The transverse mode selection in the probing laser was accomplished using the aperture, placed in the cavity, and the longitudinal mode selection was provided by the Fabry-Perot cavity standard used as the output mirror. The probing radiation after the collimator *(11)* was a parallel light beam with the diameter up to 3 cm, which allowed observation of the steam-plasma cloud development.

The interferometer was attached to the SFR-1 M high-speed recording camera *(12)*, in which the plane of the film was conjugate with the meridian section of the laser beam, acting on the sample, by means of the objective *(13)*. The high-speed camera operated in the time magnifier regime. The described setup allowed recording of time-resolved holograms of the focused image of the laser plasma plume. Separate holographic frames provided temporal resolution no worse than 0.8 μs (the single frame exposure time) and the spatial resolution in the object field ~ 50 μm. The error in the determination of the electron density was ~ 10% and it was governed by the precision with which the shifts of the fringes could be determined in the photographically developed interference patterns.

The diffraction efficiency of the holograms allowed one to reconstruct and record interference and shadow pictures of the studied process under the stationary conditions. The shadow method was most sensitive to grad n, so that the nature of the motion of the front of a shock wave outside the laser plasma and of the motion of the plasma jet could be determined from the reconstructed shadow patterns. This gave information on the motion of the shock front and the laser plasma front generated at the surfaces of metal samples. It was found that the nature of the motion of the shock wave front was practically independent of the target material and was governed primarily by the average power density of the laser radiation.

The reconstructed interference patterns were used to determine the spatial and temporal distributions of the electron density in a laser plasma plume.

The reliability of the results obtained by the method of fast holographic cinematography was checked by determination of the velocity of the front of a luminous plasma jet by a traditional method using slit scans recorded with a second SFR- 1M streak camera.

To study the surface shape of the crater that appears on the plate, we used the fringe projection method, which in the present case appeared to be more efficient than holographic methods of surface relief imaging and the stereophotogrammetric method, since, already at the stage of fringe projecting, it allowed obtaining a picture with controllable sensitivity of measurements and sufficiently good visibility of fringes, controlled visually. The sensitivity of measurements (relative fringe displacement) was set by changing the period of the projected fringes, and the good visibility was provided by changing the angle of illumination of the studied surface till removing the light flares from the crater surface. The present method is thoroughly described and successfully used in [2].

The optical scheme of the apparatus used to visualize the topography of crater is shown on Figure 2. Radiation of helium-neon laser *1* LGN-215 collimated by telescopic system *2* was used to illuminate Mach-Zehnder interferometer *3*. The interference picture was then projected to the sample *4* being studied. During the above procedure a system of dark and light strips was observed on the treated surface, and besides the configuration of strips was connected synonymously with the depth of the crater in the point of interference

$$h = d \cdot \Delta k / (tg\beta + tg\gamma).$$

Here d is a period of interference strips on the flat (nonirradiated) zone of the target, Δk - the displacement of a dark strip, β - an angle between the perpendicular to the surface of the irradiated sample and the projected interference surface, γ - an angle between the normal to the surface of the irradiated sample and the optical axis of the photographic camera.

The sample surface was optically mated with the picture surface 5 with the help of the objective "Helios-44-2". Contour stripes on the picture surface were fixed on the photographic film.

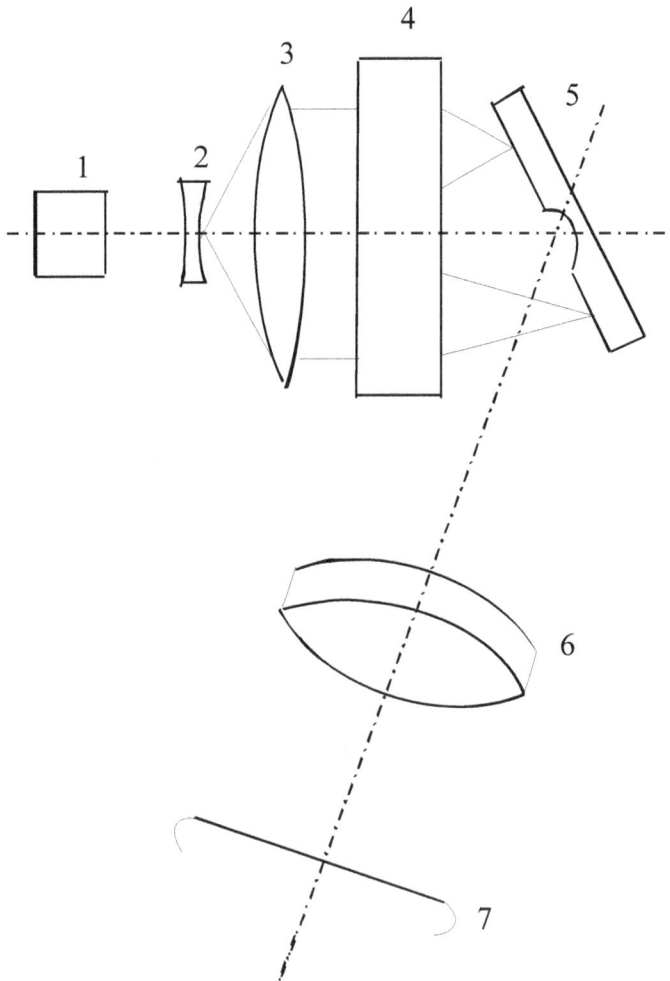

Figure 3. Schematic diagram of the visualization apparatus.

2. Experimental results and discussion

The experimental results have shown that at any polarity of the applied voltage [with positive or negative potential at the irradiated sample with respect to the electrode (3) the topography of the crater is practically identical and is determined by the energy distribution over the focusing spot of the laser radiation (Figs 4, 5).

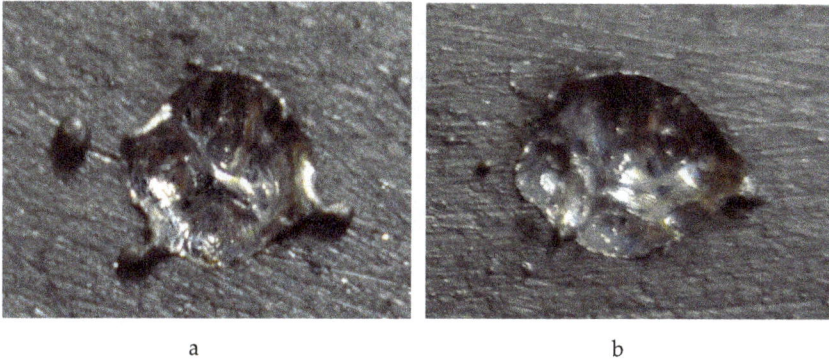

a   b

Figure 4. Photographs of the craters obtained under the action of laser pulses on the target in the absence of the external electric field (a) and in the presence of the field (b).

Figures 6 a - c display the interferograms, reconstructed from the holograms recorded at different instants in the course of high-speed holographic motion-picture shooting. The figure clearly illustrates both the initial stage of the laser torch development and the plasma flow around the electrode (3) at different directions of the external electric field strength vector.

Figures 6 d - f represent the data on the distribution of concentration of free electrons in the plasma of an evaporated metal at different instants, obtained by processing the interferograms [1]. Although the energy distribution over the laser radiation focusing spot is not uniform, the lines of equal concentration are practically smooth, which is an evidence of relatively uniform ionization of the eroded substance steams. It is essential that, despite a substantial increase in the plasma formation over time, the mean electron concentration in the torch remains practically unchanged and even slightly grows, which may be associated both with a constant increase in the mass of emitted substance and with secondary ionization of the plasma by the laser radiation. Note, that the presence of an external electric field weakly affects the concentration of electrons in the laser plasma plume.

When the interelectrode separation was 2 cm, the maximal transverse size of the steam-plasma cloud at the surface of the electrode (3) for negative voltage on the target was 2 cm and in the absence of the external electric field it was 1.5 cm. This may be observed both in the interferograms and by the burn on the polyethylene film protecting the second electrode. As seen from interferograms, after reaching the second electrode in 56, 64, and 72 μs, respectively, the steam-plasma cloud practically does not grow in the transverse dimensions. Probably it is due to the flowing out of the plasma from the interelectrode gap through the hole in the electrode *(3)*, which is used for passing the laser radiation to the target (the hole diameter being 1 cm).

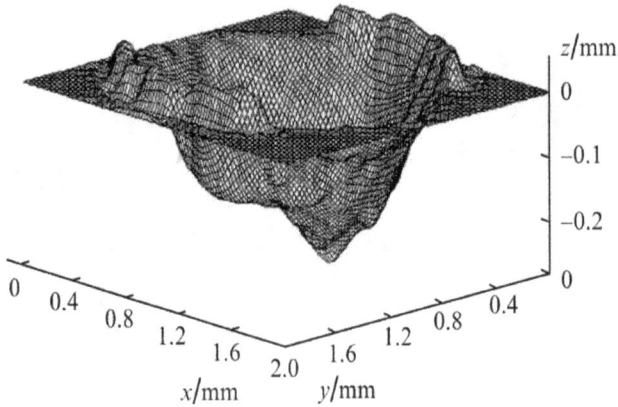

a

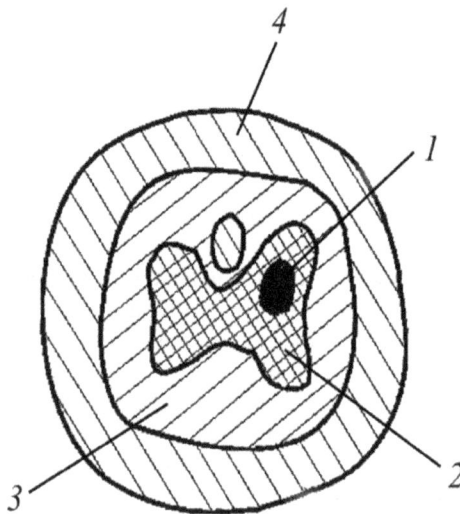

b

Figure 5. Volume topogram of a crater (a) and the distribution of light energy density over the transverse cross-section of the laser beam (b): 4.5 *(1)*, 3.5 *(2)*, 1.2 *(3)*, and 0.8 J mm^{-2} *(4)*.

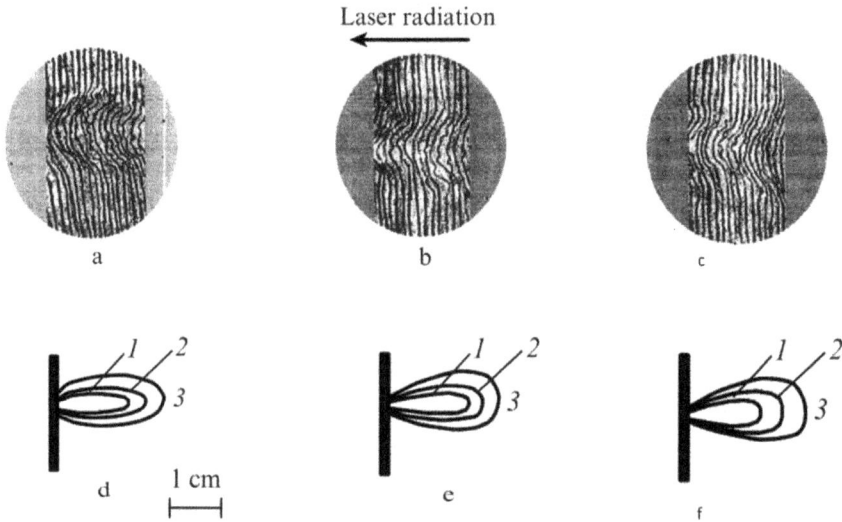

Figure 6. Interferograms of laser plasma torches (a, b, c) and electron concentration isolines in them (d, e, f) at the negative target potential **(b,** e) and at the positive target potential (c, f) at the instants 72 μs after the onset of the laser action; curve *(1)* corresponds to the electron concentration 5×10^{18}, curve *(2)* to 2.5×10^{18}, and curve *(3)* to 10^{18} cm^{-3}.

Figure 7 presents the time dependences of the plasma torch front motion velocity at different directions of the external electric field strength vector, calculated by using the information, obtained by analyzing the temporal variation of the interferograms. It is seen that even when the plasma front reaches the electrode (3), its velocity not only does not decrease (which is typical for late stages of the laser plasma torch existence [3]), but even increases; this happens both in the presence of the external electric field of any orientation and in the absence of the field. As already mentioned, this is due to the permanent and significant increase in the mass of the material, carried out under the action of laser radiation on the irradiated sample, as well as to the secondary ionization of plasma by laser radiation.

The maximal expansion velocity of the plasma torch amounted to 350 m/s for the negative voltage at the target, 310 m/s in the absence of the external electric field, and 270 m/s for the positive voltage applied to the target.

Our investigation showed that the time evolution of the leading edge of a luminous plasma moving away from the surface of a sample, deduced from the slit scans, differed from the time evolution of the front of the plasma jet, which was recorded by the shadow method. This allowed us to conclude that the concentration of the heavy particles, responsible for the radiation emitted by the plasma, was low at the front of the laser plasma jet, whereas the electron density was sufficient for reliable determination of the contribution of electrons to refraction in a hologram.

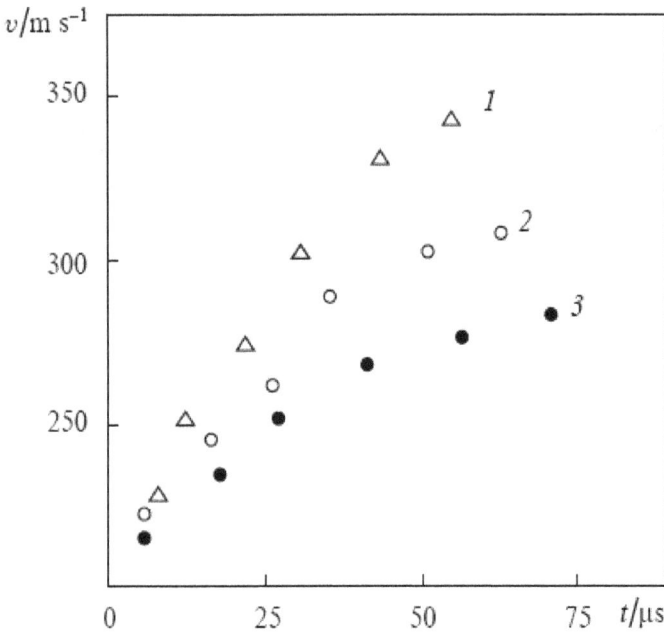

Figure 7. Time dependences of the velocity of the plasma torch front motion at the negative target potential *(1)*, in the absence of the field *(2)*, and at the positive target potential *(3)*.

The distribution of the density of cold air was determined and the electron density distribution was refined by two-wavelength holographic cinematography. We supplemented the system shown schematically in Fig. 1 with a second probe laser and an SFR-1M camera which recorded holograms at the wavelength of the radiation emitted by this laser. The second source of probe radiation was a laser utilizing a rhodamine 6G solution excited by a coaxial flashlamp. The use of a standard power supply system from a GOR-100M laser made it possible to generate output radiation pulses of 30 – 40 μs duration. The line width was reduced employing a plane-parallel Fabry-Perot interferometer. This made it possible to obtain scan holograms of the process at $\lambda_1 = 0.69\ \mu m$ and $\lambda_2 = 0.58\ \mu m$, and to separate the contribution of electrons from that of heavy particles to the refraction of a plasma jet.

This two-wavelength holographic cinematography method was used to determine the radial distributions of the electron density and of the heavy-particle concentration at different moments in time and for different sections of laser plasma near the irradiated surface of a irradiated sample

At distances of 10 - 15 mm from the surface of a sample it was found that heavy particles ("hot" atoms and ions) of metals and molecules of atmospheric gases made only a small contribution to refraction. At large distances (where there were no "heated" luminous particles) the contribution of the cold dense air became significant. This was due to the pushing out of air by a plasma cloud.

When either positive or negative potential is applied to the sample, many small droplets appear on its surface after the laser action (Figures 4 and 8). In particular, at the laser pulse energy 20 J, the diameter of the focusing spot 2 mm, and the electric field strength 10^6 V cm^{-1} we observed ejection of droplets having the mean characteristic size less than 0.1 mm to the distance up to 2 cm from the crater centre. The maximal characteristic size of the droplets was 0.4 mm.

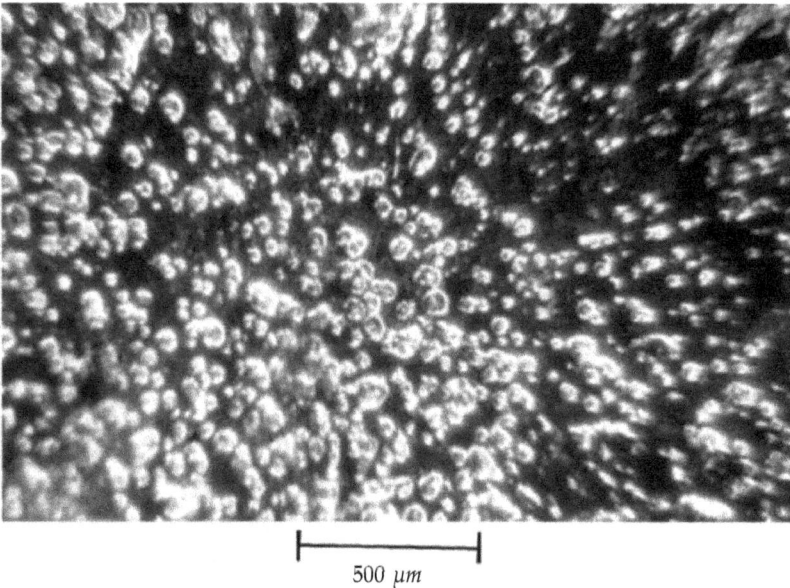

500 μm

Figure 8. Microscopic surface relief of the crater outer zone photographs.

In the absence of the external electric field the mean size of the droplets was ~ 0.4 mm. The droplets were seen at the distance up to ~ 1 cm from the crater centre.

In accordance with the results presented above, the dynamics of the processes on the surface of a sample, placed in an external electric field with the strength from 0 to 10^6 V m^{-1} and subject to the action of the pulsed laser radiation with the parameters mentioned above, is thought to be the following. The primary plasma formation and the initial stage of the laser torch development, in principle, do not differ from those observed in the absence of the external electric field. The metal is melted and evaporated. As a result of local formation of steam and plasma [4, 5], the erosion torch begins to form with the fine-dispersed liquid-droplets phase. Note, that the bulk evaporation is promoted by the gases, diluted in the metal, and by the spatiotemporal nonuniformity of the laser radiation [4]. At a radiation flux density $10^6 - 10^7$ W cm² the bulk evaporation is typical of all metals used in the experiments [5]. Obviously, the presence of the external electric field affects (increases or decreases depending on the direction of the field strength vector) the velocity of motion of the plasma front and causes some distortion of the plasma cloud shape. It is essential that the mentioned differences (at the considered parameters of laser radiation) are observed only at the initial stage of the laser plume development, because after the steam-plasma cloud reaches the electrode (3) an electric breakdown (short-circuit) occurs, and the external field in the interelectrode gap disappears.

Consider now the motion of the molten metal droplets in the steam-plasma cloud. In our opinion, the significant difference in the characteristic size of droplets, observed on the surface of the irradiated sample in the presence of the external electric field (independent of the direction of the field strength vector) and in the absence of the field, is a manifestation of the following mechanism of droplet formation [6]. It is known that at the surface of a liquid (including a liquid metal) the formation of gravity-capillary waves [7] is possible under the action of various perturbations. Undoubtedly, the examples of such perturbations are the spatially nonuniform evaporation of the target material due to nonuniform heating caused by nonuniform energy distribution over the focusing spot [8], the nonuniform primary plasma formation [6, 9, 10] caused by roughness of the irradiated sample surface [8], and, in the first place, the slop of the molten metal initiated by each spike of laser radiation, acting on the exposed sample [2].

Using the method presented in [11], one can show that at insignificant thickness of the molten metal layer (confirmed by the view of the 'outer' (directed) zone of the crater, particularly, the absence of fillets of significant height at the crater edge) the dispersion equation for the gravity-capillary waves takes the form

$$\omega^2 = \frac{\alpha k^3}{\rho} + g k - \frac{k E_0 E'}{4 \pi \rho \xi'} \Big|_{z=0},$$

(1)

where α is the surface tension coefficient of the molten metal; ρ is the metal density; g is the free fall acceleration; k is the magnitude of the wave vector of the gravity-capillary wave; E_0 is the electric field strength at the surface ($z = 0$) of the molten metal (the z axis is perpendicular to the surface of the irradiated sample, directed towards the laser radiation source and parallel to the vector E_0); $E' = -\partial\varphi/\partial z$ is the perturbation of the electric field in the space surrounding the molten metal; ξ' is the small displacement of the surface of the liquid in the z axis direction in the gravity-capillary wave.

Because for the uniform field E_0 the potential is $\varphi = -E_0 \cdot z$ (the potential at the metal surface is considered to be zero), the displacement of the mentioned surface by the small quantity ξ' leads to a small distortion of the potential:

$$\varphi' \Big|_{z=0} = E_0 \cdot \xi'.$$

(2)

It follows from Fig. 3 that the maximal concentration of electrons in the plasma formation does not exceed ~ .10¹⁸ cm³, which corresponds to the change in the dielectric constant of the medium ε by approximately 10^{-5}. Therefore, near the metal surface $\varepsilon \cong 1$ and with the boundary condition (2) taken into account

$$\varphi' = E_0 \xi' e^{-kz}.$$

In this case, the dispersion equation for gravity-capillary waves takes the form

$$\omega^2 = \frac{\alpha k^3}{\rho} + gk - \frac{k^2 E_0^2}{4\pi\rho}.$$

Because the frequency of the gravity-capillary waves ω is determined by the temporal characteristics of the abovementioned perturbations and, therefore, does not depend on the strength of the electric field E_0, the growth of the magnitude E_0 (independent of the direction of the vector $\vec{E}_0$) should cause the increase in the magnitude of the wave vector

$k = \dfrac{2\pi}{\Lambda}$ and the decrease in the wavelength Λ of the gravity-capillary wave. If we assume that the droplets are 'torn away' by the plasma flow from the 'tops' of the gravity-capillary wave and, therefore, their characteristic size is proportional to Λ, then it becomes clear why in the presence of the external electric field (of any direction) the observed mean size of the droplets becomes essentially reduced.

The escaped droplets possess the charge of the same sign as the sample. That is why the droplets begin to move with acceleration towards the second electrode. However, since the maximal initial velocity of the outgoing droplets under the analogous conditions [8] is ~ 45 m s⁻¹, i.e., an order of magnitude smaller than the velocity of steam-plasma cloud spreading, the droplets do not reach the electrode (3) before the moment of the breakdown in the interelectrode gap. In what follows (in the absence of the external electric field) the droplets move under the action of the same forces as in [8] and, therefore, in the way, described in [8]. In this case, having acquired at the stage of accelerated motion in the electric field the velocity, exceeding the initial one, the droplets may fly to a greater distance along the surface of the irradiated sample than in the absence of the electric field, which is observed in the experiment. Moreover, having moved to a greater distance from the sample surface and, therefore, being affected by the plasma for longer time before returning to the surface, the droplets may be split into finer parts than in the absence of the external field.

It should be noted that the droplets in the erosion plume may appear not only due to the molten pool surface instability, but also due to the condensation of the steams of the erosion products [12, 13]. Moreover, since the droplets produced in the course of condensation of steams may be charged [14], they, similar to those carried out from the molten pool, in the electric field may be removed from the crater to a greater distance than in the absence of the electric filed. However, this mechanism of plasma formation is dominating under somewhat different conditions of laser radiation acting on the material [12 – 15], namely, at significantly greater mean radiation flux density (~ 10^8 – 10^9 W cm⁻²) and smaller exposure duration (single pulses of laser radiation were used with the duration ~ 100 – 200 ns and with less smooth temporary shape). In the case of such a regime of laser metal processing one observes the screening of the irradiated sample by the plasma cloud, which is possible only at the concentration of the ablated material steam essentially exceeding 10^{18} cm^{-3} (see Fig. 6). In this case, one observes intense formation of droplets with the dimension ~ 200 nm and smaller, and this process is most active at the late stages of laser radiation action on the material (at decreasing intensity of laser action) [15] and even after its termination [12]. At smaller radiation flux density, characteristic of the experiment considered in the present work (~10^{18} cm^{-3}), droplet condensation from the steam of ablation products is expected to be less intense. Therefore, the essential contribution of the condensation mechanism to the process of formation of large drops (having the size ~ 0.1 – 0.4 mm, see Figures 4 and 8), especially at early stages of the process, i.e., before filling the entire interelectrode gap by the plasma cloud, seems to be hardly probable.

3. Acoustic waves generation

Investigating the acoustic emission let us use the model of a loaded zone radiating waves into elastic medium. Corresponding to this model let us consider the destructed zone as a spherical segment with the curvature radius R, depth d and diameter $2r_1$. z-ax of the coordinate system is directed along the laser beam. It is important that the parameters of the irreversibly deformed zone are changing on the time:

$$R = R(t),\ d = d(t),\ r_1 = r_1(t);$$

here t is the time.

The displacement vector in an elastic zone consists of a longitudinal and a transversal components, $\vec{A} = \vec{A}_l + \vec{A}_t$, and each of them can be described by corresponding wave equation. Because of the presence of the media board direct in the elastic wave generating zone the solution of the wave equations system we shall search as a sum of the volume and the surface components

$$\vec{A}_l = \vec{A}_{lO} + \vec{A}_{l\Pi} = \nabla \psi_O + \nabla \psi_\Pi,$$

$$\vec{A}_t = \vec{A}_{tO} + \vec{A}_{t\Pi} = \vec{A}_{t0} + rot(\vec{B}).$$

With regard to the symmetry of our problem, $\vec{A}_{t0} = 0$, scalar potential

$$\psi_O(\omega) = -\tilde{A}(\omega) \cdot \frac{\exp(-ik_l r)}{r},$$

$$B_p = B_z = 0,$$

$$\psi_\Pi(\omega) = -Z_0(k_R \rho) \cdot (\tilde{B}(\omega) \cdot \exp(-\chi_l z) + \tilde{Q}(\omega) \cdot \exp(\chi_l z)),$$

$$B_\varphi(\omega) = Z_1(k_R \rho) \cdot (\tilde{D}(\omega) \cdot \exp(-\chi_t z) + \tilde{S}(\omega) \cdot \exp(\chi_t z)).$$

Here ω is a frequency, $k_\ell = \omega/c_\ell$, c_ℓ and c_t – longitudinal and transversal sound velocities,

$\tilde{A}(\omega)$ – wave amplitude, $k_R = \omega / c_R$, c_R – surface wave velocity,

$\chi_i = (k_R^2 - k_i^2)^{1/2}$, $\tilde{B}(\omega)$, $\tilde{D}(\omega)$, $\tilde{Q}(\omega)$, $\tilde{S}(\omega)$, – os

cillation amplitudes, $Z_i(x)$ – spherical function.

When $\rho \to 0$ and $z \to \infty$ the result must remain finite, therefore $Z_i(x) = J_i(x)$ (Bessel function), $\tilde{Q}(\omega) = \tilde{S}(\omega) = 0$, and

$$\tilde{A}(\omega) = A(\omega)\frac{\vec{r}}{r^3}(1 + ik_l r)\exp(-ik_l(r - R)) + B(\omega)(\bar{\rho}_0 k_R J_1(k_R \rho) + \bar{z}_0 \chi_l J_0(k_R \rho)) \times$$
$$\times \exp(-\chi_l(z - h)) + D(\omega)(\bar{\rho}_0 \chi_t J_1(k_R \rho) + \bar{z}_0 k_R J_0(k_R \rho))\exp(-\chi_t(z - h)) \tag{3}$$

where $A(\omega) = \tilde{A}(\omega) \cdot \exp(-ik_l R)$, $B(\omega) = \tilde{B}(\omega) \cdot \exp(-\chi_l h)$, $D(\omega) = \tilde{D}(\omega) \cdot \exp(-\chi_t h)$.

Let us consider that on the surfaces $r = R$ and $z = h$ temporal dependence of pressure in the plasma cloud is

$$P/_{r=R} = p(t).$$

On the surface of spherical segment $R = R(t)$

$$\sigma_{rr} = -p(t), \quad \sigma_{r\theta} = \sigma_{r\phi} = 0;$$

on the surface $z = h$

$$\sigma_{zz} = \varrho(t), \quad \sigma_{\varrho z} = 0, \quad \sigma_{z\phi} = 0.$$

Here σ_{ij} are the components of the stress tensor, r, θ, φ are the coordinates of spherical system.

Out of the spherical segment $R = R(t)$ medium is elastic, and so

$$\left[\lambda\left(\frac{\partial A_r}{\partial r} + 2\frac{A_r}{r} + \frac{1}{r}\frac{\partial A_\theta}{\partial \theta} + \frac{A_\theta}{r}ctg\theta\right) + 2\mu\frac{\partial A_r}{\partial r}\right]_{r=R(t)} = -p(t),$$

$$\left(\frac{\partial A_\rho}{\partial z} + \frac{\partial A_z}{\partial \rho}\right)_{z=h(t),\rho=\rho_1(t)} = 0, \tag{4}$$

$$\left[\lambda\left(\frac{\partial A_\rho}{\partial \rho} + \frac{A_\rho}{\rho} + \frac{\partial A_z}{\partial z}\right) + 2\mu\frac{\partial A_z}{\partial z}\right]_{z=h(t),\rho=\rho_1(t)} = -p(t).$$

Here A_i are the components of the displacement vector $\vec{A}$, λ, μ are the Lame coefficients.

Substituting A_i from the equation (3) to the system (4) we can calculate for each temporal moment $A(\omega,R,d,\rho_1)$, $B(\omega,R,d,\rho_1)$, $C(\omega,R,d,\rho_1)$, $\sigma_{zz}(\omega,R,d)$ and

$$\sigma_{zz}(t) = \int_{-\infty}^{+\infty} \sigma_{zz}(\omega,R,d)Exp[i\omega t]d\omega = \int_{-\infty}^{+\infty} \sigma_{zz}(\omega,R(t),d(t))Exp[i\omega t]d\omega = \int_{-\infty}^{+\infty} \sigma_{zz}(\omega,t)Exp[i\omega t]d\omega.$$

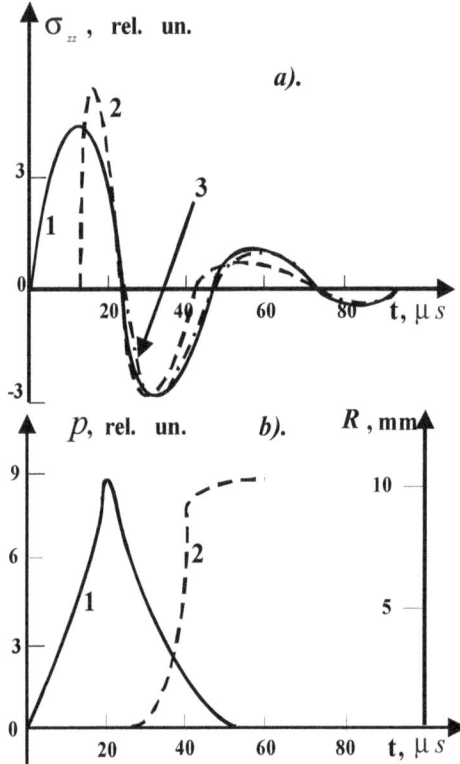

Figure 9. *a)*. Temporal dependences of pressure of acoustic wave on the action of laser pulse with duration of 20 μs on cupper sample: 1 – experimental dependence; 2 – dependence calculated without account of crater growth; 3 – dependence calculated with account of crater growth. *b)*. 1 – temporal dependences of pressure of plasma cloud on the border of irreversibly deformed zone; 2 – temporal dependences of curvature radius.

The results of the calculations with $R(t) = R_{max} \cdot \exp(\dfrac{t^2}{\tau_0^2})$ for $t<0$ and $R(t) = R_{max}$ for $t>0$ are presented on the figure 9; $\tau_0 = 40$ μs.

Evidently that at action on a surface of the copper sample of a rhodamine laser impulse duration ~ 20 μs time of growth of a zone of destruction makes approximately 40 мкс, that will well be coordinated with time of existence of plasma formation at a surface of the target exposed to laser-plasma processing (~ 50 μs). Use of model of the loaded area with the moving borders radiating acoustic waves in the elastic medium allows us to solve the important practical problem – the definition of a law of time growth of a zone of irreversible deformations on a surface of the sample exposed to pulse laser-plasma processing.

4. Laser treating of transparent insulators

Now let us to investigate the dynamics of crater growth and of changes in the density of an inelastically deformed material on the surface of a transparent insulator when it interacts with a millisecond light pulse that has a complex temporal profile. The experimental setup used in the study was similar to presented in Fig. 1. Radiation was provided by a GOR-100M ruby laser operating in the free-running regime. This made it possible to generate pulses of $\tau \sim 1.2$ ms duration with an energy E that could be varied within the limits 5-50 J. The radiation passed through focusing system and was directed to sample. Both single-lens and double-lens focusing systems were used; they formed an image of stop on the surface of the sample. The diameter D of the focusing spot obtained in this way was varied in the course of our experiments from 1 to 2 mm.

Part of the laser radiation was directed by the front face of the glass wedge to an IMO-2N energy meter whose entry pupil was located in the focal plane of lens. The radiation reflected by the rear face of the wedge was directed to FEK-14 coaxial photocell and the photocell signal was applied to the input of an S8-13 oscilloscope, which recorded the temporal profile of the laser pulse.

The sample was placed in a window in one of the arms of a Mach-Zehnder holographic interferometer which was illuminated by radiation from a second ruby laser operating in the free-running regime. The longitudinal modes emitted by this probe laser were selected by a Fabry -Perot etalon, used as the exit mirror, and the transverse modes were selected by a stop placed inside a resonator. The probe radiation was directed to collimator which produced a parallel beam with a diameter of ~ 3 cm. A field of view of this kind was sufficient to study the growth of a crater, the changes in the density of the sample in the inelastically deformed zone, the formation and propagation of elastic waves in the sample, and the processes that occurred in the gas and in the plasma cloud near the sample.

The interferometer was in contact with an SFR-1M high-speed camera. The plane of a photographic film in the camera was made to coincide, with the aid of objective, with the conjugate plane of a meridional cross section of the light beam acting on the sample. The camera was operated in the cine mode.

Fig. 10 shows the holographic interferograms at different moments from the beginning of the interaction of a laser pulse of energy $E \approx 35$ J with a sample made of polymethylmethacrylate (PMMA) on which the pulse formed a focusing spot $D \approx 1$ mm in diameter. The temporal profile of this pulse had the form shown in Fig. 11. We used these interferograms and solved the Abel equation to find the fields representing the distributions of the refractive index n (z, r, t) in space and time, and we then applied the Lorentz - Lorenz expression

$$\rho = \frac{(n^2 - 1)}{(n^2 + 2)} \frac{\mu}{R_m}.$$

to find the density fields $\rho\,(z,\ r,\ t)$ shown in Fig. 12. Here, μ is the molar mass of the material of the sample; $R_m = 4\,/\,3\pi N$, N is the number of particles responsible for refraction in a unit volume. The ratio $\mu\,/\,\rho$ was found from the parameters of the unperturbed medium.

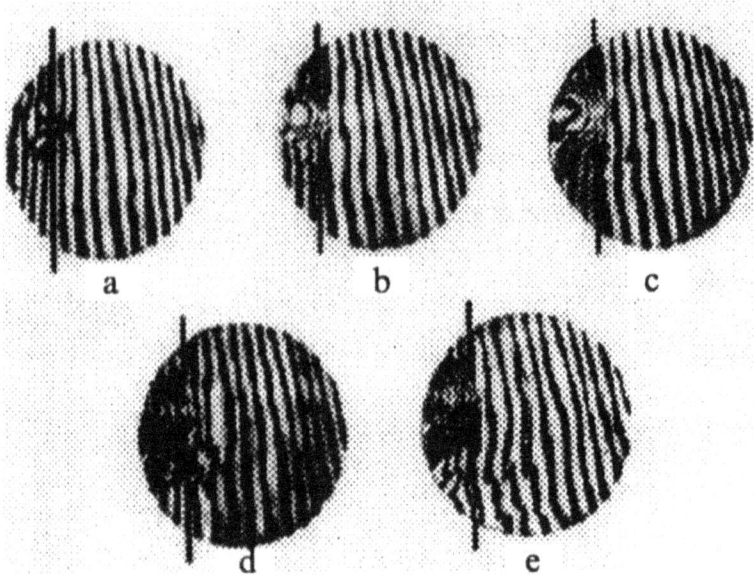

Figure 10. Interferograms recorded at the moments $t = 9.6\ \mu s$ (a), $16.0\ \mu s$ (b), $25.6\ \mu s$ (c), $48.0\mu s$ (d), and $70.4\ \mu s$ (e) relative to the beginning of the interaction.

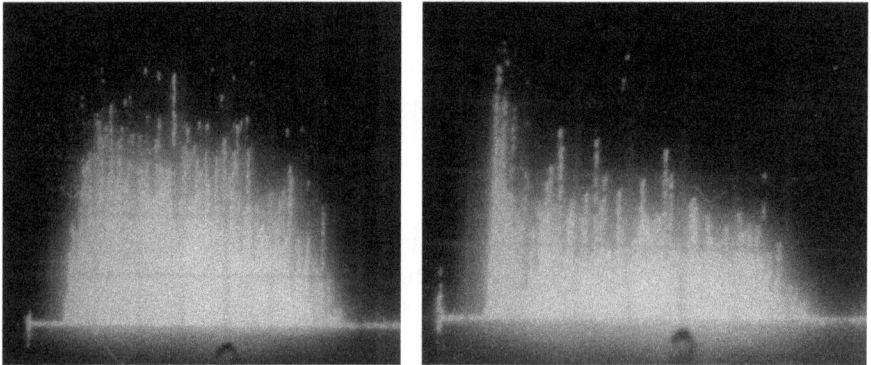

Figure 11. Temporal profile of a laser pulse before and after passing of a sample

Fig. 12 also shows the profiles of a crater formed on the surface of PMMA and recorded at different moments. These profiles yielded the time dependences of the diameter d, depth h, and volume V of the crater (Fig. 13). Fig. 14 shows the dependences of the final diameter, depth and volume of the crater on the energy of the light pulse interacting with the surface of PMMA.

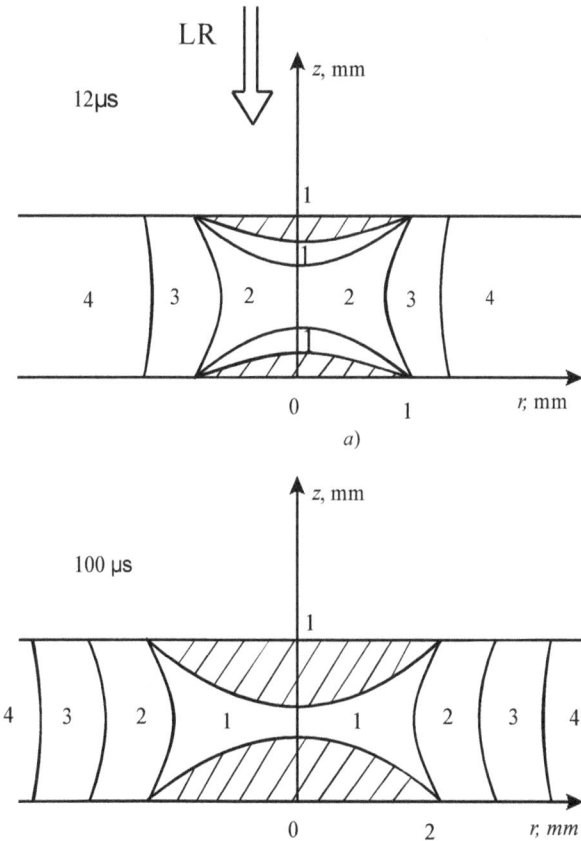

Figure 12. Profile of a crater and of a constant-density field at the moments $t = 12$ μs (a) and 100 μs relative to the beginning of the interaction.

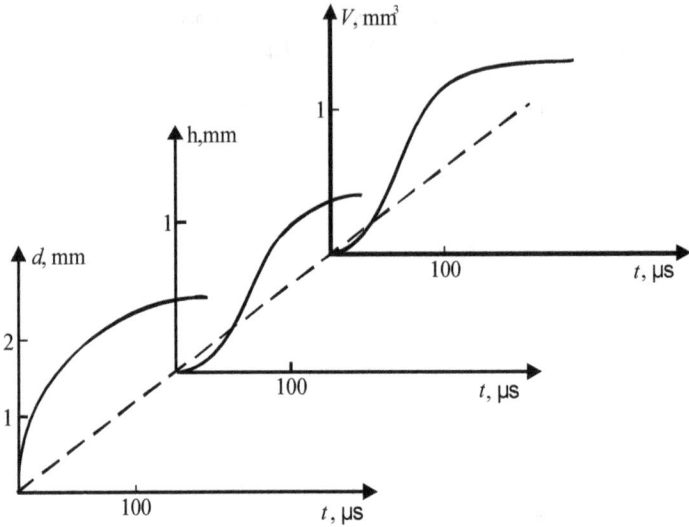

Figure 13. Crater diameter, depth, and volume time dependences.

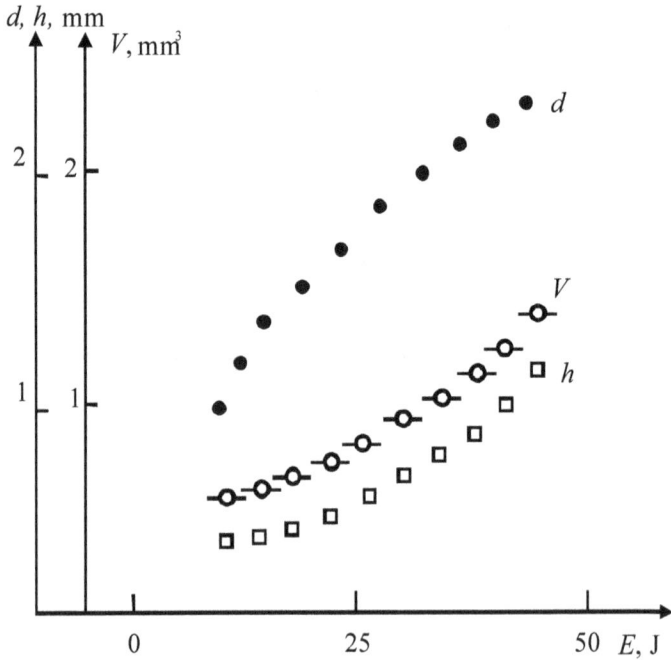

Figure 14. Dependences of the final diameter, depth and volume of a crater on the energy of a laser pulse.

5. Discussion of results

First of all, it is worth noting that the time dependences of the diameter d, depth h, and volume V of the crater (Fig. 13) were smooth and did not react in any way to the separate spikes of the laser pulse profile E (f). The dependences $d(t)$, $h(t)$, and $V(t)$ were found to be similar for the same value of

$$E(t) = 2\pi \int_0^t \int_0^{\tau_0} q(t)\, d\tau\, dr$$

irrespective of the profile $q(t)$ (which varied in a random manner from one pulse to the next). The constant-density surfaces changed equally smoothly (Fig. 4).

Obviously, even in the case of a very high power density ($\sim$ 10 MW cm^{-2}) the energy of a single spike ($\sim$ 0.1 J) was insufficient to stimulate any significant ejection of the mass of the target.

The growth of the crater and the temperature rise in the region adjoining the target surface (it was the temperature rise that altered the density of the target material) were in this case determined by the normal evaporation process, in spite of the appearance of a laser-plasma jet (Fig. 10; the presence of a plasma was detected in an analysis of the field n (z, r) by a method similar to that described in [3]). This was deduced from the observation that the mass of the evaporated material m (I) was governed entirely by the energy E_t (more exactly, by the absorbed energy which was related to E_t).

It is also worth noting the observation (Fig. 14) that an increase in the energy E of a laser pulse first resulted in an increase in the final size of the crater because of the preferential increase in its diameter d. When d reached $\sim$ 1.5D, the three-dimensional crater growth began. In the first stage the volume V varied in accordance with the law (Fig. 14)

$$V = D\, exp[C(E-E_0)] \tag{5}$$

where $E \approx 7$ J, $C \approx 0.2$ J^{-1}, and $D = 0.05$ mm^3, whereas in the second stage it obeyed the law

$$V = V_b + A(E - E_0)^\alpha, \tag{6}$$

where $\alpha \approx 1.5$, $A \approx 2$ mm^3 J $^{-3/2}$, and $V_b \sim 0.5$ mm^3 (governed by the volume V at which the change in the volume became different).

It is important to note that when the energy E was sufficient for the observation of both stages, the crater varied with time in exactly the same manner as before: first its diameter d increased and the three-dimensional growth was observed only from $d \sim 1.5$ D.

The nature of the crater growth was, in our opinion, governed by the temperature distribution T (z, r) in the sample being irradiated (we recall that T (z, r) governs the density distribution ρ (z, r); see Fig. 12). In fact, along the crater perimeter the density and temperature gradients can exceed considerably | grad ρ | and | grad T\ on the axis of the system (the z axis), which is to be expected because a transparent sample is exposed to a light beam with sharp boundaries. This sharpness of the boundaries is responsible for a high value of | grad T | at the perimeter of the focusing spot. On the other hand, the transparency of the medium results in a practically homogeneous (and weak) absorption of the energy in layers with different values of z and it is responsible for the low value of | grad T| in this direction.

Since the thermal energy flux is $q \sim -$ grad (T), the peripheral (relative to the axis of the system) part of the sample surface is heated strongly. For this reason the evaporation is strong in the same region. Since the mass of the evaporated material is

$$\Delta m = \rho h \, \Delta S = \Delta g_t / \beta \sim \Delta ES / L_b,$$

where β is the heat of evaporation, we find that

$$S \sim \exp E / (\rho h L_b),$$

in good agreement with equation (5) when h is almost constant (because the component of grad T directed along the z axis is small).

When the boundaries of the evaporation zone spread beyond the focusing spot, the temperature gradient at the perimeter of this spot decreases, resulting finally in the equalization of the gradient over the whole crater profile. The heat flux also becomes equalized in all directions and the evaporation process is then three-dimensional. We are now in the second stage of the process when $\Delta d \sim \Delta h$.

If we assume that the crater is a spherical segment (in the case of a shallow crater we have $h \ll d$ and since the crater were shallow in our experiments, this is justified as a first approximation), then

$$\Delta V = 4\pi R^2 \Delta R,$$

where R is the radius of curvature of the crater.

In the case of the processes exhibiting axial symmetry (which was true of the process we investigated) we have $R \sim E^{1/2}$ [16] and therefore $V \sim E^{3/2}$, which is in good agreement with equation (6).

The crater growth practically stopped after $t_{max} \sim 100$ μs from the beginning of the interaction with the target (we recall that the pulse duration was $\tau \approx 1.2$ ms) and t_{max} depended weakly on E. Obviously, at the power densities of the light flux used in our investigation the energy absorbed at $t > t_{max}$ was no longer sufficient to maintain the vaporization temperature on the surface of the grown crater. The plasma cloud was still maintained near the target surface and, consequently, local (and resulting in a much smaller loss of mass) vaporization still continued.

6. Conclusions

The studies performed have shown that under the action of laser radiation with the mean radiation flux density ~ 10^6 – 10^7 W/cm^2 t the surface of metals in the external electric field with different polarity and the strength up to 10^6 V m^{-1} the characteristic size of the target substance droplets, carried out of the irradiated zone, decreases by several times with increasing external electric field strength. Probably, this is due to a change in the wavelength of the gravity-capillary wave, excited on the molten metal surface. The observed effect offers the possibility to control the size of the metallic droplets in the course of laser deposition of thin films.

At action on a surface of the copper sample of a rhodamine laser impulse duration ~ 20 μs time of growth of a zone of destruction makes approximately 40 мкс, that will well be coordinated with time of existence of plasma formation at a surface of the target exposed to laser-plasma processing (~ 50 μs). Use of model of the loaded area with the moving borders radiating acoustic waves in the elastic medium allows to solve the important practical problem – the definition of a law of time growth of a zone of irreversible deformations on a surface of the sample exposed to pulse laser-plasma processing.

The effective growth of a crater formed as a result of the interaction of a laser pulse of duration of at least ~ 1 ms with the surface of a transparent insulator did not last more than 100 μs. Initially, the area of the crater increased and then, but only after it was twice as large as the focusing spot, the damage region began to grow in three dimensions. The crater growth dynamics was governed primarily by the evaporation mechanism: the individual spikes of a laser pulse interacting with the target did not influence the growth process.

Author details

A. Yu. Ivanov and S. V. Vasiliev
Grodno State University, Grodno, Belarus

7. References

[1] Ivanov A.Yu. Acoustic diagnostics of materials laser treating process. Grodno: GrSU, 2007.

[2] Bosak NA., Vasil'ev S.V., Ivanov A.Yu., Min'ko L.Ya., Nedolugov VI., Chumakov A.N. *Kvantovaya Elektron.*, 27, 69 (1999) *[Quantum Elelctron.*, 29, 69 (1999)].

[3] Barikhin B.A., Ivanov A.Yu., Nedolugov VI. *Kvantovaya Elektron.*, 17, 1477 (1990) *[Soy. J. Quantum Electron.*, 20, 1386 (1990)].

[4] Goncharov V.K., Kontsovoy V.L., Puzyrev MV. *Inzhenerno JIzicheskii zhurnal*, 66, 662 (1994) *[Journal of Engineering Physics and Thermophysics*, 66, 588 (1994)].

[5] Goncharov V.K., Kontsevoy V.L., Puzyrev MV. *Kvantovaya Elektron.*, 22, 249 (1995) *[Quantum Elelctron.*, 25, 232 (1995)].

[6] Zaykin A.E., Levin A.V., Petrov AL., Stranin S.A. *Kvantovaya Elektron.*, 18, 708 (1991) *[Soy. J. Quantum Elelctron.*, 21, 643 (1991)].

[7] Zaykin A.E., Levin A.V., Petrov AL. *Kvantovaya Elektron.*, 21, 486 (1994) *[Quantum Elelctron.*, 24, 449 (1994)].

[8] Vasil'ev S.V., Ivanov A.Yu., Lyalikov AM. *Kvantovaya Elektron.*, 22, 830 (1995) *[Quantum Elelctron.*, 25, 799 (1995)].

[9] Dorofeev IA., Libenson MN. *Opt. Spektrosk.*, 76,73(1994) *[Opt. Spectrosc.*, 76, 66 (1994)].

[10] Chumakov AN. et al. *Kvantovaya Elektron.*, 21, 773 (1994) *[Quantum Electron.*, 24, 718 (1994)].

[11] Rabinovich MI., Trubetskov DI. *Vvedenie v teoriyu kolebanii i voln* (Introduction to the Theory of Oscillations and Waves) (Moscow: Nauka, 1984).

[12] Goncharov V.K., Kozadayev K.V. *Inzhenerno-fizicheskii zhurnal*, 83, 80 (2010) *[Journal of Engineering Physics and Thermophysics*, 83, 90 (2010)].

[13] Chumakov AN., Bereza NA., Hu Dz.D., Bosak NA., Guo Z.H., Hie K.K. *Jnzhenerno-fizicheskiizhurnal*, 84, 524 (2011)*[Journalof Engineering Physics and Thermophysics*, 84, 567 (2011)].

[14] Klimentov SM. et al. *Laser Phys.*, 8 (6), 1(2008).

[15] Goncharov V.K., Kozadayev K.V., Shchegrikovich DV. *Inzhenerno-flzicheskiizhurnal*, 84, 781 (2011) *[Journalof Engineering Physics and Thermophysics*, 84, (2011)].

[16] Ashmarin I.I, Bykovskii Yu A, Gridin V A, et al. *KvantovayaElektron. (Moscow)* 6 1730 (1979) *[Sov. J. Quantum Electron.* 9 1019(1990)].

Chapter 8

Kinetics and Dynamics of Phase Transformations in Metals Under Action of Ultra-Short High-Power Laser Pulses

V.I. Mazhukin

Additional information is available at the end of the chapter

1. Introduction

Action of super-power and ultra-short laser pulses on highly absorbing condensed media is investigated in the past two decades [1] - [5]. The urgency of this problem is primarily determined by a variety of practical applications of pulsed laser irradiation. In a short period of time scientists have mastered such operations as ablation of elemental materials by femtosecond lasers (100 fs) [6], femtosecond nanostructuring [7], generation of metallic nanoparticles and nanostructures by laser ablation of massive targets by 40 fs and 2ps pulses [8], etc.

The use of super-power $G\sim10^{12}\div10^{15}$ W/cm^2 and ultra-short $\tau_L\approx10^{-12}\div10^{-15}$ s laser pulses for dimensional processing of materials, such as cutting of various materials by pico-femtosecond laser pulses [9], micro-drilling by a femtosecond laser [10], surface etching of metals (Al, Cu, Mo, Ni) and semiconductors (Si) [11], is accompanied by realization of unique physical conditions. In particular, the duration of action becomes comparable with the characteristic times of thermalization and phase transitions in matter. This leads to the need to address complex fundamental problems, including the heating of the material and the kinetics of phase transitions in a strong deviation from local thermodynamic equilibrium.

It should be noted that the majority of laser technologies is associated with the beginning of phase transformations in the material. In particular, the action of pico - femtosecond laser pulses of high intensity on solid targets is one of the ways to create individual particles with unique characteristics, or to form their streams, consisting of a cluster, liquid or solid fragments of the target. The formation of the particle flux in the pulsed laser ablation is observed for a wide range of materials: metals, semiconductors and insulators. The possibility of usage of laser ablation products in practical applications was the impetus for a number of experimental and theoretical studies [12], aimed at studying the conditions and mechanisms of formation of particles of nano - and micro- sizes during the laser exposure.

The accumulation of knowledge of the experimental nature in the first place in this dynamic field leads not only to the variety of practical applications of pulsed laser action, but also to a number of issues of independent physical interest.

The main features of ultrashort action on metals are associated with high speed and voluminous nature of the energy release of the laser pulse. The high rate of heating of a condensed medium is associated with rapid phase transformations of matter, characterized by the transfer of superheated phase boundaries of high-power fluxes of mass and energy. Overheated metastable states at the interface are characterized by temperatures, whose values can be hundreds of degrees higher than the equilibrium values of the melting point or boiling point. Removal of the energy by the flow of matter in conjunction with volume mechanism of energy release of laser radiation contribute to the formation of metastable superheated regions in the volume of solid and liquid phases with near-surface temperature maximum. Calculations [13] showed that the maximum speed of melting front are comparable to the speed of sound ~ (0.5 - 6) km / s, and the phase velocity of the solidification front are (10 - 200) m / s. Accordingly, the maximum superheating / undercooling can reach several thousand / hundred degrees. The achievement of such overheating and overcooling leads to large gradients of the Gibbs energy, which actually determine the driving force for high-speed phase transformations.

Physics of supercooled states in metallic and nonmetallic systems, because of their widespread use in the production of new materials technology, is relatively well-studied [14]. Superheated states received little attention until recently, largely because of the difficulty of experimental investigation and non-obviousness of their application. The situation changed with the advent of femtosecond laser pulses and their application for the production of nanoparticles and nanomaterials [15] - [17].

The purpose of this chapter is the theoretical study of nonequilibrium pulsed laser heating of metals, the kinetics and dynamics of phase transitions in a deep deviation from the local thermodynamic equilibrium.

The main tool for studying the processes initiated by laser pulses of picosecond and femtosecond duration are the methods of mathematical modeling and computational experiment (CE). The possibilities of experimental approaches in this area are very limited due to the large transience of the processes. Computational experiments are preferred in the cases where the natural experiment is not possible, is very difficult or very expensive. The statement of CE is especially convincing in the studies of the kinetics and dynamics of fast processes. For its statement, the computational experiment, which is an important link connecting theory and field experiments, requires the development of appropriate models, determining the properties of all substances studied, the development of computational algorithms and the creation of program codes.

Construction of a theory that covers most of the features of pulsed laser action on materials is very difficult. The methods of mathematical modeling in theoretical constructs have that advantage that you can use the phenomenological and the experimental data.

The description of the kinetics of fast phase transitions of the first kind is carried out in two classes of mathematical models: continuum and atomistic. Continuous models are based on the equations of continuum mechanics, as a rule, are represented as partial differential equations with appropriate boundary conditions and equations of state. Continuum model is used to describe the macro-level processes, heterogeneous kinetics and dynamics of phase transitions of the 1st kind [18]. Atomistic approach is based on a model of molecular dynamics is used to describe the kinetics of homogeneous phase transitions [19].

2. Brief theory of phase transformations of the 1st type

The basis of first-order phase transformations - melting-solidification and evaporation-condensation are two qualitatively different mechanisms: heterogeneous and homogeneous ones. The heterogeneous mechanism is characterized by a sharp interface between the phases (phase front of zero thickness) and determines the dynamics of phase transformations of the 1st type. The homogeneous mechanism of phase transitions that arise usually under the influence of the volume heating or cooling is associated with the processes of volume melting and boiling or spontaneous crystallization and condensation.

Historically, the theory of phase transformations of the 1st type of melting is based on two fundamental sections of classical physics, thermodynamics and kinetics.

2.1. Thermodynamical approach

Thermodynamics is a macroscopic theory [20], which from the energy point of view, considers the properties of macroscopic bodies in equilibrium. This allows to have a great community for the conclusions of thermodynamics. However, thermodynamics does not take into account the internal structure of the considered bodies and some of its conclusions and regulations do not have physical clarity. One way to describe the equilibrium processes in the equilibrium thermodynamics is a theory of thermodynamic potentials.

2.1.1. Phase equilibrium and phase transformations

The properties of a thermodynamic system are determined by thermodynamic parameters. The energy state of a thermodynamic system in equilibrium is uniquely determined by the parameters of the system. There is a unique relationship between the parameters of the system, which is mathematically represented by the state function. From a mathematical point of view, this means that the function has a total differential.

The basis of the method of thermodynamic potentials is just the possibility of introducing the state functions for the equilibrium processes, with total differentials describing the change in the state of a thermodynamic system.

The main identity of thermodynamics of equilibrium processes is usually represented as

$$TdS = d\varepsilon + pdV \text{ or } d\varepsilon = TdS - pdV \tag{1}$$

Depending on the choice of the two independent parameters, one can introduce thermodynamic potentials, which differentiation allows determining the other unknown parameters of the state. In general, a thermodynamic potential may be a function of various parameters. In this notation, the internal energy is given as a function of entropy and volume: $\varepsilon = \varepsilon(S,V)$. It is the function of the state and it has a total differential with respect to its variables. The total differential $d\varepsilon$ can be used to determine the temperature T and pressure p. However, the usage of entropy S and volume V as two independent variables is inconvenient because they are difficult to control in experiment. It is more convenient to use pressure p and temperature T as two independent variables.

After consecutive transformation of internal energy U first into enthalpy $H = H(S,p)$

$$H = \varepsilon + pV, \tag{2}$$

which total differential, taking into account the basic identity of thermodynamics has the form

$$dH = d\varepsilon + pdV + Vdp = TdS + Vdp \tag{3}$$

and then into Gibbs energy $G = G(T,p)$ (Gibbs thermodynamic potential)

$$G = H - TS = \varepsilon + pV - TS \tag{4}$$

one can obtain the total differential of Gibbs energy, which with account of the basic identity of thermodynamics (1), takes the form

$$dG = -SdT + Vdp \tag{5}$$

It is convenient in the fact that the independent variables T and p are easily to modify and to control in experiment.

After reaching the equilibrium state of the system, Gibbs potential takes its minimum value and becomes constant: $dG = 0$. This allows to use the condition of minimum of the Gibbs potential for the description of equilibrium states in which $T = const$ and $p = const$.

Thermodynamic systems in equilibrium state do not necessarily have to be a homogeneous medium. A system in equilibrium may be composed of several phases, different in their physical and chemical properties, separated by the phase boundaries not changing over time.

The multiphase thermodynamic systems are most simply described, components of which are in equilibrium states, and there is no transfer of matter, energy and momentum through interphase boundaries. In this case, such thermodynamic system is in equilibrium and methods of equilibrium thermodynamics apply to describe it.

Given that during phase transformations, each of the phases is a system with variable mass, the notion of chemical potential $\mu(p,T)$ is introduced into the thermodynamic description. It is used to take into account not only exchange of energy but also the exchange of mass (particles). To determine the chemical potential, the term which takes into account the possibility of changing the number of particles in a homogeneous system (the same can be done with the other potentials) is formally introduced into the expression for the thermodynamic Gibbs potential:

$$dG = -SdT + Vdp + \mu dN$$

dN determines the change of the number of particles in the system.

If the thermodynamic potential dG is given as a function of temperature and pressure, then the value of the chemical potential is written as: $\mu = \left(\dfrac{\partial G}{\partial N}\right)_{p,T}$

The chemical potential can be written through other thermodynamical functions but in this case it will be written in terms of other state parameters:

$$\mu = \left(\frac{\partial G}{\partial N}\right)_{p,T} = \left(\frac{\partial U}{\partial N}\right)_{V,S} = \left(\frac{\partial H}{\partial N}\right)_{p,S} \tag{6}$$

From the relation (6) it follows, that the chemical potential is a physical quantity that is equal to the value of some thermodynamical potential (with constant certain parameters) that is required to add to the system to change its number of particles by unity.

2.1.2. The conditions of equilibrium for two-phase single-component system

If macroscopic transport does not occur through the interphase boundaries, and the phases themselves are in a state of thermodynamic equilibrium, such thermodynamic system, in spite of its heterogeneity, will be in a state of thermodynamic equilibrium.

For the phase equilibrium in the one-component two-phase system, the following three conditions must be fulfilled:

1. the condition of thermal equilibrium, that means the equality of temperatures at both sides of the interphase boundary: $T_1 = T_2 = T_{eq}$,

2. the condition of mechanical equilibrium, consisting of the equality of pressure at the both sides of the interphase boundary: $p_1 = p_2 = p_{eq}$,

3. the condition of the equality of the Gibbs energy per particle, consisting of the requirement of the absence of the macroscopic transfer of molecules (atoms) of this material from one phase to another: $\mu_1(p,T) = \mu_2(p,T)$

In principle, we can use not the Gibbs energy, but any thermodynamic potential, which has a minimum in equilibrium. It is not difficult to show the validity of these conditions. We shall use the total differential for the internal energy, taking into account changes in the number of particles:

$$d\varepsilon = TdS - pdV + \mu dN$$

We write this expression for each phase of the closed-loop system:

$$de_1 = T_1 dS_1 - p_1 dV_1 + \mu_1 dN_1$$
$$de_2 = T_2 dS_2 - p_2 dV_2 + \mu_2 dN_2$$

(7)

The closedness of the system automatically gives the following equation:

$$de_1 = de_2, \quad dS_1 = dS_2, \quad dV_1 = -dV_2, \quad dN_1 = -dN_2 \ .$$

Consider an equilibrium two-phase system under some simplifying assumptions. Assume that the phases do not change the volume and do not exchange particles, i.e. $dV_1 = dV_2 = 0$, $dN_1 = dN_2 = 0$. Combining the equations (7) for this case, we obtain the expression: $de_1 + de_2 = dS_1(T_1 - T_2) = 0$, which gives the condition of thermal equilibrium $T_1 = T_2 = T_{eq}$. Assuming the constancy of the entropy and the number of particles in phases, i.e. $dS_1 = dS_2 = 0, dN_1 = dN_2 = 0$ we obtain the condition of mechanical equilibrium: $p_1 = p_2 = p_{eq}$.

Given the conditions of thermal and mechanical equilibrium of (7) we obtain the equality of chemical potentials in the different phases $\mu_1(p,T) = \mu_2(p,T)$. This equation can be solved for the variables T and p and may represent the equilibrium curves of the two phases in the form $p_{eq} = p(T_{eq})$ or $T_{eq} = T(p_{eq})$. If we consider the boundary between liquid and solid, the equilibrium melting curve is obtained $T_m = T(p_m)$.

When describing the interface between liquid and gas, the equilibrium vaporization curve is obtained $p_{sat} = p(T_b)$.

It should be noted that the processes at the interface are static in nature as in the case of equilibrium of different phases and also during phase transitions. There is a constant process of transition of particles from one phase to another at the interface. In equilibrium, these opposing processes compensate each other, and during supply or withdrawal of heat to one of the phases one of these processes begins to dominate leading to a change in the amount of matter in various states of aggregation.

If the components of the thermodynamic system are not in equilibrium with each other, then there are thermodynamic flows through their interface. This will be a process of transformation of matter from one state to another, i.e. phase transformation. Assuming that the occurring processes are quasi-static and the flows are small, one can use the methods of equilibrium thermodynamics to describe such non-equilibrium system. In this case we assume an infinitely small difference between the thermodynamic parameters in different parts of the system.

2.1.3. The driving force of phase transformations

In the process of phase transitions of the 1st type, a number of quantities undergo abrupt changes at the interface, so in the following text in the thermodynamic equations, the sign of the differential will be replace by the corresponding value of the difference $dF \approx \Delta F$ for the temperature of the phase transition T_{eq}.

The driving force of phase transitions of the first type is determined by the difference of Gibbs energy (or the magnitude of overheating/overcooling ΔT) for two phases at the interface, defined in (4) and (5) and can be written in two forms

$$\Delta G = \left(\varepsilon_1 - \varepsilon_2\right) + p_{eq}\left(V_1 - V_2\right) - T_{eq}\left(S_1 - S_2\right) = \Delta\varepsilon + p_{eq}\Delta V - T_{eq}\Delta S \tag{8}$$

$$\Delta G = V\Delta p - S\Delta T \tag{9}$$

where $\Delta p = p - p_{eq}$, $\Delta T = T - T_{eq}$, $\Delta\varepsilon = \varepsilon_1 - \varepsilon_2$, $\Delta V = V_1 - V_2$, $\Delta S = S_1 - S_2$

Equilibrium. In equilibrium, $\Delta G = 0$ and the equality (8) takes the form:

$$\Delta\varepsilon + p_{eq}\Delta V - T_{eq}\Delta S = \Delta H - T_{eq}\Delta S = 0$$

where the difference of enthalpy $\Delta H = \Delta\varepsilon + p_{eq}\Delta V$, is known as the equilibrium latent heat of transformation

$$L_{eq} = \Delta H = T_{eq}\Delta S. \tag{10}$$

From equation (9) we can obtain the dependence of the equilibrium pressure P_{eq} on the temperature T_{eq}, that is known as the curve of Clausius-Clapeyron:

$$\frac{dp_{eq}}{dT_{eq}} = \frac{L_{eq}}{T_{eq}}\frac{1}{\Delta V} \tag{11}$$

If one take as ΔV the difference of the volumes of vapor V_{vap} and condensed V_{sol} phases, then since $V_{vap} \gg V_{sol}$ for ideal gas $\Delta V = V_{vapor} - V_{solid} \approx \dfrac{RT}{P}$ one obtain the expression

$$\frac{dp_{eq}}{p_{eq}} = \frac{L_v}{R}\frac{dT_{eq}}{T_{eq}^2}$$

After integration, we find the temperature dependence of the equilibrium vapor pressure that is widely used for many materials:

$$p_{eq} = p_b \exp\left[\frac{L_v}{R}\left(\frac{1}{T_b} - \frac{1}{T_{eq}}\right)\right]$$

The vapor pressure of the material in equilibrium with solid or liquid phase is called the saturated vapor pressure and with the notation $p_{eq} = p_{sat}$, $T_{eq} = T_{sur}$ is usually written as

$$p_{sat} = p_b \exp\left[\frac{L_v}{R}\left(\frac{1}{T_b} - \frac{1}{T_{sur}}\right)\right] \tag{12}$$

where T_{sur} is the temperature of the surface of the condensed phase, p_b, T_b are the equilibrium values of pressure and the boiling point under normal conditions, L_v is the latent heat of evaporation, R is the gas constant.

As it follows from equation (9), at constant pressure, the difference of Gibbs energy ΔG is linearly proportional to the overheating/ overcooling

$$\Delta G = -S \cdot \Delta T = \frac{L_{eq}\Delta T}{T_{eq}} \tag{13}$$

In the future, assuming the difference of energy ΔG to be equal to the rate of the phase transformation, one can obtain, that in the thermodynamic approach, the rate of conversion at constant pressure for small deviations from equilibrium is linearly proportional to the overheating / overcooling ΔT

$$v \approx K\Delta T, \tag{14}$$

where K is the constant of proportionality between the normal speed limits and its overcooling. The constant K does not have any clear physical sense and is chosen experimentally for each material. The dependence (14) by its form coincides with the well known relation for the determination of the linear crystal growth rate obtained on the basis of classical molecular-kinetic models in which the constant of proportionality is called the kinetic coefficient [21]. The main application of the relation (14) and its various modification [22, 23] is found in the description of different processes of melting - solidification. Comparison with experiment showed that the equation (14) gives good agreement mainly at small overcoolings [24, 25]. The kinetic coefficient K is the main parameter, characterizing the mobility of the boundary crystal -melt. Despite the great importance of this characteristic, there are only a few experiments to successfully measure the kinetic coefficient in metals and alloys [26]. The main difficulties of experimental determination are associated with the great complexity of measuring the overcooling at the solidification front. Currently, the main approach to the determination of the quantitative evaluation and qualitative understanding of the physics of the processes behind the coefficient K are the methods of molecular dynamics [27] - [30].

2.2. Kinetic approach

The structure particles of matter are in continuous motion and appear in the main provisions of the molecular-kinetic theory in which all processes are considered at the atomic or molecular level, the particles obey the Boltzmann statistics, and the speed of processes is given by

$$R = R_0 \exp(-Q / k_B T) \tag{15}$$

where Q is the activation energy, k_B is the Boltzmann's constant, $k_B T$ is the average thermal energy for one atom, R_0 is the pre-exponential factor that affects the process rate. The exponential term $\exp(-Q / k_B T)$ is known as Boltzmann's factor, that determines the part of atoms or molecules in the system that have the energy above Q at the temperature T.

2.2.1. Melt kinetics

For the first time, the conditions of crystal growth from liquids were formulated by Wilson [31], who suggested that the atoms have to overcome the diffusion barrier in order to make the transition from liquid to solid phase. The rate of accession of atoms to the crystal lattice is expressed by the relation analogous to (15)

$$R_c(T) = av \exp(-Q / k_B T) \tag{16}$$

where Q is the activation energy for overcoming the diffusion barrier, v is the frequency of attempts of transitions, a is the atom diameter.

The rate of accession of atoms to the crystal was estimated in [31] using the diffusion coefficient of liquid $D = \dfrac{a^2}{6} v_D \exp(-Q / k_B T)$, that allowed to estimate the velocity of the crystallization front as

$$v_{s\ell} = \frac{6D}{a} \frac{L_m}{k_B T_m} \frac{\Delta T}{T_{s\ell}} \tag{17}$$

where $\Delta T = T_{s\ell} - T_m$, $T_{s\ell}$ is the temperature of the interphase boundary.

Later, Frenkel [32], using the Stokes-Einstein relation $D = \dfrac{k_B T}{3\pi a \eta}$ between the coefficients of diffusion D and viscosity η expressed the crystal growth rate in terms of viscosity

$$v_{s\ell} = \frac{2k_B T}{\pi a^2 \eta} \frac{L_m}{R T_m} \frac{\Delta T}{T_{s\ell}} \tag{18}$$

The expressions (17), (18) allowed establishing a linear relationship between the rates of crystal growth and overcooling. Wilson-Frenkel theory [31-33] contributed to a better understanding of the microscopic processes associated with the growth of crystals from the melt. A generalization of the obtained results allowed to obtain that, except for very large deviations from equilibrium, where the homogeneous nucleation mechanism can dominate, the process of melting-solidification proceeds heterogeneously. The heterogeneous nucleation mechanism involves the inclusion of the motion of the liquid–solid interface into the consideration. The velocity of this interface $v_{s\ell}$, as a function of the deviation from the equilibrium melting temperature T_m is called the response function of the interface and is the main value characterizing the processes of crystallization and melting. As with the similar equation (14) obtained in the thermodynamic approach, the equations of Wilson-Frenkel showed a good agreement with experiments for very small overcooling of the boundary, i.e. for small deviations from equilibrium.

Theoretical studies of the velocity of the interface are based on some modifications and generalizations of Wilson-Frenkel theory. Their meaning is reduced to taking into account several factors, such as the latent heat of melting L_m, interatomic distance a, efficiency coefficient f, showing the proportion of atoms that remain in the solid phase at the border crossing. In the modifications [24,34,35], it was considered that the crystal has always lower enthalpy than the melt. This is the amount of energy needed for atoms of the crystal to make the transition from crystal to melt. Escape rate of atoms of the crystal followed by the addition to the active points of the liquid, contains this energy difference in the form of the Boltzmann factor $\exp\left(-L_m / k_B T_{s\ell}\right)$

$$R_m\left(T_{s\ell}\right) = C_1 \exp\left(-Q / k_B T_{s\ell}\right) \exp\left(-L_m / k_B T_{s\ell}\right) \tag{19}$$

The rate of the reverse flow of atoms into the crystal from the melt depends only on the diffusion process in liquid

$$R_c\left(T_{s\ell}\right) = C_2 \exp\left(-Q / k_B T_{s\ell}\right) \tag{20}$$

where C_1, C_2 are some constants that should be determined.

In the equilibrium point $T_{s\ell} = T_m$, the rates $R_m\left(T_m\right)$ and $R_c\left(T_m\right)$ are equal, which gives $C_1 \exp\left(-Q / k_B T_m\right) \exp\left(-L_m / k_B T_m\right) = C_2 \exp\left(-Q / k_B T_m\right)$, and $C_1 = C_2 \exp\left(L_m / k_B T_m\right)$.

The velocity of the interphase boundary $v_{s\ell}$ is equal to the difference of the rates $R_c\left(T\right)$ и $R_m\left(T\right)$

$$
\begin{aligned}
v_{s\ell}\left(T_{s\ell}\right) &= R_c\left(T_{s\ell}\right) - R_m\left(T_{s\ell}\right) = C_2 \exp\left(-Q / k_B T_{s\ell}\right) - C_1 \exp\left(-Q / k_B T_{s\ell}\right) \exp\left(-L_m / k_B T_{s\ell}\right) = \\
&= \exp\left(-Q / k_B T_{s\ell}\right)\left[C_2 - C_1 \exp\left(-L_m / k_B T_{s\ell}\right)\right] = C_2 \exp\left(-Q / k_B T_{s\ell}\right)\left[1 - \exp\left(\frac{L_m}{k_B}\left(\frac{1}{T_m} - \frac{1}{T_{s\ell}}\right)\right)\right] = \\
&= C_2 \exp\left(-Q / k_B T_{s\ell}\right)\left[1 - \exp\left(\frac{L_m}{k_B}\left(\frac{T_{s\ell} - T_m}{T_{s\ell}}\right)\right)\right] = C_2 \exp\left(-Q / k_B T_{s\ell}\right)\left[1 - \exp\left(\frac{L_m}{k_B}\left(\frac{\Delta T}{T_{s\ell}}\right)\right)\right]
\end{aligned}
\tag{21}
$$

$\Delta T = T_{s\ell} - T_m$, for overheating $\Delta T > 0$, for overcooling $\Delta T < 0$.

The constant C_2 is associated with other physical constants using the following relation: $C_2 = afv$. The studies have shown that the resulting equation (21) well predicts the velocity of the melting-solidification front of silicon [36] in a fairly wide temperature range.

The problems of melting-crystallization of monatomic metals in the modes of rapid heating/cooling, typical of ultra-short laser irradiation, use a different expression for the velocity $v_{s\ell}$. It is based on the assumption [37, 38], that crystallization of single-atom metals, (that are characterized by high velocities $v_{s\ell} \approx 50 \div 100$ m/c [35, 39]) is not diffusion-limited, but is limited only by the collision frequency during transition from liquid to crystal surface. The modification of the equation (21) consists of replacement of the diffusion term with the thermal velocity $v_T = \left(3k_B T / m\right)^{1/2}$. The velocity of interphase boundary $v_{s\ell}$ is written as:

$$v_{s\ell}\left(T_{s\ell}\right) = \frac{a f}{\lambda}\left(3k_B T_{s\ell} / m\right)^{1/2}\left[1-\exp\left(\frac{L_m}{k_B}\left(\frac{\Delta T}{T_{s\ell}}\right)\right)\right] = C \cdot \left(T_{s\ell}\right)^{1/2}\left[1-\exp\left(\frac{L_m}{k_B}\left(\frac{\Delta T}{T_{s\ell}}\right)\right)\right] \quad (22)$$

where $C = \frac{a f}{\lambda}\left(3k_B / m\right)^{1/2}$, λ is the mean free path, m is the atomic mass.

Thus, the kinetic approach allows us to obtain an expression for the response function $v_{s\ell}\left(T_{s\ell}\right)$ without fitting coefficients and suitable for a wide range of overheating/overcooling. The kinetic dependence (22) is asymmetric for the processes of melting and solidification. However, large deviations from equilibrium require additional modification of the kinetic dependences (21) (22), because they do not take into account the dynamic effects associated with the occurrence of high pressures generated by the high velocity of propagation of phase fronts.

2.2.2. Kinetics of vapor

One of the key processes in the zone of laser irradiation is the transition of condensed matter to the gaseous state. Evaporation process is characterized by high power consumption and large increase in the specific volume of the substance.

Investigation of evaporation process began in the 19th century [40, 41] and continues to this day [42 -]. This fact is defined by practical importance and not fully clarified features of non-equilibrium behavior of matter when it evaporates.

2.2.2.1. The simplest model of kinetics of evaporation in vacuum

The thermodynamic relation (9) gives that at a constant temperature the difference between the Gibbs energy between the two phases, one of which is an ideal gas, is linearly proportional to the pressure difference:

$$\Delta G = V \Delta p = k_B T \frac{\Delta p}{p} \quad (23)$$

On the basis of the formula (23), it can be assumed that the rate of phase transformation at constant temperature should be linearly proportional to the pressure difference:

$$v = v_0 \frac{\Delta p}{p} \quad (24)$$

The simplest model for the growth kinetics of vapor phase was developed by Hertz [40] and Knudsen [41] about 100 years ago. Its formulation is qualitatively the same as the thermodynamic model (23), (24)

$$j_m = j_m^+ - j_m^- = \frac{1}{\sqrt{2\pi m k_B}} \left(\frac{p_{sat}}{\sqrt{T_{sur}}} - \frac{p}{\sqrt{T}} \right) \tag{25}$$

where j_m is non-equilibrium flow of atoms at the surface of evaporation, j_m^- is the flux of atoms that collide with the surface under the assumption that the adhesion coefficient is 1. Formally, this flow can be determined using the relation connecting the vapor pressure with the equilibrium particle flux directed to the condensed surface: $p = \sqrt{2\pi} \, m \upsilon j_m^-$, where υ is the average velocity in one direction, $\upsilon = \sqrt{\dfrac{2 k_B T}{m}}$, what gives $j_m^- = \dfrac{p}{\sqrt{2\pi m k_B T}}$. Since the nature of the flux $j_m^{(-)}$ remained undetermined and the values of T and correspondingly p unknown, then during the formulation of the boundary conditions in the problem of evaporation into vacuum it is supposed [20, p.281] to use a single-term version of the Hertz-Knudsen formula to determine the evaporation rate:

$$j_m \approx j_m^+ = \frac{p_{sat}}{\sqrt{2\pi m k_B T_{sur}}}, \tag{26}$$

that takes into account the connection between the pressure of saturated vapor p_{sat} and flux of evaporated atoms at the surface temperature T_{sur}. The equilibrium vapor pressure p_{sat} depends on the temperature T_{sur}, as it is shown in the equation (12).

However, the representation of the process of surface evaporation in the form of a simple model, which does not take into account the reverse influence of evaporated atoms, does not remove the internal contradictions inherent in the model of Hertz - Knudsen. Let us write the expressions for the fluxes of momentum and j_i and energy j_e of the particles, moving away from the evaporation surface

$$\rho < V_z^2 >= j_i^{(+)} + j_i^{(-)} = j_i = \rho_\upsilon R T_\upsilon + \rho_\upsilon u^2$$
$$\rho < V^2 V_z >= j_e^{(+)} + j_e^{(-)} = j_e = \rho_\upsilon u \left(\frac{u^2}{2} + C_p T_\upsilon \right),$$

It is easy to see that these fluxes, which completely describe this one-dimensional flow, are impossible to be characterized by any temperature. If we equate these fluxes to the corresponding thermodynamic expressions containing the velocity, temperature and density and find these values, we will see that the system of equations has two distinct complex solutions, which have no physical meaning.

The ambiguity of the solution for real values of the thermodynamic parameters is associated with the possibility of discontinuous solutions of the type of shock wave. The complexity of the solution in this case is due to the thermodynamic non-equilibrium of the evaporation flow of Hertz-Knudsen, which can not be described in terms of thermodynamic concepts. When taking into account collisions in the non-equilibrium layer, the evaporative flux is thermalized, but the temperature on the outer side of this layer no longer coincides with the surface temperature.

2.2.2.2. Approximation of the Knudsen layer

Intense surface evaporation is essentially non-equilibrium process. In addition to the thermodynamic equilibrium, this process also has a gas-kinetic non-equilibrium in a thin (Knudsen) layer of vapor, directly adjacent to the interface. Gas-kinetic non-equilibrium is due to the flow of material through the phase boundary. The mass flow increases with the growth of the evaporation rate and, consequently, the degree of non-equilibrium of the process increases. From the physical considerations, the maximum velocity of material flow on the outside of the Knudsen layer is limited to the local speed of sound $u \leq u_{sound} = \left(\gamma R T_v \right)^{1/2}$, or $M = u / u_{sound} \leq 1$, where u is the gas-dynamic velocity, M is the Mach number. The maximum deviation from equilibrium is determined by the maximum value of mass flow, which is known to be achieved at $M = 1$.

Under the conditions of phase equilibrium, when the saturated vapor pressure p_{sat} is equal to the external pressure, the flow of vaporized material is balanced by the return flow of particles and the total mass flux through the boundary is zero. The distribution of particle velocities in vapor is in equilibrium and can be described by the Maxwell function with zero average velocity. In cases where the vapor pressure above the surface is less than the saturated vapor pressure, in the system condensed matter-vapor, the directed movement is formed with $u > 0$ and is characterized by non-zero material flow through the phase boundary. The decreasing reverse flow leads to a deviation from the equilibrium in the distribution of the particles. The magnitude of the flux of returning particles decreases with the increase of the rate of evaporation, and the distribution function at the evaporation surface becomes increasingly different from Maxwellian one.

In general, the non-equilibrium distribution function is found by solving the Boltzmann equation in a region with a characteristic size of a few mean free paths. This area is adjacent to the evaporation surface, where the kinetic boundary conditions are set, taking into account the interaction of individual particles with the interface. Similar problem was solved by various methods in many studies, taking into account, in particular, the differences from unity and variability of the coefficient of condensation, which determines the probability of attachment of the particle in its collision with the evaporation surface. (See, for example. [42 - 46]).

Methods of non-equilibrium thermodynamics are used to describe the evaporation process together with other approximate phenomenological approaches [47 -52]. A more general and fundamental approach is to use molecular dynamics method, which was used in [53, 54] to analyze the evaporation process. A recent review on the issue of non-equilibrium boundary conditions at the liquid-vapor boundary is given in [55].

For the equations of continuum mechanics, thin Knudsen layer is a gas-dynamic discontinuity. The knowledge of the relations at this break, connecting the parameters of the condensed medium and the evaporated material, is needed to deal with the full gas-hydrodynamic problem that arises, for example, during the description of laser ablation, taking into account the variability of the Mach number and instability of the evaporation front [56]. The use of kinetic approaches, which explicitly consider the structure of the Knudsen layer, in such cases is difficult because of the emerging problem of significant difference of space-time scales. The solution of these problems is associated with the additional computational difficulties and is not always possible. Therefore, usually another approach is used that allows to determine the matching conditions with certain assumptions about the form of the non-equilibrium distribution function inside the break [57 - 60] without solving the kinetic problem. Approximation of the distribution function in the Knudsen layer was carried out in different models. But to obtain physically reasonable boundary conditions it is necessary to formulate criteria which these conditions must meet. In addition, attention is paid to the peculiarities of the behavior of the fluxes of mass, momentum and energy as the Mach number tends to unity that allows to use the requirement of extremum of total fluxes of mass j_m, momentum j_i and energy j_e as one of the criteria for $M = 1$

"β" - model. In "β" – model [61, 62], a compound distribution function is used to describe one-dimensional non-equilibrium flow of particles on the inner side of the planar Knudsen layer: $f = f^{(+)} + f^{(-)}$. Here, the distribution $f^{(+)}$ for particles, flying out of the surface is given as a Maxwell-shaped function $f^{(+)} = f(\rho_{sat}, T_{sur}, 0)$ with density of saturated vapor ρ_{sat} for the surface temperature T_{sur}. The distribution $f^{(-)}$ characterizes the flow of particles returning to the surface and is supposed to be proportional to a "shifted" Maxwell function $f^{(-)} = f(\rho_v, T_v, u)$ with density ρ_v, temperature T_v and mean velocity u steady flow of vapor on the outside of the Knudsen layer:

$$f^{(+)} = f(\rho_{sat}, T_{sur}, 0), \quad V_z > 0$$

$$f^{(-)} = f(\rho_v, T_v, u) = \beta \left(\frac{m}{2\pi k_B T_v}\right)^{3/2} \exp\left(\frac{V_x^2 + V_y^2 + \left(V_z^2 - u\right)^2}{2k_B T_v}\right) \quad V_z < 0$$

The flows of mass, momentum and energy $j_k = j_k^{(+)} + j_k^{(-)}$, $k = m, i, e$ calculated using $f^{(+)}, f^{(-)}$ must be equal to their gas-dynamic values j_k that are determined by function f :

$$\rho < V_z > = j_m^{(+)} + j_m^{(-)} = j_m = \rho_v u$$

$$\rho < V_z^2 > = j_i^{(+)} + j_i^{(-)} = j_i = \rho_v RT_v + \rho_v u^2 \qquad (27)$$

$$\rho < V^2 V_z > = j_e^{(+)} + j_e^{(-)} = j_e = \rho_v u \left(\frac{u^2}{2} + C_p T_v\right),$$

where $C_p = \dfrac{\gamma}{\gamma+1} k_B$ is the vapor heat capacity per particle at constant pressure, for single-

atom gas $\gamma = \dfrac{5}{3}$. From the solution of equations (27), it is possible to obtain gas-dynamic

conditions on the break, allowing to determine the magnitude of ρ_v, T_v и β in terms of ρ_{sat}, T_{sur} and M. The calculations showed that total fluxes j_m, j_i, j_e have extrema depending on M at $M = 0.88$, 1.18, 1.22 correspondingly.

Thus, the requirement of the extrema of all flows in the selected point $M = 1$ in the model [61,62] does not met, that can serve as evidence of a bad choice of the distribution function $f^{(-)}$.

"$\varepsilon - \delta$" - model. The drawbacks of the "β" – model can be eliminated by selection of a more general form of the distribution function for the reverse flow of particles $f^{(-)}$ [63]:

$$f^{(-)} = \beta f + \alpha f_1 \quad \alpha + \beta = 1$$

All 3 flows .. will have extrema at $M = 1$, if the function f_1 is set to be equal to $f_1 = f(\rho_1, T_1, u_1)$, where the values of ρ_1, T_1, u_1 are written in terms $\rho_{sat}, T_{sur}, \rho_u, T_v$ using additional fitting parameters ε and δ

$$\rho_1, T_1^{1/2} = \varepsilon \rho_{sat} T_{sur}^{1/2} + (1-\varepsilon) \rho_v T_v^{1/2},$$
$$\rho_1, T_1^{1/2} = \varepsilon T_{sur}^{1/2} + (1-\varepsilon) T_v^{1/2},$$
$$u_1 = \delta M (\gamma R T_1)^{1/2} = \delta (T_1/T_v)^{1/2} u$$

For example, at $\varepsilon = 0.70$ and $\delta = 0.32$ all three flows j_k will have extrema at $M = 1$ with values $j_m = 0.853, j_i = 0.557, j_e = 0.892$. It is clear, that this version of selection of fittings coefficients is not the only one.

"α" - model. It is possible to suggest another phenomenological model [63], where strict localization of extrema $j_k(M = 1)$ is achieved without usage of fitting coefficients for such distribution function $f^{(-)}$, that do not depend on gas-dynamic values. An example of such function is function

$$f^{(-)} = \alpha^7 f_0 \left(\rho_{sat}, \alpha^2 T_{sur} \right),$$

It takes into account the decrease of temperature $T_\alpha = \alpha^2 T_{sur}$ of the reverse flow of particles as compared to the surface temperature T_{sur}. Due to this change of T_α, the ratio of the normalized fluxes j_e/j_m ceases to be a constant and takes the form $(1 - \alpha^8)(1 - \alpha^{10})$, which ensures that the correct limiting value is equal to 1.25 in the equilibrium case for $\alpha = 1$. The equation for α, that is obtained from the equity of fluxes (27) has a relatively simple form:

$$\frac{\left(1-\alpha^8\right)\left(1-\alpha^{10}\right)}{\left(1-\alpha^9\right)^2} = \frac{\pi}{8}\frac{\gamma^2 M^2\left[(\gamma-1)M^2+2\right]}{(\gamma-1)\left(1+\gamma M^2\right)^2} \tag{28}$$

The right-hand side of the equation (28) has a maximum at $M=1$, that determines localization of the extrema of α and j_k. The values of $j_k(M=1)$ are equal correspondingly to $j_m = 0.85, j_i = 0.56, j_e = 0.90$.

Modified Crout model. The property of localization of the flows $j_k(M=1)$ also present in the model suggested by D. Crout [57]. It uses non-equilibrium function of distribution of particles, written in analytical form with temperature that is anisotropic by directions: T_L and T_T.

$$f\left(\rho,T_L,T_T,u,v,w\right) = \left\{\rho\left[\frac{\pi^3}{h_T m^3}(2k_B T_L)\cdot(2k_B T_T)\right]^{1/2}\cdot \exp\left[-m\left(\frac{(u-u_0)^2}{2k_B T_L}+\frac{v^2}{2k_B T_T}+\frac{w^2}{2k_B T_T}\right)\right]\right\}$$

where T_L is the lateral temperature along x axis, T_T is the transversal temperature along y,z axis; u,v,ω are the components of the velocity vector along corresponding axes x,y,z, u_0 is the drift velocity.

The modification of the Crout model [64] consists of explicit introduction of the Mach number into the main relations that allows to obtain:

$$T_v = \alpha_T(M)T_{sur}, \quad \rho_v = \alpha_\rho(M)\rho_{sat}, \tag{29}$$

$$\alpha_T(M) = \frac{2\gamma M^2\left(m^2+0.5\right)^2}{\left(1+\gamma M^2\right)^2 m^2 t^2}, \quad \alpha_\rho(M) = \frac{1}{\exp\left(-m^2\right)+\pi^{1/2}m\left(1+\text{erf}(m)\right)}\cdot\frac{\left(1+\gamma M^2\right)m^2}{\gamma M^2\left(m^2+0.5\right)^2},$$

The value of m is determined from non-linear equation

$$F(M)\left(m^2+0.5\right)^2 - m^2\left(m^2+a+1.5\right) = 0, \tag{30}$$

where $F(M) = 1+\dfrac{3\gamma M^2-1}{\left(\gamma M^2-1\right)^2}$, $a = 2t^2 - 0.5\pi^{1/2}mt - 1$, .

$$t = \frac{2m}{\pi^{1/2}} + \frac{1+\text{erf}(m)}{\exp\left(-m^2\right)+\pi^{1/2}m\left(1+\text{erf}(m)\right)}, \quad \text{erf}(m) = \frac{2}{\sqrt{\pi}}\int_0^m e^{-y^2}dy.$$

For $M=1$: $T_v = 0,633 T_{sur}$, $\rho_v = 0,328\rho_{sat}$. $p_v = 0.208 p_{sat}$,

$M=0$: $T_v = T_{sur}$, $\rho_v = \rho_{sat}$, $p_v = p_{sat}$.

For the numerical solution of the nonlinear equation (30), one can use the Newton's iterative procedure. All fluxes j_m, j_i, j_e have extrema at $M=1$. Calculations using an anisotropic non-equilibrium particle distribution function give the corresponding extreme values of the fluxes: $j_m = 0.84, j_i = 0.55, j_e = 0.88$.

The calculations show that specific choice of the model has relatively little effect on the magnitude of the momentum flux j_2, but significantly affects the flux of mass j_m and energy j_e. Fig.1 shows the dependencies of the ratio of the normalized fluxes $\bar{j}_3 / \bar{j}_1 = j_3 j_1^{(+)} / j_1 j_3^{(+)}$ on M for all discussed models. From the comparison of the extreme values of the fluxes $j_k(M=1)$ and behavior of the curves $j_m(M), j_e(M)$ it follows, that due to bad choice of $f^{(-)}$, the less favorable for the description of evaporation kinetics is "β" – model.

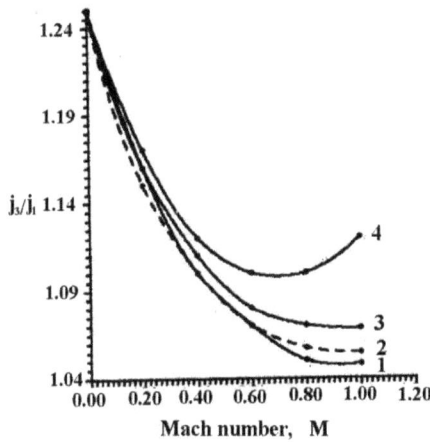

Figure 1. Dependency of the normalized ratio of energy and mass fluxes j_3 / j_1 on Mach number for different models: 1 – "ε-δ"-model, 2 – Crout model, 3 – "α"-model, 4 – "β"-model (Knight).

Table 1 for all models shows numerical values of ρ_v and T_v (normalized by ρ_{sat} and T_{sat} correspondingly) depending on the Mach number that changes from zero to unity. Comparative analysis of tabular data, as well as the behavior of the curves j_m, j_e, show a marked difference of the values of ρ_v and T_v, obtained using "β" model, from the values, obtained using other models. The values of ρ_v turned out to be underestimated, and T_v overestimated as compared to their real values at the outer side of the Knuden layer.

	"β"-model (Knight)		"ε-δ"-model		"α"-model		Crout model	
M	T	N	T	N	T	n	T	n
0.0	1.000	1.000	1.000	1.000	1.000	1.000	1.000	1.000
0.1	0.960	0.861	0.958	0.864	0.960	0.869	0.953	0.861
0.2	0.922	0.748	0.916	0.753	0.920	0.758	0.910	0.749
0.3	0.866	0.654	0.876	0.662	0.881	0.666	0.870	0.658

	"β"-model (Knight)		"ε-δ"-model		"α"-model		Crout model	
M	T	N	T	N	T	n	T	n
0.4	0.851	0.576	0.837	0.587	0.844	0.588	0.833	0.582
0.5	0.817	0.511	0.799	0.525	0.808	0.524	0.798	0.519
0.6	0.785	0.457	0.763	0.472	0.773	0.470	0.763	0.466
0.7	0.754	0.410	0.727	0.428	0.740	0.424	0.730	0.421
0.8	0.724	0.371	0.693	0.391	0.705	0.386	0.697	0.384
0.9	0.696	0.337	0.660	0.360	0.672	0.355	0.665	0.352
1.0	0.669	0.308	0.628	0.333	0.640	0.328	0.633	0.326

Table 1. Normalized values ρ_v and T_v depending on the Mach number for different models

The modified Crout model, $"\varepsilon - \delta"$ – model, $"\alpha"$ – model fulfill the requirement of the extremum of the flows j_k at $M = 1$. The difference of the results that were obtained using these models does not exceed 1.5%. Any of these models can be used to describe the kinetics of the process of non-equilibrium surface evaporation.

2.3. Conclusions

The performed brief analysis of the kinetics of phase transitions is an introduction to the construction of the models that combine mathematical description of kinetics of high-speed phase transformations with dynamics of the macro-processes (heat and mass transfer) under conditions of a strong deviation from local thermodynamic equilibrium that are typical for ultra-short super-power laser action on metals.

3. Thermodynamic and thermo-physical properties of phonon and electron Fermi gas

Determination of physical characteristics of a medium, including equations of state under conditions of local thermodynamic equilibrium can be carried out either experimentally or by means of calculation using distribution functions - a Maxwell-Boltzmann function for ideal gas and ideal plasma and Fermi one for degenerate electron gas, and for the phonon gas – Bose function. In case of violation of the conditions of local thermodynamic equilibrium distribution functions are determined by solving the classical kinetic Boltzmann equation or quantum-kinetic equations. The presence of distribution function is just the required minimum of information that can be used to describe nonequilibrium processes with reasonable accuracy.

The influence of ultrashort high-energy laser on a strongly absorbing media (metals, semiconductors) is in a very short temporal and spatial scales and leads to disturbance of their general local-thermodynamic equilibrium. Irradiated targets in these conditions are presented in the form of two subsystems - electron and phonon each of which is in local thermodynamic equilibrium and are characterized by their temperatures and equations of state. As a consequence, all processes are described in the two-temperature approximation [65], [66]. The target at pico- and femtosecond influence may be heated to very high temperatures and pressures at which the thermal and mechanical properties of matter are not known in general. One of the most important problems for the mathematical modeling is the necessity to determine thermophysical, optical and thermodynamic properties in a wide (tens and hundreds of electronvolts) temperature and frequency ranges for each of the subsystems.

3.1. Electron subsystem

The most important thermophysical and thermodynamic characteristics of the electron Fermi gas within the scope of heat-conducting mechanism of energy transfer are: heat capacity C_e, thermal diffusivity χ_e and thermal conductivity λ_e. For its determination using fundamental physical quantities, which include the electron mean free paths l_{ee}, l_{eph} and the characteristic times (frequency) of interaction for two scattering mechanisms: the electron-electron τ_{ee} and electron-phonon τ_{eph}.

3.1.1. Fermi-Dirac integral and its approximation

For a quantitative description of electrical, thermodynamic and thermophysical properties of degenerate electron gas with distribution function

$$f_e = \frac{1}{\exp\left(\dfrac{\varepsilon_e - \mu}{k_B T_e}\right) + 1}, \tag{31}$$

widely used Fermi-Dirac functions $F_{k+1/2}(\mu(T_e)/T_e)$ expressed in terms of integrals of the form

$$F_{k+1/2}(\mu(T_e)/T_e) = \int_0^\infty \frac{\left(\dfrac{\varepsilon_e}{k_B T_e}\right)^{k+1/2} d\left(\dfrac{\varepsilon_e}{k_B T_e}\right)}{\exp\left(\dfrac{\varepsilon_e - \mu}{k_B T_e}\right) + 1} = \int_0^\infty \frac{E^{k+1/2} dE}{\exp\left(E - \dfrac{\mu}{k_B T_e}\right) + 1}, \tag{32}$$

where T_e, ε_e - temperature and energy of electron, $\mu(T_e)$ - chemical potential, $\mu(0) = \varepsilon_F$ - Fermi energy.

In the future, the Fermi integrals will be represented as a function of dimensionless energy of the chemical potential:

$$F_{k+1/2}(\eta) = \int_0^\infty \frac{E^{k+1/2} dE}{\exp(E - \eta) + 1}, \tag{33}$$

where $E = \varepsilon_e/k_B T_e$, $\eta(T_e) = \mu(T_e)/k_B T_e$ - dimensionless energy and chemical potential of electrons.

The integral of form $F_{1/2}(\eta) = \int\limits_0^\infty \dfrac{E^{1/2}dE}{\exp(E-\eta)+1}$ is used to determine the electron density

$$N_e = \int\limits_0^\infty g(E)f(E)dE = \int\limits_0^\infty \dfrac{g(E)dE}{\exp(E-\eta)+1} = \int\limits_0^\infty F(E)dE,$$

where $g(E) = \dfrac{2^{1/2}m^{3/2}}{\pi^2\hbar^3}(k_BT_e)^{3/2}E^{1/2} = g_0\cdot(k_BT_e)^{3/2}E^{1/2}$ is a density of states,

$F(E) = g(E)f(E) = g_0(k_BT_e)^{3/2}\dfrac{E^{1/2}}{\exp(E-\eta)+1}$ is distribution function of the free electron

energy, $g_0 = \dfrac{2^{1/2}m^{3/2}}{\pi^2\hbar^3}$. Then

$$N_e = g_0\cdot(k_BT_e)^{3/2}\int\limits_0^\infty\dfrac{E^{1/2}dE}{\exp(E-\eta)+1} = g_0\dfrac{2}{3}\varepsilon_F^{3/2} = g_0\cdot(k_BT_e)^{3/2}F_{1/2}(\eta) \qquad (34)$$

From the known distribution function of particle energy $F(E)$, using the ratio

$$<x(E)>= \dfrac{\int\limits_0^\infty x(E)F(E)dE}{\int\limits_0^\infty F(E)dE}$$ the average value of any physical quantity depends on energy can

be found. Since the average energy of the electron gas is defined as the ratio of Fermi integrals

$$<\varepsilon_e> = \dfrac{\int\limits_0^\infty E\cdot F(E)dE}{\int\limits_0^\infty F(E)dE} = (k_BT_e)\dfrac{\int\limits_0^\infty \dfrac{E^{3/2}dE}{\exp(E-\eta)+1}}{\int\limits_0^\infty \dfrac{E^{1/2}dE}{\exp(E-\eta)+1}} = (k_BT_e)\dfrac{F_{3/2}(\eta)}{F_{1/2}(\eta)}. \qquad (35)$$

Similarly, we can determine the other physical quantities of the electron gas.

The chemical potential $\mu(T_e)$ depends on the temperature, so the Fermi integrals can be conveniently represented as a function of dimensionless temperature $\xi = \dfrac{k_BT_e}{\varepsilon_F}$. In [67] for the integrals of the form (33) has been proposed convenient approximation, which allows to express the integrals $F_{k+1/2}(\xi)$ through the transcendental gamma-functions $\Gamma(k+1/2)$ and the dimensionless temperature ξ:

$$F_{k+1/2}(\xi) = A\xi^{-3/2}\left[1+(B/\xi)^2\right]^{k/2}, \qquad (36)$$

where $A = \dfrac{2}{3}\dfrac{\Gamma\left(k+\frac{3}{2}\right)}{\Gamma\left(\frac{3}{2}\right)}$, $B = \left[A\left(k+\frac{3}{2}\right)\right]^{-1/k}$ coefficients expressed in terms of gamma-

functions. Equation (36) has correct asymptotics at $\xi \to 0$ and $\xi \to \infty$:

$$F_{k+1/2}(\eta) \approx 1/\xi^{k+3/2}\left(k+3/2\right), \qquad \xi \to 0$$

$$F_{k+1/2}(\eta) \approx \frac{2}{3}\frac{\Gamma\left(k+3/2\right)}{\Gamma(3/2)}\frac{1}{\xi^{3/2}} \qquad \xi \to \infty$$.

The integral of $k = 1/2$ order can easily be determined from the expression (34):

$$F_{1/2}(\xi) = \frac{2}{3}\frac{1}{\xi^{3/2}}.$$

The most frequently used Fermi integrals that are expressed through (36) have the form:

$$F_{-1/2}(\xi) = \frac{4}{3}\frac{\xi^{-1/2}}{\left(\xi^2 + \frac{4}{9}\right)^{1/2}}, \quad F_{3/2}(\xi) = \frac{\left(\xi^2 + \left(\frac{2}{5}\right)^2\right)^{1/2}}{\xi^{5/2}}, \quad F_{5/2}(\xi) = \left[\frac{5}{2}\xi^{-3/2} + \frac{2}{7}\xi^{-7/2}\right]. \quad (37)$$

The maximum error compared with the exact solution [68] for integrals at $k = -1$ and $k = 1$ does not exceed 8%, but increases slightly with increasing of k.

Approximation (36) allows to obtain simple analytical expressions for the physical quantities of the electron gas at arbitrary temperatures.

3.1.2. Equations of state

Using the approximating expressions (36) and (37) equations of state for degenerate electron gas can be written as simple analytical expressions at arbitrary temperatures. Since the average electron energy $<\varepsilon_e>$ and its pressure can be represented as

$$<\varepsilon_e> = T_e\frac{F_{3/2}}{F_{1/2}} = \frac{3}{2}\varepsilon_F(\xi^2 + 0.16)^{1/2}, \qquad (38)$$

$$p = \frac{2}{3}N_e<\varepsilon_e> = \frac{2}{3}N_eT_e\frac{F_{3/2}}{F_{1/2}} = N_e\varepsilon_F(\xi^2 + 0.16)^{1/2}, \qquad (39)$$

3.1.3. Electron heat capacity $C_e(T_e)$

The expression for the heat capacity of the electron gas $C_e(T_e) = \dfrac{\partial}{\partial T}(N_e\langle\varepsilon_e\rangle)$ can be obtained from the relations

$$\frac{\partial}{\partial \eta} F_{k+1/2} = \left(k+1/2\right) F_{k-1/2} , \quad \frac{\partial \eta}{\partial T} = \frac{3}{T} \frac{F_{1/2}}{F_{-1/2}} , \quad C_e\left(T_e\right) = \frac{N_e}{F_{1/2}} \left(\frac{5}{2} F_{3/2} - \frac{9}{2} F_{-1/2}\right).$$

Using the approximating expressions (37), heat capacity of electron gas can be represented with an error not exceeding 5% as the following function

$$C_e\left(T_e\right) \cong \frac{3}{2} \frac{N_e k_B^2 T_e \left[K\right]}{\left[T_e^2 + \left(\frac{3\mathcal{E}_F}{\pi^2}\right)^2\right]^{\frac{1}{2}}} , \qquad (40)$$

where $N_e = z N_a$, z - the number of valence electrons, N_a - the concentration of atoms (ions) of lattice.

The resulting expression gives the classical linear dependence of heat capacity of a degenerate electron gas vs. temperature [69] in low temperature region $T_e \ll \varepsilon_F$,

$C_e = \frac{\pi^2}{2} \frac{k_B^2 T_e N_e}{\varepsilon_F}$ and constant value at $T_e \gg \varepsilon_F$ equal to heat capacity of gas with Maxwell distribution $C_e = 3/2 k_B N_e$. Dependences for copper and aluminum are shown in Fig. 2.

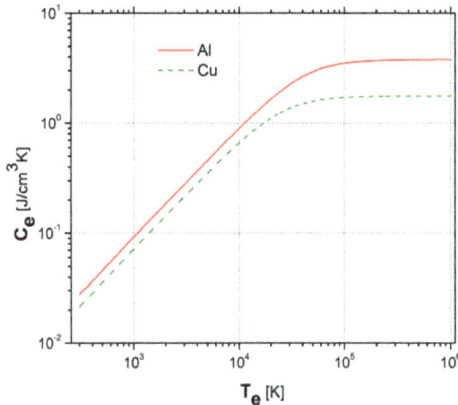

Figure 2. Temperature dependence of electron heat capacity $C_e(T_e)$.

3.1.4. The thermal diffusivity of a degenerate electron gas χ_e.

The thermal diffusivity of electron gas is proportional to the product of the mean free path l_e and average velocity of the electron $<v_e>$:

$$\chi_e = \frac{1}{3} l_e <v_e> , \qquad (41)$$

In metals electron mean free path due to several mechanisms: pair of electron-electron collisions, electron-phonon collisions and scattering by plasmons.

Electron-electron collisions dominate at temperatures comparable to the Fermi energy $T_e/\varepsilon_F \approx 1$. Electron-phonon interaction is dominant at low temperatures $T_e/\varepsilon_F \ll 1$. The interaction associated with the excitation of plasmons occurs at high temperatures, exceeding the plasma frequency energy $T_e > \hbar\omega_r$ ($\hbar\omega_r \cong 10-20$ eV). Taking into account high temperature region of occurrence and limitation of experimental data about reducing of the mean free path of electrons due to plasmon excitation (it is known only for some metals), electron scattering by plasmons will not be considered.

3.1.4.1. The electron-electron thermal diffusivity $\chi_{ee}(T_e)$

The mean free path of an electron in pair electron-electron collisions l_{ee} is determined from the known gas-dynamic formula

$$l_{ee} = \frac{1}{N_e \sigma_{ee}}, \qquad (42)$$

where σ_{ee} - scattering cross section with energy $\Delta\varepsilon$ transfer for electrons with energies $\varepsilon_1, \varepsilon_2$. The cross section σ_{ee} is expressed through the transport cross section of the collision of two isolated electrons $\sigma_{\ell\ell}^{tr}$ and Fermi integrals $F_{-1/2}, F_{1/2}$. In turn, the transport cross section of collision of two isolated electrons $\sigma_{\ell\ell}^{tr}$ in a field of screened Coulomb potential $U = \frac{e^2}{r}\exp\left(-\frac{r}{d}\right)$ is expressed through the differential scattering cross section $d\sigma$ determined in the Born approximation [70, p.560]. The final cross sections will be written in form

$$\sigma_{ee}^{tr} = \frac{2\pi}{9}\left(\frac{4}{9\pi}\right)^{4/3} z^{-4/3}\frac{r^4}{r_B^2}\frac{\left[\ln(t+1)-\dfrac{t}{(t+1)}\right]}{\left(\xi^2+0.16\right)},$$

$$\sigma_{ee} = \sigma_{ee}^{tr}\left(\frac{F_{-1/2}}{2F_{1/2}}\right)^2 = \frac{2\pi}{9}\left(\frac{4}{9\pi}\right)^{4/3} z^{-4/3}\frac{r^4}{r_B^2}\frac{\xi^2\left[\ln(t+1)-\dfrac{t}{(t+1)}\right]}{\left(\xi^2+\frac{4}{9}\right)\left(\xi^2+0.16\right)}, \qquad (43)$$

where $t = 4\langle k\rangle^2 d^2 = \left(\frac{9\pi}{4}\right)^{4/3} z^{1/3}\frac{r_B}{r}\cdot\left[\left(\xi^2+0.16\right)\cdot\left(\xi^2+\frac{4}{9}\right)\right]^{1/2}$, $r_B = \dfrac{\hbar^2}{me^2} = 0.529 \cdot 10^{-8}$cm $-$ Bohr radius, $r = \left(\dfrac{3}{4\pi N_a}\right)^{1/3}$ - the average distance between atoms, e - the electron charge, d – field acting radius (Debye) $U = \dfrac{e^2}{r}\exp\left(-\dfrac{r}{d}\right)$.

At low temperatures $T_e \ll \varepsilon_F$, the effective cross section σ_{ee} is small and amounts to

$$\sigma_{ee} = \sigma_{ee}^{tr} \frac{\xi^2}{\left(\xi^2 + 4/9\right)} \approx \sigma_{ee}^{tr} \cdot \left(\frac{T_e}{\varepsilon_F}\right)^2 .$$ Maximum of cross section is achieved at $T_e \geq \varepsilon_F$, Fig.3,

when the degeneracy is passed and the electron-electron collisions with large energy transfer become possible. At very high temperatures $T_e \gg \varepsilon_F$ cross section becomes the Coulomb one, Fig.3, and decreases logarithmically. The mean free path of electrons l_{ee} is determined by the formula (42):

$$l_{ee} = \frac{1}{N_e \sigma_{ee}} = \left[N_e \cdot \frac{2\pi}{9} \left(\frac{4}{9\pi}\right)^{4/3} z^{-4/3} \frac{r^4}{r_B^2} \frac{\xi^2 \left[\ln(t+1) - \frac{t}{(t+1)}\right]}{\left(\xi^2 + \frac{4}{9}\right)\left(\xi^2 + 0.16\right)} \right]^{-1} =$$

$$= \frac{2}{\pi^2} \left(\frac{4}{9\pi}\right)^{1/3} \frac{1}{r^2} \frac{1}{z^{1/3} N_a} \cdot \frac{t^2}{\xi^2} \left[\ln(1+t) - \frac{t}{1+t}\right]^{-1}$$

(44)

Calculations indicate that the mean free path for Al and Cu change in a wide range (~$10^{-2} \div 10^{-7}$) cm and have a minimum at $T_e \approx \varepsilon_F$.

The average thermal velocity of electron $<v_e>$ is expressed through its average energy $<\varepsilon_e>$:

$$<v_e> = \left(\frac{2}{m} <\varepsilon_e>\right)^{1/2} = \left(\frac{3\varepsilon_F}{m} \left(\xi^2 + 0.16\right)^{1/2}\right)^{1/2} = \left(\frac{3}{2}\right)^{1/2} \left(\frac{9\pi}{4}\right)^{1/3} \frac{z^{1/3}}{r} \frac{\hbar}{m_e} \left(\xi^2 + 0.16\right)^{1/4}, \quad (45)$$

Taking into account (44) and (45) electronic thermal diffusivity χ_{ee} takes the form:

$$\chi_{ee} = \frac{1}{3} l_{ee} <v_e> = \frac{1}{3} \frac{2}{\pi^2} \left(\frac{4}{9\pi}\right)^{1/3} \frac{1}{r^2} \frac{1}{z^{1/3} N_a} \cdot \frac{t^2}{\xi^2} \left[\ln(1+t) - \frac{t}{1+t}\right]^{-1} \cdot \left(\frac{3}{2}\right)^{1/2} \left(\frac{9\pi}{4}\right)^{1/3}$$

$$, \quad (46)$$

$$\cdot \frac{z^{1/3}}{r} \frac{\hbar}{m_e} \left(\xi^2 + 0.16\right)^{1/4} = \left(\frac{2}{3}\right)^{1/2} \frac{4}{3\pi} \frac{\hbar}{m_e} \frac{t^2}{\xi^2} \frac{\left(\xi^2 + 0.16\right)^{1/4}}{\left[\ln(1+t) - \frac{t}{1+t}\right]} = \chi_0 \cdot \Phi_{ee}(\xi)$$

where $\Phi_{ee}(\xi) = \dfrac{t^2}{\xi^2} \dfrac{\left(\xi^2 + 0.16\right)^{1/4}}{\left[\ln(1+t) - \dfrac{t}{1+t}\right]}$ is dimensionless function,

$$\chi_0 = \sqrt{\frac{2}{3}} \frac{4}{3\pi} \frac{\hbar}{m_e} = 0.402, \left[\frac{cm^2}{s}\right].$$

Fig. 4 shows the temperature dependence of $\chi_{ee}(T_e)$ for *Al* and *Cu*. Temperature dependence of electron-electron thermal diffusivity $\chi_{ee}(T_e)$ of both metals has a deep minimum at $T_e \approx \varepsilon_F$. Its value reaches ~ 20÷30 [cm²/s], Fig.4. At removal of degeneracy, when $T_e > \varepsilon_F$ the thermal diffusivity increases due to decreasing of the effective cross section σ_{ee}. With further increase of temperature $T_e \gg \varepsilon_F$ the thermal diffusivity continues to increase and its dependence $\chi_{ee} \sim \xi^{5/2} / \ln \xi^2$ coincides with temperature dependence of thermal diffusivity of Maxwell electron plasma. At low temperature $T_e \ll \varepsilon_F$ dependence of thermal diffusivity is inversely proportional to square of temperature $\chi_{ee} \sim T_e^{-2}$. Strong growth of $\chi_{ee}(T_e)$ with decreasing of T_e leads to the fact that χ_{ee} reach $(1-5) \cdot 10^5 \dfrac{cm^2}{s}$ values which is 3-4 orders higher than the actual electron thermal diffusivity of metals at room temperatures. Thus, the resulting expression (46) is a good approximation only for high temperatures. Under normal conditions ($T_e \sim 300K$) it is necessary to consider the scattering of electrons of metals by phonons to determine thermal diffusivity, this interaction is dominant at low temperatures.

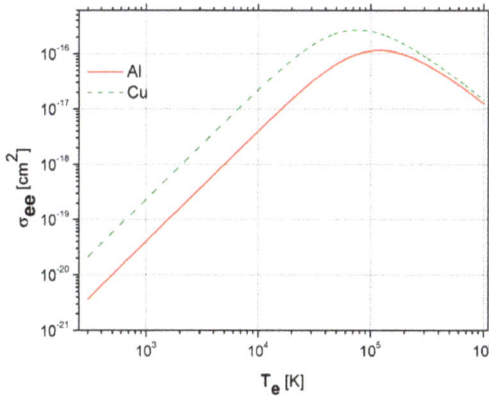

Figure 3. Temperature dependences of electron-electron scattering cross section $\sigma_{ee}(T_e)$.

3.1.4.2. The electron-phonon thermal diffusivity

The mean free path, defined by electron-phonon interaction is described by the assumption of elastic scattering of conduction electrons of metal on lattice oscillations. To determine it is convenient to use the phenomenological approach [70], in which the crystal is considered as an elastic continuum. Lattice oscillations at the same time considered as a wave of elastic deformations. To simplify the density fluctuations are presented as deviations of each atom (ion) from the average, which square of amplitude is directly proportional to temperature T_{ph}. According to the macroscopic theory of elasticity we can obtain an expression for the mean free path, expressed in terms of macroscopic quantities [71] by expressing the force tending to return the atom (ion) to the equilibrium state through the Young's modulus E:

$$\ell_{eph} = \frac{E \cdot r}{N_e k_B T_{ph}} \tag{47}$$

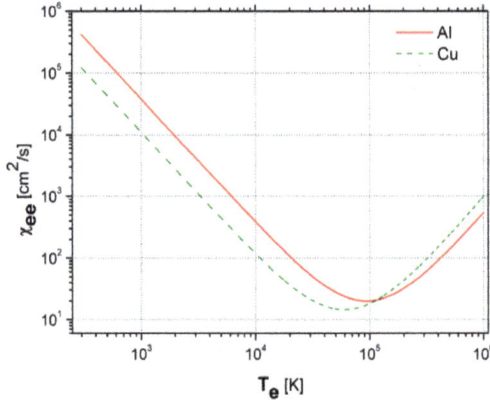

Figure 4. Temperature dependence of electron-electron thermal diffusivity $\chi_{ee}(T_e)$.

From the expression (47) it follows that ℓ_{eph} is inversely proportional to the lattice temperature T_{ph} in electron-phonon interaction.

During melting of a metal the number of collectivized electrons remains practically unchanged. The modulus of elasticity is only one value (excluding the jump in specific volume, which usually does not exceed 10%) that changes. The melting of most metals is accompanied by decrease in elastic modulus by 2-3 times [72]. This decrease causes a corresponding increase in density fluctuation and, consequently, an abrupt decrease in the mean free path ℓ_{eph}:

$$\left(\ell_{eph}\right)_k = \left(\frac{E \cdot r}{z N_a k_B T_{ph}}\right)_k , \tag{48}$$

where $k = s,l$ subscripts denoting membership in the solid and liquid phases, respectively,

$r = \left(\dfrac{3}{4\pi} \dfrac{1}{N_a}\right)^{1/3} = 0.6204 N_a^{-1/3}$ is a distance between the atoms. From below l_{eph} is limited by

Bohr diameter value. The mean free path of electrons taking into account the scattering by phonons , calculated for aluminum and copper from the relation (48) showed that, compared with the mean free path of the electron-electron scattering values for both metals decreased by several orders of magnitude: for the high-temperature region of $1.5 \div 2$ orders of magnitude, while at low temperatures for $4 \div 5$ orders of magnitude.

The thermal diffusivity $\chi_{eph}(T_e, T_{ph})$, defined by the mechanism in electron-phonon interaction is calculated by the formula:

$$\left(\chi_{eph}\left(T_e, T_{ph}\right)\right)_k = \left(\frac{1}{3} \ell_{eph}\left(T_{ph}\right) < v_e\left(T_e\right) >\right)_k, \tag{49}$$

Dependences $\chi_{eph}\left(T_e, T_p\right)$ for Al and Cu calculated from (49) at $T_e = T_{ph}$ are shown in Fig.5. The calculations show that at room temperature values of the electron thermal diffusivity χ_{eph}, for both materials are 3-4 orders of magnitude less than the thermal diffusivity χ_{ee} and reach values of 10^2 cm²/s typical for metals under normal conditions. Under equilibrium conditions with temperature T_{ph} increasing, thermal diffusivity χ_{eph}, undergoing break at the phase transition decreases rapidly to a value of ~ 1 cm²/s at a temperature of ~ 5 000K. Function $\chi_{eph}\left(T_e, T_p\right)$ increase by several orders of magnitude with electron temperature increase.

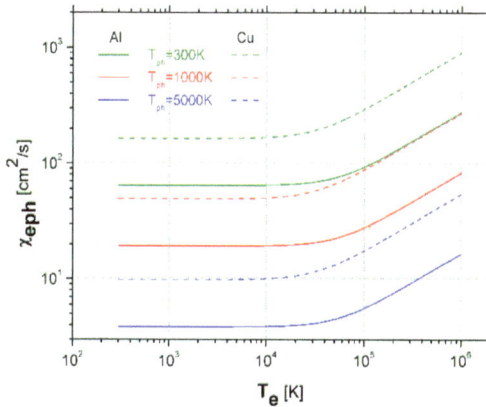

Figure 5. Temperature dependences of thermal diffusivity $\chi_{eph}(T_e, T_{ph})$ for Al and Cu with fixed $T = T_e = T_{ph}$.

3.1.4.3. The resulting thermal diffusivity of electron Fermi gas $\chi_e\left(T_e, T_{ph}\right)$

Calculations have shown that taking into account only a pair electron-electron collisions leads to a strong (by several orders of magnitude) overestimation of the thermal diffusivity of the electron gas at low temperatures $T_e \ll \varepsilon_F$. Accounting for electron-phonon collisions gives a more realistic values of $\chi_e(T_e)$ at low temperatures.

By averaging mean free paths ℓ_{ee}, ℓ_{eph} we obtain an expression for resulting thermal diffusivity, at arbitrary temperature

$$\left(\chi_e\left(T_e,T_{ph}\right)\right)_k = \left(\frac{\chi_{ee}\chi_{eph}}{\chi_{ee}+\chi_{eph}}\right)_k = \left(\frac{1}{3}\ell_e <v_e>\right)_k = <v_e> \left(\frac{\ell_{ee}\,\ell_{eph}}{3\left(\ell_{ee}+\ell_{eph}\right)}\right)_{k=s,\ell}, \qquad (50)$$

Fig.6 shows temperature dependences of total electron thermal diffusivity $\chi_e(T_e,T_{ph})$ for *Al* and *Cu* in equilibrium $T_e = T_p$.

3.1.5. *The thermal conductivity of the electron gas*

According to elementary kinetic theory, the thermal conductivity of the gas is

$$\left(\lambda_e(T_e,T_{ph})\right)_k = \left(\frac{1}{3}C_e\,\ell_e(T_e,T_{ph}) <v_e>\right)_k = \left(C_e\chi_e(T_e,T_{ph})\right)_k, \quad \left[\frac{W}{m\cdot K}\right] \qquad (51)$$

Thus, the thermal conductivity λ_e can be determined through the heat capacity C_e and averaged thermal diffusivity of electron gas $\chi_e\left(T_e,T_{ph}\right)$. The temperature dependences of λ_e for *Al* and *Cu* calculated for the equilibrium case when $T_e = T_{ph}$ are shown in Fig.6. In accordance with the results, the equilibrium electron thermal conductivity $\lambda_e(T_e,T_{ph})$ at temperatures not exceeding the boiling temperature of the equilibrium is practically independent of temperature. In high temperature region $T_e > 1eV$ thermal conductivity increases rapidly due to the dominance of electron-electron scattering. It is natural that in this region the thermal conductivity of the electron gas depends on electron density and λ_e increases with increasing of electron concentration. For this reason, electron thermal conductivity of aluminum is higher than the same one of copper at the high-temperature. In low temperature region where electron-phonon interaction is dominated, the ratio is inverse. Because of the greater mean free path ℓ_{eph}, electron thermal conductivity of copper is higher than that of aluminum.

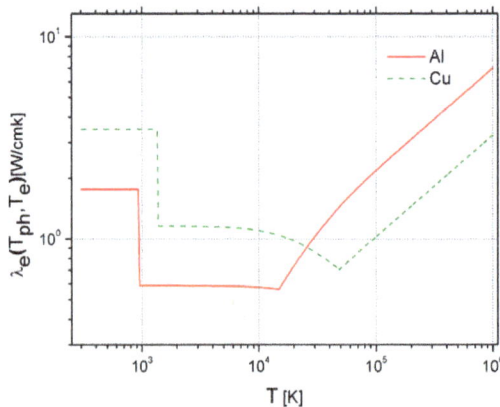

Figure 6. Temperature dependence of total electron thermal conductivity $\lambda_e(T_e,T_{ph})$.

3.2. Phonon gas

3.2.1. Heat capacity of phonons

The main consequence of the existence of lattice oscillations is the possibility of its thermal excitation, which is appeared as a contribution to the heat capacity of solid.

Taking into account the process of melting, heat capacity of phonon gas can be written as [69], [73]:

$$C_{ph,k} = \begin{cases} 3k_B N_{a,k}, & k=s,l, \quad T>T_D \\ \dfrac{12\pi^4}{5} N_a k_B \left(\dfrac{T_{ph}}{T_D}\right)^3, & T<T_D \end{cases}, \quad \left[\dfrac{J}{m^3 K}\right] \tag{52}$$

T_D is a Debye temperature.

3.2.2. The thermal diffusivity of phonon gas χ_{ph}

Phonons are considered as a gas of particles. From elementary kinetic theory, thermal diffusivity of a gas is given by

$$\chi_{ph} = \frac{1}{3}\ell_{ph}\,\upsilon_{sound} \tag{53}$$

where ℓ_{ph} - phonon mean free path. It is assumed that the phonons move with velocity of sound υ_{sound}.

The mean free path of phonons ℓ_{ph} is determined from the description of thermal motion in a solid by means of notions of the phonon gas. The interaction between the phonons can be characterized by some effective cross section which is proportional to the mean square of thermal expansion of the body or the mean square of density fluctuation ρ [69]: $\Delta^2 = N_a k_B T_p \beta$, where β is a coefficient of compressibility. Taking into account $\beta = (\rho \upsilon_{sound}^2)^{-1}$:

$$\Delta^2 = N_a k_B T_{ph}\beta = \frac{N_a k_B T_{ph}}{\rho \upsilon_{sound}^2} = \frac{k_B T_{ph}}{M \upsilon_{sound}^2}$$

Assigning to phonons radius equal to thermal oscillations amplitude, we can taken into account scattering of sound waves and determine the mean free path ℓ_{ph}

$$\ell_{ph} = \frac{r}{\Delta^2 \gamma^2} = \frac{M \upsilon_s^2}{k_B T_{ph}} \frac{r}{\gamma^2}, \tag{54}$$

where M - the mass of the atom, γ - Griuneyzen constant.

Accounting for the expressions for the mean free path ℓ_{ph} (54) and the velocity of sound υ_{sound} [73] expressed through the Fermi velocity

$$\upsilon_{sound} = \left(\frac{m}{3M}\right)^{1/2} \upsilon_F = \left(\frac{m}{3M}\right)^{1/2} \left(\frac{9\pi}{4} z\right)^{1/3} \frac{\hbar}{m} \frac{1}{r}, \quad M = M_0 A,$$

thermal diffusivity of phonons can be written as:

$$\chi_{ph} = \frac{1}{3} \ell_{ph} \upsilon_s = \frac{M_0 A \upsilon_S^3}{3 k_B T_{ph}} \frac{r}{\gamma^2} = \left(\frac{3}{4\pi}\right)^{1/3} \frac{\pi^2}{(3m)^{3/2}} \frac{\hbar^3}{(M_0 A)^{1/2}} \frac{z}{\gamma^2} \frac{N_a^{2/3}}{k_B T_{ph}}$$

Taking into account the melting process:

$$\left(\chi_{ph}\right)_k = \left(2.831 \times 10^{-13} \frac{z}{A^{1/2} \gamma^2} \frac{N_{a,\kappa}^{2/3}}{T_{ph}[K]}\right)_k \quad \left[\frac{cm^2}{s}\right], \quad k = s, l \tag{55}$$

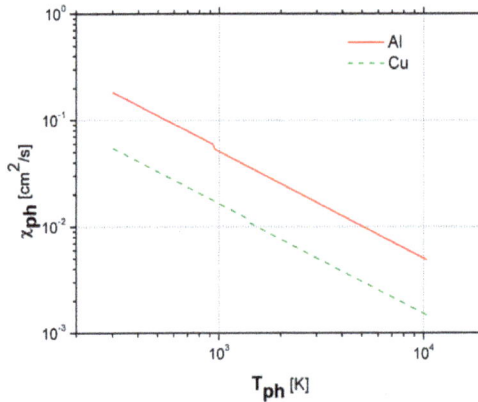

Figure 7. Temperature dependences of phonon thermal diffusivity $\chi_{ph}\left(T_{ph}\right)$.

3.2.3. Thermal conductivity of phonon gas λ_{ph}

The thermal conductivity of phonon gas can be expressed through its heat capacity C_{ph} and thermal conductivity. Taking into account expressions for C_{ph} and χ_{ph} in the high-temperature region $T_{ph} > T_D$ thermal conductivity takes the form:

$$\left(\lambda_{ph}\right)_\kappa = C_{ph} \chi_{ph} = \left(\frac{3\pi^2}{(3m)^{3/2}} \frac{\hbar^3}{M_0^{1/2}} \frac{z}{A^{1/2} \gamma^2} \frac{N_a^{5/3}}{T_{ph}[K]}\right)_k = \left(1.17 \times 10^{-35} \frac{z}{A^{1/2} \gamma^2} \frac{N_{a\kappa}^{5/3}}{T_{ph}[K]}\right)_\kappa, \tag{56}$$

Graphic dependences of λ_{ph} for *Al* and *Cu* are represented in Fig. 8.

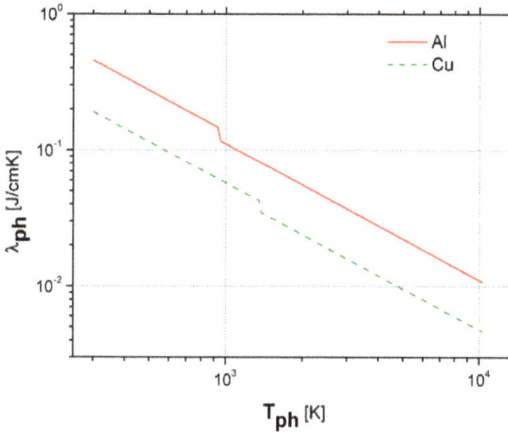

Figure 8. Temperature dependence of the phonon thermal conductivity.

3.2.4. Equilibrium heat capacity and thermal conductivity of metal

The temperature dependences of the total heat capacity and thermal conductivity of a metal can be represented as corresponding sums, which consist of two parts – electron (40), (51) and phonon (52), (55):

$$C(T) = C_e(T) + C_{ph}(T),$$
$$\lambda(T) = \lambda_e(T) + \lambda_{ph}(T) \tag{57}$$

Temperature dependences of $\lambda(T)$ and $C(T)$ are shown in Figs 9, 10. For comparison, reference data from [72], [74] - [77] are shown by markers.

According to the equilibrium theory of metals [69], [73] in the temperature range $T_0 < T < T_m$ heat transport by electrons is dominated. As predicted by the Wiedemann-Franz law, equilibrium electron thermal conductivity is practically independent of temperature. Its contribution to the total thermal conductivity of the metal with temperature increasing, is much greater than the contribution of the phonon component $\lambda_e > \lambda_{ph}$. The phonon part of thermal conductivity is inversely proportional to the temperature, and its contribution is noticeable only at low temperatures. As a result, the resulting thermal conductivity of solid-state becomes linearly decreasing temperature dependence. At the phase transition solid - liquid thermal conductivity of both metals decreases abruptly in 2 ÷ 3 times and decreases slowly with increasing temperature.

At very low temperatures, much smaller than Debye temperature $T < T_D$, these relations become invalid, which is associated with the change of interaction mechanisms. Since the thermal conductivity of metal bodies at low temperatures is determined by the transfer of thermal energy by sound waves, rather than electrons. The heat capacity and the number of phonons in this case are proportional to ~ T^3, and the electrical resistance, defined by the scattering of electrons by phonons is inversely proportional to absolute temperature.

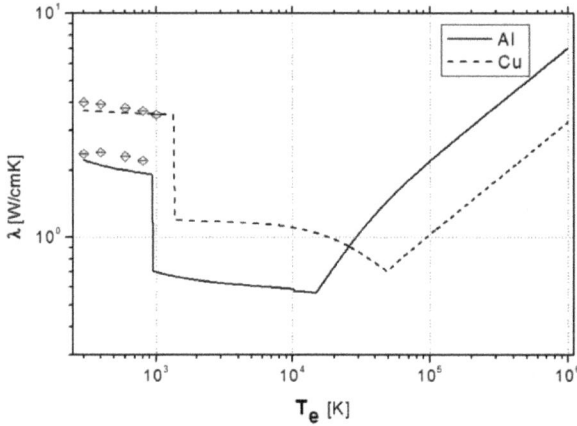

Figure 9. Temperature dependences of total thermal conductivity $\lambda(T)$.

Calculations showed that at temperatures close to the plasma $T \geq 1$ eV, the thermal conductivity of metals increases rapidly, and the thermal conductivity of aluminum, due to three times exceeded electron density becomes higher than the thermal conductivity of copper. However, this situation can have meaning only in the case of strongly nonequilibrium with hot electrons and cold lattice. Under equilibrium conditions, the use of calculated data should be restricted to the region of the critical point, in which neighborhood, as known, all physical properties of metals vary sharply.

Comparison of the obtained theoretical data with reference data was carried out in a relatively narrow temperature range, since the physical properties of most metals were measured in the temperature range from several tens to several thousand degrees, and usually not exceeding the boiling temperature of the equilibrium. In the range 300 ÷ 1500 K, the comparison has shown a complete qualitative agreement of results with a good quantitative correlation, Figs.9, 10.

3.3. Electron-phonon interaction

During laser action on metals, entire pulse energy is transmitted directly to the electrons. The result is a highly nonequilibrium region in the solid with hot electrons and cold lattice. Cooling of electron subsystem realized by two mechanisms: energy transfer by electron thermal conductivity, leveling the gradients of electron temperature and electron-phonon interaction, which leads to heating of the lattice.

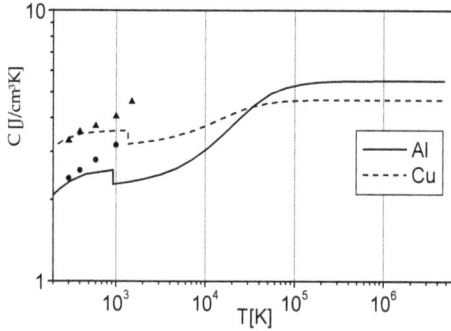

Figure 10. Temperature dependences of total heat capacity $C(T)$.

3.3.1. The average frequency of electron-phonon collisions $\langle v_{eph} \rangle$.

Per time unit, the electron undergoes an average of $\sim 1/2\,\tau$ collisions with the emission of phonons, and roughly the same, but slightly smaller, number of collisions with the absorption of phonon. Estimate the time of inelastic collisions by using the following equation [70]

$$\frac{1}{2}\tau_{eph} = \frac{1}{N_e \langle v_e \rangle \sigma_{eph}}.$$

Hence, the average frequency of electron-phonon collisions with energy transfer of electrons to crystal lattice can be represented as

$$\langle v_{eph} \rangle = 2N_e < v_e > \sigma_{eph} = 2zN_a < v_e > \sigma_{eph} \tag{58}$$

where σ_{eph} is scattering cross section of electrons by phonons.

The cross section for electron-phonon scattering, calculated under the assumption that the product of the wave number of electron and atom displacements from equilibrium [78] has the form

$$\sigma_{eph} = 4\pi \left(\frac{4}{9\pi}\right)^{4/3} \frac{m k_B T_{ph}}{\hbar^2} \left(\frac{r}{r_B}\right)^2 \frac{r^4}{z^{4/3}} \frac{\xi \cdot \left[\ln(t_1 + 1) - \dfrac{t_1}{(t_1 + 1)}\right]}{\left(\xi^2 + 0.16\right)^{1/2} \left(\xi^2 + \dfrac{4}{9}\right)^{1/2}} \tag{59}$$

The cross section has a maximum at $T_e \approx \varepsilon_F$. At $T_e \ll \varepsilon_F$ due to the Pauli principle, the cross section is proportional to temperature T_e. After removing the degeneracy $T_e \gg \varepsilon_F$, cross section of electron-phonon collisions decreases with T_e increasing. The linear lattice temperature T_{ph} dependence of the cross section of electron scattering in metals reflects the fact that the effective radius of the atoms can be identified with the amplitude of its thermal fluctuations.

Taking into account expressions for σ_{eph} (59) and $<v_e>$ (45), the average rate of energy transfer can be written as:

$$\langle v_{eph}\rangle =\left(\frac{3}{2}\right)^{1/2}\frac{8}{3\pi}\left(\frac{r}{r_B}\right)^2\frac{k_B}{\hbar}T_{ph}\Phi_{eph}\left(\xi\right)\cdot\left\{\begin{array}{l}\Phi_{eph}\left(\xi\right)=\dfrac{\xi\cdot\left[\ln\left(t_1+1\right)-\dfrac{t_1}{\left(t_1+1\right)}\right]}{\left(\xi^2+0.16\right)^{1/4}\left(\xi^2+\dfrac{4}{9}\right)^{1/2}}\quad \text{при } \xi<1\\[20pt]\Phi_{eph}\left(\xi\right)=\dfrac{\xi\cdot\left[\ln\left(t_1+1\right)-\dfrac{t_1}{\left(t_1+1\right)}\right]}{\left(\xi^2+0.16\right)^{3/4}\left(\xi^2+\dfrac{4}{9}\right)^{1/2}}\quad npu\ \xi\geq1\end{array}\right\}\quad(60)$$

The dependences of the mean frequency of transmission electron energy to phonons from the temperature T_e at T_{ph} =300 K and T_{ph} = 1000 for aluminum and copper are shown in Fig.11.

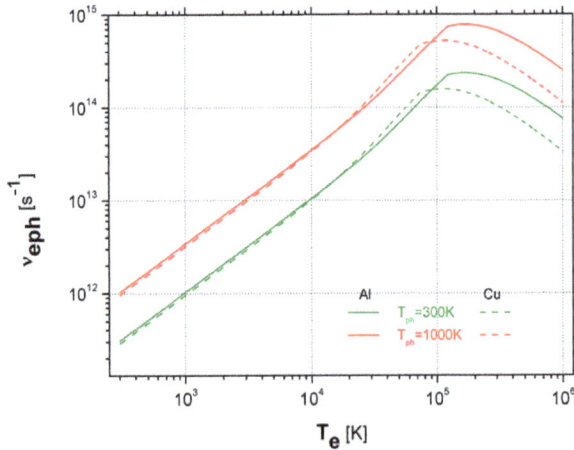

Figure 11. Temperature dependences of frequency of energy transfer of electrons to phonons at 300 K and = 1000 K.

3.3.2. Energy transfer

The mechanism of energy transfer to the lattice due to electron-phonon collisions was discussed in several papers [79-81]. The collision of an electron with the lattice phonon is absorbed or is born, for which law of conservation of energy and momentum works:

$$\Delta\varepsilon = mv_1^2/2 - mv_2^2/2 = \pm\hbar\,\Omega$$

$$p = 2mv_F \sin\frac{\varphi}{2} \cong \frac{\hbar\Omega}{v_{sound}} \tag{61}$$

where Ω - the phonon frequency, φ - the angle between the initial and final electron momentum mv_1 and mv_2, v_{sound} - velocity of sound, p - the electron momentum.

From the conservation laws (61) it follows that energy of phonon, excited by an electron with a momentum p equal to $\Delta\varepsilon = \hbar\Omega = 2pv_{sound}\sin\frac{\varphi}{2}$.

At a spherically symmetric scattering the energy transferred by one electron per unit time, averaged by φ, equals [81]:

$$<\Delta\varepsilon> = \frac{2}{3}\frac{\hbar\Omega_D}{k_B T_{ph}} pv_{sound}V_{eph}\left[1 - \frac{1}{32}\left(\frac{\hbar\Omega_D}{pv_{sound}}\right)^3\right], \text{ where } \hbar\Omega_D \text{ - the Debye energy.}$$

Taking into account velocity of sound v_{sound}, expressed in terms of the Fermi velocity [73], $<\Delta\varepsilon>$ in its final form is written:

$$<\Delta\varepsilon> = \frac{2}{3}p_F v_{sound}^2 \frac{\hbar k_D}{k_B T_{ph}}\left[1 - \frac{1}{32}\left(\frac{\hbar k_D v_{sound}}{p_F v_{sound}}\right)^3\right]\frac{V_{eph}}{k_B T_{ph}} =$$

$$= \left(\frac{9\pi}{2}\right)^{1/3}\left(\frac{3}{2}\right)^{1/2}\frac{4}{3}\frac{z}{mM}\left(\frac{r}{r_B}\right)^2\frac{\hbar^3}{r^4}\times\left[1 - \left(\frac{1}{2^4 z}\right)\right]\Phi_{eph}(\xi), \tag{62}$$

3.3.3. The electron-fonon coupling factor

The average energy transferred to phonons by electrons per volume unit per time unit is equal to the product of the average energy transferred by a single electron $<\Delta\varepsilon>$ and concentration of electrons, which may participate in the transfer of energy to phonons

$$\frac{d\varepsilon}{dt} = \begin{cases} <\Delta\varepsilon>\dfrac{k_B(T_e - T_{ph})}{\varepsilon_F}N_e & T_e < \varepsilon_F \\ <\Delta\varepsilon>z(T_e)N_e & T_e \geq \varepsilon_F \end{cases}$$

or

$$\frac{d\varepsilon}{dt} = <\Delta\varepsilon>\frac{k_B(T_e - T_{ph})}{\varepsilon_F}N_e = g(\xi)(T_e - T_{ph}), \tag{63}$$

where $g(\xi)$ is electron-fonon coupling factor, equal to

$$g(\xi) = \langle \Delta \varepsilon \rangle \frac{k_B N_e}{\varepsilon_F} = \left(\frac{3}{2}\right)^{1/2} \left(\frac{8}{9\pi}\right)^{1/3} \frac{8}{3} \cdot \frac{\hbar k_B}{r_B^2 M_0} \frac{z^2}{A} N_a \left[1 - \left(\frac{1}{2^4 z}\right)\right] \Phi_{eph}(\xi), \left[\frac{W}{m^3 K}\right]. \qquad (64)$$

The temperature dependence of the electron-fonon coupling factor $g(T_e)$ is shown in Fig.12. The results calculated for *Al* in [82] are shown to compare by markers.

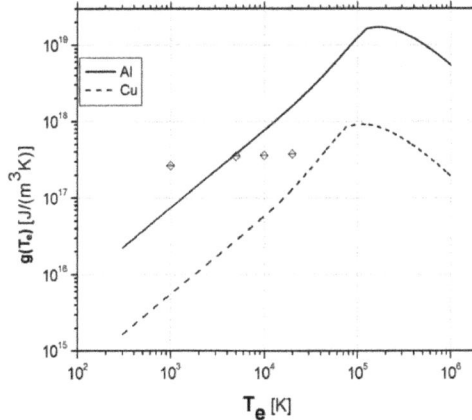

Figure 12. Temperature dependences of electron-fonon coupling factor $g(T_e)$.

3.4. Conclusions

Important thermodynamic and thermophysical properties of a degenerate electron and phonon gas: equation of state, heat capacity, the energy exchange between subsystems in an arbitrary temperature range are determined using fundamental physical quantities - the mean free paths and times (frequency) of the electron - electron and electron-phonon collisions.

Using Fermi integrals technique and its subsequent approximations have provided the temperature dependence of all characteristics in the form of simple analytical expressions.

4. Formulation and solution of the problem of laser action on metals using the method of dynamic adaptation

Action of intense laser radiation on metals may be accompanied by heating, evaporation, plasma formation. The dynamics of the processes in the condensed media and in the flow of vapor and gas environment surrounding the evaporating surface depend on several parameters of the laser pulse: the level of the absorbed intensity, wavelength, duration and spatial and temporal distribution of the energy, as well as optical, thermal and hydrodynamic characteristics of the condensed and gaseous media. It should be noted that the process of interaction is qualitatively different depending on where the target is placed: in the vacuum or the gaseous medium.

As it was already mentioned, the main features of ultra-short super-power action on metals are associated with the high speed and voluminous nature of the laser pulse energy release. The high rate of energy input leads to a strong deviation of the system from the local thermodynamic equilibrium (LTE). The basis of the mathematical description of the pulsed laser heating of metals is a phenomenological two-temperature model (TTM) of parabolic type proposed in the 50s by the authors of [65], [66]. The application of TTM involves the description of the situations with small deviations from LTE. Ultrafast heating of metal targets by super-power laser pulses causes a strong deviation from the LTE and requires appropriate adjustments of the mathematical model. In the first place, the kinetic non-equilibrium phase transitions and powerful dynamic effects associated with the movement of the phase fronts must be taken into account. The consideration of these processes requires the explicit description of the kinetics of the phase transitions and the formulation of the conservation laws at the phase fronts, which are hydrodynamic breaks.

4.1. Mathematical description of processes in the irradiation zone

Laser radiation propagates from the right to the left and strikes the surface of the metal target. Then the radiation is partially absorbed and partially reflected. The absorbed energy is consumed for heating, phase transformations and the generation of shock waves in the solid phase. Fig.13 shows spatial location of phases, moving interphase boundaries $\Gamma_{s\ell}(t)$, $\Gamma_{\ell v}(t)$ and shock wave in solid $\Gamma_{sh,s}(t)$.

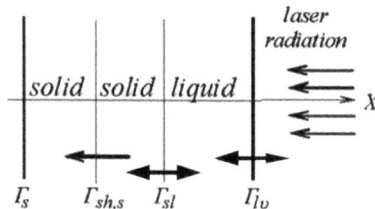

Figure 13. Spatial phase configuration.

Formulation of the problem was carried out under the following assumptions and limitations:

- The mechanisms of volume of melting and evaporation are not included into the consideration.
- The melting and evaporation fronts appear at the irradiated surface during its noticeable overheating $\Delta T_{s\ell} = T_{sur} - T_m$ and $\Delta T_{\ell v} = T_{sur} - T_b(p_{sat})$, the overheated metastable states act sustainably during the consideration.

The mathematical description and modeling of pico- and femtosecond laser action on metal target in vacuum was performed within the framework of a two-temperature and spatially one-dimensional multi-front nonequilibrium hydrodynamic Stefan problem, written for the two phases - solid and liquid.

4.1.1. System of equations

Action of high-power laser pulses on metal targets initiate in them a number of interrelated processes with nonlinear behavior. The description of their behavior is performed using a system of hydrodynamic equations, two energy balance equations for electron and phonon subsystems, and the transport equation of the laser radiation. The system of equations can be written in the domain with three moving boundaries $\Gamma_{sh,s}(t)$, $\Gamma_{s\ell}(t)$, $\Gamma_{kv}(t)$.

$$
\left(
\begin{aligned}
&\frac{\partial \rho}{\partial t} + \frac{\partial (\rho u)}{\partial x} = 0 \\
&\frac{\partial (\rho u)}{\partial t} + \frac{\partial (\rho u^2)}{\partial x} + \frac{\partial p}{\partial x} = 0 \\
&\frac{\partial (\rho_e \varepsilon_e)}{\partial t} + \frac{\partial (\rho_e u \varepsilon_e)}{\partial x} = -\left(p\frac{\partial u}{\partial x} + \frac{\partial W_e}{\partial x} + g(T_e)(T_e - T_{ph}) + \frac{\partial G}{\partial x} \right) \\
&\frac{\partial (\rho \varepsilon_{ph})}{\partial t} + \frac{\partial (\rho u \varepsilon_{ph})}{\partial x} = -\left(p\frac{\partial u}{\partial x} + \frac{\partial W_{ph}}{\partial x} - g(T_e)(T_e - T_{ph}) \right) \\
&\frac{\partial G}{\partial x} + \alpha(T_e)G = 0, \qquad \rho_e = z\frac{m}{M}\rho, \\
&P = P_e(\rho_e, T_e) + P_{ph}(\rho, T_{ph}), \quad \varepsilon_e = C_e(T_e)T_e, \quad \varepsilon_{ph} = C_{ph}(T_{ph})T_{ph} \Big)_{k=s,\ell} \\
&\quad t>0, \ \Gamma_s < x < \Gamma_{sh,s}(t) \cup \Gamma_{sh,s}(t) < x < \Gamma_{s\ell}(t) \cup \Gamma_{s\ell}(t) < x < \Gamma_{\ell v}(t) \qquad\qquad (65) \\
&W_e = -\lambda(T_e,T_{ph})\frac{\partial T_e}{\partial x}, \ W_{ph} = -\lambda(T_{ph})\frac{\partial T_{ph}}{\partial x}, \ P(\rho,T) = P(\rho_e,T_e) + P(\rho,T_{ph}),
\end{aligned}
\right.
$$

the designations: $\rho, u, \varepsilon, T, p$ are the density, gas-dynamic velocity, internal energy, temperature and pressure correspondingly, $\alpha(T_e)$, $R(T_e)$ are the coefficient of volume absorption of the laser radiation and surface reflectivity, G is the density of the laser radiation, $C_e(T_e)$, $C_{ph}(T_{ph})$, $\lambda_e(T_e, T_{ph})$, $\lambda_{ph}(T_{ph})$ are electron and phonon heat capacity and heat conductivity, $g(T_e)$ electron-phonon coupling factor. Indexes s, ℓ, v, designate solid, liquid, vapor phases, e, ph designate electron and phonon components.

4.1.2. Edge conditions

We use edge conditions in the form of initial conditions at $t = 0$ and boundary conditions at $x = \Gamma_s$, $\Gamma_{sh,s}(t)$, $\Gamma_{s\ell}(t)$, $\Gamma_{\ell v}(t)$.

Initial conditions

$$
t = 0: \quad u(0,x) = 0, \quad p = 0, \quad \rho = \rho_0, \quad T_e = T_{ph} = T_0 = 293K \qquad (66)
$$

Boundary conditions

$x = \Gamma_s$: the condition of zero mass and heat flow is used as boundary conditions at the left (fixed) boundary:

$$x = \Gamma_s : \ \rho_s u_s = 0, \quad W_T = 0; \tag{67}$$

$x = \Gamma_{s\ell}(t)$: the model of surface melting-crystallization is used as boundary conditions at the moving interphase boundary $\Gamma_{s\ell}(t)$.

The model of surface melting-crystallization is a non-equilibrium kinetic version of the Stefan problem [18], [83,84], formulated for the conditions of significant deviation from LTE and consisting of three conservation laws: mass, momentum and energy, written in the stationary (laboratory) coordinate system:

$$x = \Gamma_{s\ell}(t): \ \rho_s(u_s - v_{s\ell}) = \rho_\ell(u_\ell - v_{s\ell}) \tag{68}$$

$$P_s + \rho_s(u_s - v_{s\ell})^2 = P_\ell + \rho_\ell(u_\ell - v_{s\ell})^2, \tag{69}$$

$$\left(\lambda_{ph}\frac{\partial T_{ph}}{\partial x}\right)_s - \left(\lambda_{ph}\frac{\partial T_{ph}}{\partial x}\right)_\ell = \rho_s L_m^{ne} v_{s\ell}, \tag{70}$$

These conservation laws are accompanied by the kinetic dependence of the interphase front velocity $v_{s\ell}(\Delta T_{s\ell})$ on the overheating of the solid surface (22).

$$v_{s\ell}(T_{s\ell}) = \frac{a}{\lambda}\frac{f}{}(3k_B T_{s\ell}/m)^{1/2}\left[1 - \exp\left(\frac{L_m}{k_B}\left(\frac{\Delta T}{T_{s\ell}}\right)\right)\right], \tag{71}$$

Additional account of hydrodynamic effects was carried out using the expressions for the curve of equilibrium melting $T_m(P_s)$, temperature dependence of the equilibrium latent melting heat $L_m^{eq}(T_m(P_s))$ and non-equilibrium latent melting heat L_m^{ne}

$$L_m^{ne} = L_m^{eq}(T_m(P_s)) + \Delta C_{ps\ell}\Delta T_{s\ell} + \frac{\rho_s + \rho_\ell}{\rho_s - \rho_\ell}\frac{(u_s - u_\ell)^2}{2}, \quad T_m = T_m(P_s) = T_{m,0} + kP_s,$$

where $\Delta C_{ps\ell} = (C_{p\ell} - C_{ps})$, $\Delta T_{s\ell} = (T_{s\ell} - T_m(P_s))$.

The electron component is assumed continuous with respect to the electron density N_e and temperature T_e during transition through the phase front:

$$\left(\lambda_e\frac{\partial T_e}{\partial x}\right)_s = \left(\lambda_e\frac{\partial T_e}{\partial x}\right)_\ell, \qquad T_{e,s} = T_{e,\ell}. \tag{72}$$

$x = \Gamma_{kv}(t)$: the model of surface evaporation is used as boundary conditions at the moving interphase boundary $\Gamma_{kv}(t)$. The process of surface evaporation within the approximation of the Knudsen layer is described by three conservation laws and three additional parameters at the outer side of the Knudsen layer (temperature T_v, density ρ_v and velocity u_v). In the general case, two of these parameters (usually T_v and ρ_v) are determined using certain approximation relations, e.g. (30), while the third one (usually the Mach number $M = u / u_c$) is found from the solution of the gas-dynamic equations.

Three conservation laws are written in the stationary (laboratory) coordinate system and have the form:

$$x = \Gamma_{kv}(t): \rho_k\left(u_k - v_{kv}\right) = \rho_v(u_v - v_{kv}), \tag{73}$$

$$p_k + \rho_k\left(u_k - v_{kv}\right)^2 = p_v + \rho_v\left(u_v - v_{kv}\right)^2, \tag{74}$$

$$\left(\lambda_{ph}\frac{\partial T_{ph}}{\partial x}\right)_k - \left(\lambda_v\frac{\partial T_v}{\partial x}\right)_v = \rho_k\left(u_k - v_{kv}\right)L_v^{ne}, \tag{75}$$

$$\rho_v = \alpha_\rho(M)\rho_{sat}, T_v = \alpha_T(M)T_{sur}, \tag{76}$$

$$p_{sat}(T_{sur}) = p_b \exp\left[\frac{L_v^{ne}}{RT_b}\left(\frac{\Delta T_{sur}}{T_{sur}}\right)\right], \rho_{sat} = \frac{p_{sat}(T_{sur})}{RT_{sur}}, \Delta T_{sur} = T_{sur} - T_b\left(p_{sat}(T_{sur})\right)$$

$$L_v^{ne} = L_v^{eq}(T_{sur}) + C_{pv}(T_v - T_{sur}) + \frac{p_k + p_v}{p_k - p_v}\frac{\left(u_k - u_v\right)^2}{2},$$

The boundary conditions for the electron component and the laser radiation are written as:

$$-\lambda_e\frac{\partial T_e}{\partial x} = \sigma T_e^4, \tag{77}$$

$$G(t) = \left(1 - R_k(T_e)\right) \cdot G_0 \exp\left[-\left(\frac{t}{\tau_L}\right)^2\right]. \tag{78}$$

σ is the Stefan-Boltzmann constant.

$x = \Gamma_{sh,s}(t)$: three conservation laws are written at the moving front of the shock wave, so-called Rankine-Hugoniot relations, that have the following form in the laboratory coordinate system:

$$j_{sh,s}^m = \rho_1\left(u_1 - v_{sh,s}\right) = \rho_0\left(u_0 - v_{sh,s}\right), \tag{79}$$

$$j_{sh,s}^i = p_1 + \rho_1\left(u_1 - v_{sh,s}\right)^2 = p_0 + \rho_0\left(u_0 - v_{sh,s}\right)^2, \tag{80}$$

$$j_{sh,s}^e = -W_{T,1} + j_{sh,s}^m\left[\varepsilon_1 + \frac{\left(u_1 - v_{sh,s}\right)^2}{2}\right] = -W_{T,0} + j_{sh,s}^m\left[\varepsilon_1 + \frac{\left(u_0 - v_{sh,s}\right)^2}{2}\right] \tag{81}$$

The index "0" signifies the values at the background side, "1" relates to the shock wave side.

4.2. Numerical algorithm and finite difference schemes

Mathematical feature of Stefan problems is the lack of explicit expressions for the interconnected quantities at the interphase boundaries: temperature $T_{s\ell}$, T_{sur} and velocities of the phase fronts $v_{s\ell}$ and v_{kv}. So the problem (65) – (81) will be nonlinear even for constant values of thermo-physical and optical properties. For low velocities of phase transformations ($v_{s\ell}$, $v_{kv} \ll v_{sound}$), the processes behave in a quasi-equilibrium way, and their description can be performed within the framework of equilibrium models. For example, in the problem of melting-crystallization, a phenomenological condition of the temperature equity $T_{s\ell} = T_s = T_\ell = T_m$ is used instead of the kinetic condition (71), and the differential Stefan condition is completely omitted. The influence of the phase transition is taken into account using a singular adding of the latent heat L_m^{eq} to the heat capacity (equation of state) in the point of the phase transition, so-called enthalpy method [85 - 87]. To solve the problems in this statement, the methods of the type of "pass-through" or uniform schemes have found wide application [85], where the velocity and location of the phase transition are not determined explicitly.

Fast phase transitions ($v_{s\ell}$, $v_{kv} \ll v_{sound}$), that are typical for powerful pulsed laser action, occur under the conditions of high non-equilibrium caused by the powerful flow of material [88] over the interphase boundary. Using the smoothing procedure of the enthalpy reduces the class of solutions of the phase transformations problems in the material; in particular, it excludes from consideration the effects of overheating and overcooling of the condensed matter [89, 90]. High-speed phase transitions require explicit tracking of the phase boundaries [91]. Typically, the velocity of phase boundaries is determined numerically. In computational respect, the presence of moving boundaries leads to a significant complication of the numerical solution [92].

4.2.1. Method of dynamic adaptation

Finite-difference method of dynamic adaptation [93], [94],was used to numerically solve the discussed problem. This method allows performing explicit tracking of any number of interphase boundaries and shock waves [95].

The method of dynamic adaptation is based on the procedure of transition to an arbitrary nonstationary coordinate system. Its usage allows to formulate the problem of grid generation and adaptation on a differential level, i.e. part of differential equations in the obtained mathematical model describes physical processes, and other part describes the behavior of the grid nodes [96].

4.3. Arbitrary nonstationary coordinate system

The transition to an arbitrary nonstationary coordinate system is performed using an automatic coordinate transformation using the sought solution. Formally, it is possible to use a variable change of the common form $x=f(q,\tau)$, $t = \tau$ to perform a transition from the physical space $\Omega_{x,t}$ with Euler variables (x,t) to some computational space with arbitrary nonstationary coordinate system $\Omega_{q,\tau}$ and variables (q,τ). It is assumed that this transformation corresponds to a univalent nondegenerate reverse transformation $q=\phi(x,t)$, $\tau=t$. During the transition between coordinate systems, the partial derivatives of the dependent variables are connected using the following expressions:

$$\frac{\partial}{\partial t}=\frac{\partial}{\partial\tau}+Q\frac{\partial}{\partial q},\frac{\partial}{\partial x}=\frac{1}{\psi}\frac{\partial}{\partial q},\frac{\partial^2}{\partial x^2}=\frac{1}{\psi}\frac{\partial}{\partial q}\frac{1}{\psi}\frac{\partial}{\partial q}, \tag{82}$$

where $\psi=\partial x/\partial q$ is the Jacobian of the reverse transformation, Q is the transformation function that should be determined.

In the arbitrary coordinate system, moving with velocity Q, where $\dfrac{\partial x}{\partial\tau}=-Q$, or $\dfrac{\partial\psi}{\partial\tau}=-\dfrac{\partial Q}{\partial q}$, the differential model (65) in the new variables (q,τ) takes the form:

$$\frac{\partial\psi}{\partial\tau}=-\frac{\partial Q}{\partial q},\psi=\frac{\partial x}{\partial q} \tag{83}$$

$$\frac{\partial}{\partial\tau}(\psi\rho)+\frac{\partial}{\partial q}(\rho(u+Q))=0 \tag{84}$$

$$\frac{\partial}{\partial\tau}(\psi\rho u)+\frac{\partial}{\partial q}(p+\rho u(u+Q))=0 \tag{85}$$

$$\frac{\partial}{\partial\tau}(\psi\rho_e\varepsilon_e)+\frac{\partial}{\partial q}(\rho_e\varepsilon_e(u+Q))=-\left(p\frac{\partial u}{\partial q}+\frac{\partial W_e}{\partial q}+g(T_e)(T_e-T_{ph})+\frac{\partial G}{\partial q}\right) \tag{86}$$

$$\frac{\partial(\psi\rho\varepsilon_{ph})}{\partial\tau}+\frac{\partial}{\partial q}(\rho\varepsilon_{ph}(u+Q))=-\left(p\frac{\partial u}{\partial q}+\frac{\partial W_{ph}}{\partial q}-g(T_e)(T_e-T_{ph})\right) \tag{87}$$

$$\frac{\partial G}{\partial q}+\psi\alpha(T_e)G=0 \tag{88}$$

$$p=p_e(\rho_e,T_e)+p_{ph}(\rho,T_{ph}),\ \varepsilon_e=C_e(T_e)T_e,\ \varepsilon_{ph}=C_{ph}(T_{ph})T_{ph},\ W=-\frac{\lambda(T)}{\psi}\frac{\partial T}{\partial q},$$

Thus, usage of an arbitrary nonstationary coordinate system is accompanied by the transformation of the initial differential model (65) into the extended model (83) – (88), in which the equation (83) is the equation of the inverse transform. Equations (84) – (88) describes the physical processes. The type, characteristics and kind of boundary conditions for the inverse transformation equations depend on the type conversion function Q. Because of this equation, the initial (66) and boundary conditions (67) – (81) are changed accordingly.

Thus, the unknown functions in the computational space are not only the functions of the physical fields but also the coordinates of the grid nodes. The equation of the reverse transformation is used after the determination of the function Q to construct the adaptive grid in the physical space. Its differential analogue describes the dynamics of the grid nodes, and the function Q performs controlled movement of the nodes in agreement with the dynamics of the sought solution. In the computational space, the nodes of the grid and all discontinuous and interphase boundaries are steady. The value of the function Q at the boundaries of the domain are determined from the boundary conditions.

Selection of the function Q. The function Q that is in agreement with the sought solution can be determined from the quasi-steady principle [96], which states that we should search for such nonstationary coordinate system where all processes occur in a steady way. Since the energy balance equation includes all hydrodynamic variables, to determine the function Q, we can use only two equations (86), (87), assuming that $\dfrac{\partial \varepsilon_e}{\partial \tau} = \dfrac{\partial \varepsilon_{ph}}{\partial \tau} = 0$. Then the function Q will have the form

$$Q = -u - \frac{2p\dfrac{\partial u}{\partial q} + \left(+\dfrac{\partial W_e}{\partial q} + \dfrac{\partial W_{ph}}{\partial q} + \dfrac{\partial G}{\partial q} \right)}{\left(\rho_e \dfrac{\partial \varepsilon_e}{\partial q} + \rho\dfrac{\partial \varepsilon_{ph}}{\partial q} + re \right)} \qquad (89)$$

where $W_e = -\dfrac{\lambda_e}{\Psi}\dfrac{\partial T_e}{\partial q}$, $W_{ph} = -\dfrac{\lambda_{ph}}{\Psi}\dfrac{\partial T_{ph}}{\partial q}$.

Differential schemes

The differential model (83) – (88) was approximated by a family of conservative differential schemes obtained using integration-interpolation method [97]. Computational grids $\left(\omega_m^j\right)_s$ and $\left(\omega_m^j\right)_\ell$ are introduced in the computational space $\Omega_{q,\tau}$ in each subdomain with integer q_i and half-integer $q_{i+1/2}$ nodes for the spatial variable q and step $\Delta\tau^j$ for the variable τ.

$$\left(\omega_m^j\right)_k = \left\{ \begin{array}{l} \left(q_m,\tau^j\right),\left(q_{m+1/2},\tau^j\right); \quad q_{m+1}=q_m+h, \quad q_{m+1/2}=q_m+0.5h, \\ \tau^{j+1}=\tau^j+\Delta\tau^j, \quad i=0,..,N-1, \quad j=0,1,... \end{array} \right\}_{k=s,\ell}$$

The flow variables W_e, W_{ph}, G, and functions u, Q, x, correspond to the integer nodes during approximation, while functions ε_e, ε_{ph}, T_e, T_{ph}, ρ, ψ correspond to half-integers ones $\left(q_{m+1/2},\tau^j\right)$.

The family of conservative differential schemes has the form:

$$\frac{\psi_{m-1/2}^{j+1} - \psi_{m-1/2}^{j}}{\Delta\tau^j} = \frac{Q_m^{\sigma_1} - Q_{m-1}^{\sigma_1}}{h}$$

$$\frac{\psi_{m-1/2}^{j+1}\rho_{m-1/2}^{j+1} - \psi_{m-1/2}^{j}\rho_{m-1/2}^{j}}{\Delta\tau^j} = \frac{\rho_m^{\sigma_2}\left(Q_m^{\sigma_1} - u_m^{\sigma_3}\right) - \rho_{m-1}^{\sigma_2}\left(Q_{m-1}^{\sigma_1} - u_{m-1}^{\sigma_3}\right)}{h}$$

$$\frac{u_m^{j+1}\left(\psi_{m-1/2}^{j+1}\rho_{m-1/2}^{j+1} + \psi_{m+1/2}^{j+1}\rho_{m+1/2}^{j+1}\right) - u_m^{j}\left(\psi_{m-1/2}^{j}\rho_{m-1/2}^{j} + \psi_{m+1/2}^{j}\rho_{m+1/2}^{j}\right)}{2\Delta\tau^j} = \frac{p_{m-1/2}^{\sigma_4} - p_{m+1/2}^{\sigma_4}}{h} +$$

$$+\frac{\rho_m^{\sigma_2}\left(Q_m^{\sigma_1} - u_m^{\sigma_3}\right)u_{m+1}^{\sigma_3} + \rho_{m+1}^{\sigma_2}\left(Q_{m+1}^{\sigma_1} - u_m^{\sigma_3}\right)u_m^{\sigma_3}}{2h} - \frac{\rho_m^{\sigma_2}\left(Q_m^{\sigma_1} - u_m^{\sigma_3}\right)u_{m-1}^{\sigma_3} + \rho_{m-1}^{\sigma_2}\left(Q_{m-1}^{\sigma_1} - u_{m-1}^{\sigma_3}\right)u_m^{\sigma_3}}{2h}$$

(90)

$$\frac{\left(\psi_{m-1/2}^{j+1}\rho_{e,m-1/2}^{j+1}\varepsilon_{e,m-1/2}^{j+1}\right) - \left(\psi_{m-1/2}^{j}\rho_{e,m-1/2}^{j}\varepsilon_{e,m-1/2}^{j}\right)}{\Delta\tau^j} = -\left\{\frac{\left[\rho_{e,m}^{\sigma_2}\varepsilon_{e,m}^{\sigma_5}\left(Q_m^{\sigma_1} - u_m^{\sigma_3}\right) - \rho_{e,m-1}^{\sigma_2}\varepsilon_{e,m-1}^{\sigma_5}\left(Q_{m-1}^{\sigma_1} - u_{m-1}^{\sigma_3}\right)\right]}{h} +\right.$$

$$+\frac{p_{m-1/2}^{\sigma_4}\left(u_{m-1}^{\sigma_3} - u_m^{\sigma_3}\right)}{h} + \frac{2\left[\lambda_m^{\sigma_6}\dfrac{T_{m+1/2}^{\sigma_7} - T_{m-1/2}^{\sigma_7}}{\psi_{m+1/2}^{\sigma_8} + \psi_{m-1/2}^{\sigma_8}} - \lambda_{m-1}^{\sigma_6}\dfrac{T_{m-1/2}^{\sigma_7} - T_{m-3/2}^{\sigma_7}}{\psi_{m-1/2}^{\sigma_8} + \psi_{m-3/2}^{\sigma_8}}\right]}{h} + \left[g(T_e)^{\sigma_9}(T_e - T_{ph})^{\sigma_{10}}\right]_{m-1/2} +$$

$$\left.+\frac{\left(G_m^{\sigma_{11}} - G_{m-1}^{\sigma_{11}}\right)}{h} + \frac{\left(W_{e,m}^{\sigma_{12}} - W_{e,m-1}^{\sigma_{12}}\right)}{h}\right\},$$

$$\frac{\left(\psi_{m-1/2}^{j+1}\rho_{m-1/2}^{j+1}\varepsilon_{ph,m-1/2}^{j+1}\right) - \left(\psi_{m-1/2}^{j}\rho_{m-1/2}^{j}\varepsilon_{ph,m-1/2}^{j}\right)}{\Delta\tau^j} =$$

$$-\left\{\frac{\left[\rho_m^{\sigma_2}\varepsilon_{ph,m}^{\sigma_{13}}\left(Q_m^{\sigma_1} - u_m^{\sigma_3}\right) - \rho_{m-1}^{\sigma_2}\varepsilon_{ph,m-1}^{\sigma_{13}}\left(Q_{m-1}^{\sigma_1} - u_{m-1}^{\sigma_3}\right)\right]}{h} + \frac{p_{m-1/2}^{\sigma_4}\left(u_{m-1}^{\sigma_3} - u_m^{\sigma_3}\right)}{h} +\right.$$

$$+\frac{2\left[\lambda_{ph,m}^{\sigma_{14}}\dfrac{T_{ph,m+1/2}^{\sigma_{15}} - T_{ph,m-1/2}^{\sigma_{15}}}{\psi_{m+1/2}^{\sigma_8} + \psi_{m-1/2}^{\sigma_8}} - \lambda_{ph,m-1}^{\sigma_{14}}\dfrac{T_{ph,m-1/2}^{\sigma_{15}} - T_{ph,m-3/2}^{\sigma_{15}}}{\psi_{m-1/2}^{\sigma_8} + \psi_{m-3/2}^{\sigma_8}}\right]}{h} +$$

$$\left.+\left[g(T_e)^{\sigma_9}(T_e - T_{ph})^{\sigma_{10}}\right]_{m-1/2} + \frac{\left(W_{e,m}^{\sigma_{16}} - W_{e,m-1}^{\sigma_{16}}\right)}{h}\right\},$$

Here $f^{\sigma_r} = \sigma_r f^{j+1} + (1-\sigma_r)f^j$, and $\sigma_r = \sigma_1, \sigma_2, \dots \square$ are the weight coefficients, determining the degree of implicitness of the differential schemes.

To solve the obtained system of nonlinear partial-differential equations we use a computational algorithm with enclosed iterative cycles consisting of one external and two internal cycles [98]. Each of the internal cycles uses the Newton iterative procedure.

5. Modeling results

We consider two modes of pulse with pico - and femtosecond laser fluence with a wavelength $\lambda_L = 0.8$ μm on two metal target of aluminum (Al) and copper (Cu). The influence on each of the targets was carried out with the same energy density F, respectively $F = 0.7 \, J \, cm^{-2}$ for Al and $F = 2.0 \, J \, cm^{-2}$ for Cu.

Temperature dependences of the reflectivity $R(T_e)$ and volumetric absorption coefficient $\alpha(T_e)$ were calculated according to [99, 100]. Fig. 14 shows the time profiles of the incident and absorbed laser pulse with a Gaussian profile $\tau_L = 10^{-12}$ s and $\tau_L = 10^{-15}$ s.

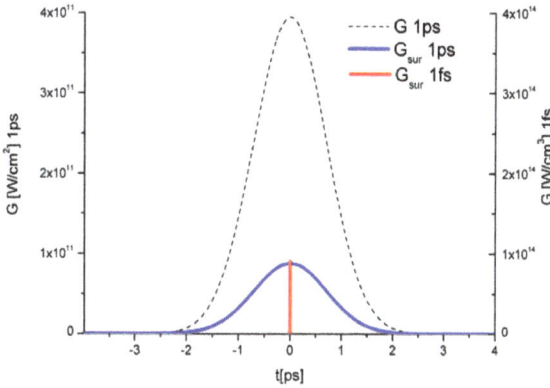

Figure 14. Time profile of pico- and femtosecond pulses on the surface of Al.

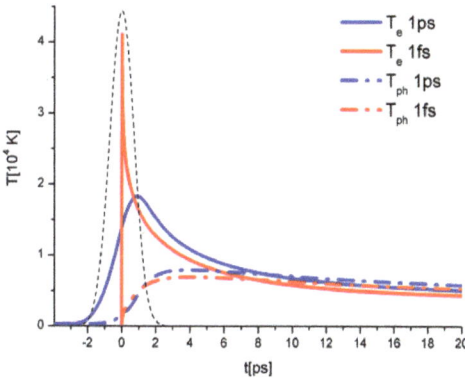

Figure 15. Time dependence of the surface temperature of Al.

The incident laser radiation is absorbed in the electron component, causing it to warm up fast. The main feature of the nonequilibrium heating is that it occurs under conditions of intense competition between the rapid release of the laser energy and the factors limiting the heat - strongly varying electron thermal conductivity coefficient $\lambda_e(T_e, T_{ph})$ and electron-fonon coupling factor $g(T_e)$.

At picosecond influence $\tau_L = 10^{-12}$, the maximum of electron temperature of the surface $T_{e,max} \approx 1.8 \cdot 10^4 K$ is achieved at the descending branch of the laser pulse at the moment $t \cong \tau_L$, Fig. 15, and the values of the coefficients vary in the range $\lambda_{e,max} / \lambda_{Al}^{eq} \approx 10$ and $g_{max} / g_{Al}^{eq} \approx 70$. The maximum value of the lattice temperature is achieved after the pulse $t \approx 5\tau_L$ and amounts $T_{ph,max} \approx 7.5 \cdot 10^3 K$. The maximum gap between the temperatures reaches the value $\Delta T = T_e - T_{ph} \approx 10^4 K$. Temperature equalization $T_e \approx T_{ph}$ occurs within a time ~ 10 ps.

Femtosecond influence $\tau_L = 10^{-15} s$ differs from picosecond that the same energy is released during the 3 orders of magnitude less than that causes more of a deviation from local thermodynamic equilibrium. The values of the coefficients $\lambda_e(T_e, T_{ph})$ and $g(T_e)$ limiting the heating of electronic components, and also increase $\lambda_{e,max} / \lambda_{Al}^{eq} \approx 27$ and $g_{max} / g_{Al}^{eq} \approx 200$, but not as much as $G(t)$. Time profile of the electron temperature of the surface $T_{e,surf}(t)$ significantly shifted relative to laser pulse time profile $G(t)$. The maximum value of the electron temperature is achieved at the end of the pulse $t \cong 2\tau_L$ and is $T_{e,max} \approx 4.1 \cdot 10^4 K$, Fig. 15. A second difference of a femtosecond influence, despite the rapid energy exchange between the subsystems is a slow heating of the lattice, which differs little from the effects of heating at picosecond influence. The maximum value of the lattice temperature is achieved at the time $t \approx 5$ ps and is $T_{ph,max} \approx 7.2 \cdot 10^3 K$. The maximum gap between the temperatures reaches a value $\Delta T = T_e - T_{ph} \approx 3.3 \cdot 10^4 K$. Temperature equalization $T_e \approx T_{ph}$ occurs within the time ~ 9 ps.

The high heating rate determines high rate of phase transformations. Fig. 16 shows the time dependences of the melting front propagation velocity $v_{sl}(t)$ for two modes of influence. Since the rate of heating of the lattice by pico - and femtosecond pulses differ slightly, then the rate v_{sl} is almost the same. The only difference is in the initial stage of the process. At the picosecond influence, the birth of the melting front starts at the front of the laser pulse $t \approx -0.1 ps$, while at the femtosecond with a considerable delay after the pulse, $t \approx 30 fs$. The maximum values of $v_{sl\,max}(t) \approx 3.75$ km/s are achieved in the time interval $t \approx 2 \div 2.5 ps$. Such a high rate of propagation of the front leads to a pressure jump at the melting surface of the solid phase $p_s = 0.16 Mbar$ for $\tau_L = 10^{-12}$ s ($p_s = 0.14 Mbar$ for $\tau_L = 10^{-15}$ s) and the formation of a shock wave moving ahead of the front of melting, Fig. 17.

Appearance of high dynamic pressure causes an increase in the equilibrium melting temperature, the curve $T_m(p_s)$ shown in Fig. 18. Its maximum value reaches $T_m(p_s) \cong 1900 K$. Due to the dynamic pressure of about one order of magnitude as compared to the equilibrium value, increases the value of non-equilibrium heat of melting $L_m^{ne}(p_s)$.

Comparison of the $T_{sl}(t)$ and $T_m(p_s(t))$ curves indicates a significant overheating of surface of the phase boundary $T_{sl}(t) > T_m(p_s(t))$. The degree of nonequilibrium of melting process is conveniently characterized by the response function $\Delta T_{sl}(v_{sl})$, which is shown in Fig.19.

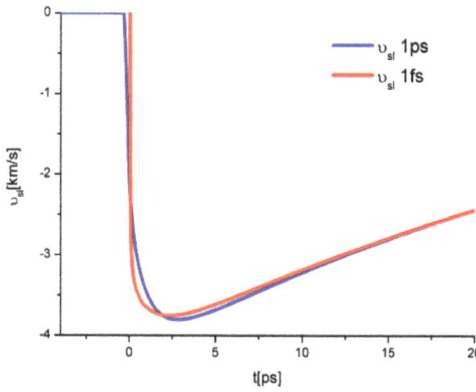

Figure 16. The time profile of the velocity of melting of aluminum.

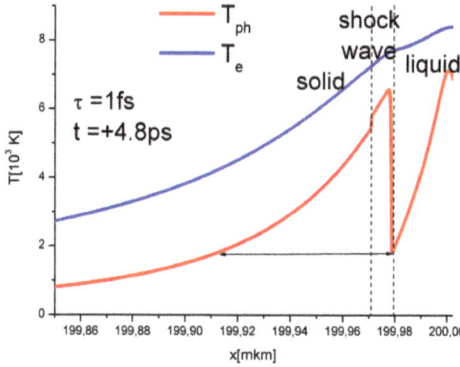

Figure 17. The spatial profiles of temperature of aluminum.

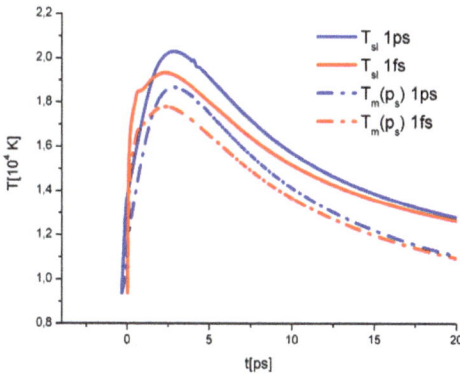

Figure 18. The time profiles of the equilibrium melting temperature and temperature of the melting front of Al.

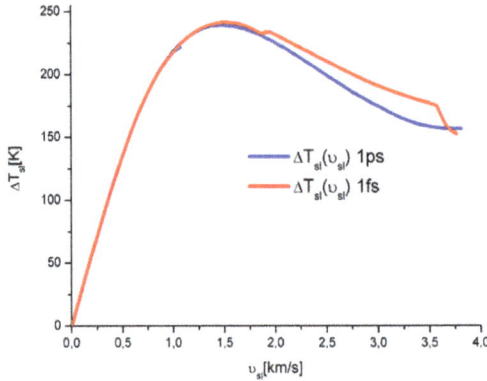

Figure 19. Response functions $\Delta T_{s\ell}\left(v_{sl}\right)$ for Al.

At low overheating (supercooling), the function $\Delta T_{s\ell}$ increases linearly with growth v_{sl}. In the range of $v_{sl} \sim 1$ km/s, at which the shock wave is formed in the solid phase, $\Delta T_{s\ell}$ function has a smooth maximum indicating the change of the mechanism of phase transformation. The thermal conductivity mechanism is dominant at low velocities $v_{sl} \leq 1 km / s$ will be displaced by hydrodynamic at large values $v_{sl} \leq v_{sound}$.

Fig. 20 shows fragment of dependence of $v_{sl}(t) > 0$ which characterizes the hardening process which takes place in the nanosecond time range. Maximum velocity of the solidification front does not exceed 45 m/s, respectively, the maximum undercooling is $10 \div 15 K$. High velocity of propagation of melting fronts are associated with strong cross-flows of matter and energy across the phase boundary.

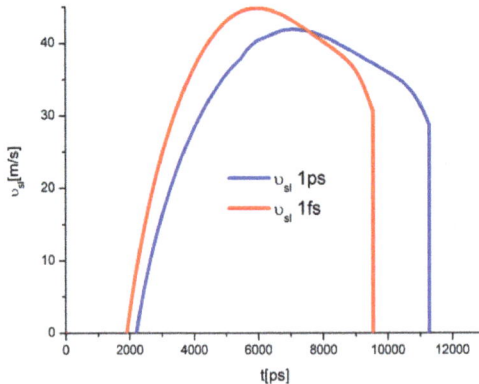

Figure 20. The time profiles of the velocity of solidification of aluminum.

In aggregate with volumetric nature of the energy transfer from the electron gas to the lattice it leads to the formation of highly superheated metastable state, characterized by the appearance of the surface temperature maximum $\Delta T_{ph,\max} = T_{ph} - T_m(p_s) \approx 4.5 \cdot 10^3 K$ in the solid phase, Fig. 17. In the liquid phase is also formed near-surface temperature maximum, Fig.20, but since the maximum velocity of evaporation front $v_{\ell v,\max} \approx 40 \ m/s$ is two orders of magnitude lower than the velocity v_{sl}, magnitude and its role are insignificant.

Action of ultrashort laser pulses on metal target with different thermo-physical properties – copper – has several differences and many in common with the action of the aluminum target. The main difference of the non-equilibrium heating and non-equilibrium phase transformation in the condensed media is in the first place connected with the rate of energy exchange between the electron and phonon subsystems. The coefficient of energy exchange $g(T_e)$ in copper is by about 1.5 orders lower than that of aluminum, so that during laser energy release, the action of one of the main factors that limit the heating of the electron subsystem turns out to be significantly lower. As a result, the maximum values of the electron temperature of the surface are several times higher than that for aluminum: $T_{e,\max} \approx 4.\cdot 10^4 K$ for picosecond pulse $T_{e,\max} \approx 1.2 \cdot 10^5 K$ for femtosecond pulse, fig. 21.

Heating and melting of the lattice take place in the picosecond range $t_m \approx +0.1 \ ps$, fig.22. For copper, the rate of energy transfer to the lattice is significantly lower, so that the velocity of the melting front turns out to be several times lower than that for aluminum $v_{s\ell,\max}(t) \approx 1.3$ km/s, fig.23. The maximum value of the pressure in the solid phase is also lower, $p_{s,\max} = 0.14 \ Mbar$. Despite the fact that the velocity $v_{s\ell}$ for copper is significantly lower than that for Al, the value of overheating of the interphase surface turns out to be significantly higher due to decrease of the hydrodynamic effects: $\Delta T_{s\ell,\max} \approx 350 K$, fig. 24. The overheating of the sub-surface region of the solid phase reaches the value of $4 \cdot 10^3 K$.

The solidification of the liquid phase takes place in the nanosecond rage – the same as for aluminum. The lifetime of the melt reaches the value of $15 \div 20$ ns, fig.25, the maximum velocity of the crystallization front - $v_{s\ell} \approx 35$ m/s. The maximum value of the overcooling of the front ~ 22 K. The overcooling of the volume of the solid reaches the value of *800 K*.

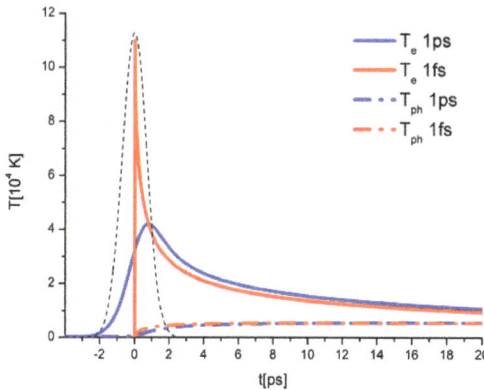

Figure 21. Time dependence of the surface temperature of Cu.

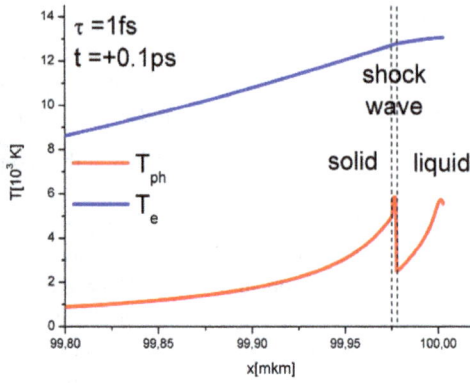

Figure 22. The spatial profiles of temperature of copper.

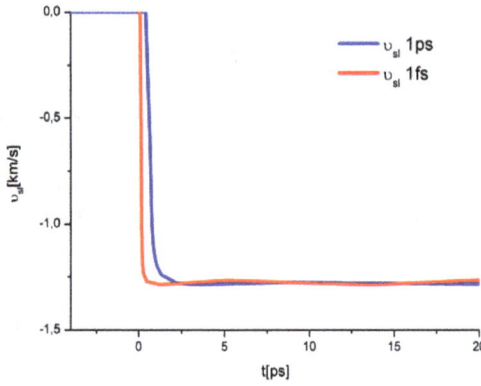

Figure 23. The time profile of the velocity of melting of copper.

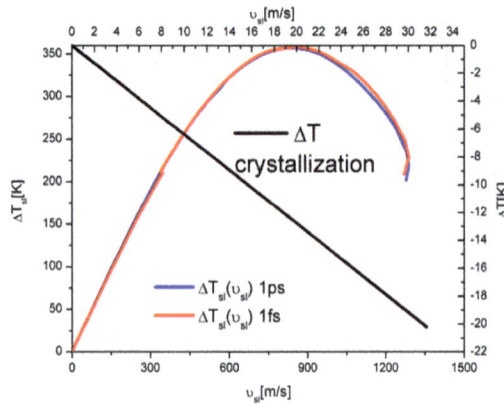

Figure 24. Response functions $\Delta T_{s\ell}\left(v_{sl}\right)$ for Cu.

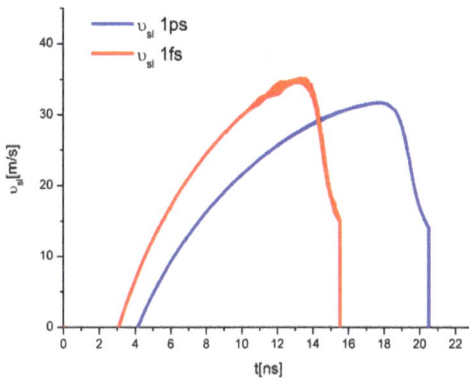

Figure 25. The time profiles of the velocity of solidification of copper.

Modern experimental methods of the investigation of the processes of melting-crystallization of the metal systems determine the velocities of the solid-melt front to be within the range of $10 \div 100$ m/s with corresponding overcooling of liquid within the range of $10 \div 400\,K$. Thus, the data from the modeling of the metastable overcooled states in the liquid phase of metals is in relatively good agreement with the experimental measurements. It is not possible to perform the same comparison with the overcooled states in the solid phase because of the lack of reliable experimental data which is caused by the complexity of statement of such experiments.

6. Conclusion

The detailed analysis of the physical processes of the non-equilibrium heating and the fast phase transformations in the metal systems under the action of ultra-short laser pulses was performed in this chapter. A mathematical model was built based on the brief analysis of the thermodynamical and kinetic approaches for the description of the phase transformations of the first order. This model is a non-equilibrium hydrodynamic version of the Stefan problem and describes the kinetics of high-speed heterogeneous phase transformations: melting, crystallization, evaporation. The model is supplemented by the explicit description of the moving phase fronts for correct reflection of the dynamics of the phase transitions.

The methods of mathematical modeling were used to investigate the regimes of pico- and femtosecond action on metals - Al, Cu. The kinetics and dynamics of phase transformations are analyzed. The results of the modeling confirmed the skewness of the high-speed processes of melting – solidification. In particular, the maximum velocity of melting reaches the value of several kilometers per second, while the maximum velocity of solidification does not exceed hundred of meters per second. The temperature dependence $v_{s\ell}\left(T_{s\ell}\right)$ does not have a sharp bend in the point of change of the direction of the phase transition.

Special attention is paid to the overheated and overcooled metastable states that are caused by the fast phase transitions. The deviation from the local thermodynamical equilibrium is noticed both in the bulk and at the interphase boundary. The response function $T_{s\ell}(v_{s\ell})$ has a linearly increasing character for low values of velocity $v_{s\ell}$ in the range of $\left[0 \div \pm 50\right] m/s$. As the velocity $v_{s\ell}$ rises more than $v_{s\ell} > 100 m/s$, a maximum is formed in the response function for both metals. The appearance of the maximum means that the dominating mechanism of the phase transition changes: the mechanism of heat conductivity is changed by the hydrodynamic one. The maximum value of overheating of the interphase boundary reaches the value of 200 ÷ 400 K, the one for overcooling is by one order lower than that value. The overheating and overcooling of the bulk are significantly higher - $4 \cdot 10^3 \div 5 \cdot 10^3 K$ for overheating and $6 \cdot 10^2 \div 8 \cdot 10^2 K$ for overcooling. The results for different materials are compared between each other and with experimental data.

These results do not pretend to be an exhaustive description of the variety of fast phase transitions, because they used the assumption of the thermodynamic stability of overheated / overcooled metastable states during all times considered. Refusal of such assumptions will require further research to include the kinetics of homogeneous phase transformations into the consideration.

Author details

V.I. Mazhukin
Keldysh Institute of Applied Mathematics of RAS, Moscow, Russia

Acknowledgement

This work was supported by the RFBR (projects nos. 10-07-00246-a, 12-07- 00436-a, 11-01-12086-ofi-m).

7. References

[1] Fujimoto J.G., Liu J.M., Ippen E.P., Bloembergen N (1984) Femtosecond Laser Interaction with Metallic Tungsten and Nonequilibrium Electron and Lattice Temperatures. Physical Review Letters. 53: 19/1837 – 1840.
[2] Chichkov B.N., Momma C., Nolte S., von Alvensleben F, Tunnermann A. (1996) Femtosecond, picosecond and nanosecond laser ablation of solids. Appl. Phys. A 63: 109–115.
[3] Shirk M.D., Molian P.A. (1998) A review of ultrashort pulsed laser ablation of materials. Journal of Laser Applications. 10: 1/18-28.
[4] Von der Linde D., Sokolovski-Tinten K. (2000) The physical mechanisms of short-pulse laser ablation. Appl. Surf. Sci. 154-155: 1-10.

[5] (2007) Part 2: Ultrafast Interactions. In: Claude R. Phipps, editor. Optical Sciences. Springer Series in Laser ablation and its applications. Santa Fe, New Mexico: Springer. 129: 121-280.

[6] Gonzales P., Bernath R., Duncan J., Olmstead T., Richardson M. (2004) Femtosecond ablation scaling for different materials. Proc. SPIE. 5458: 265-272.

[7] Chichkov B.N., Korte F., Koch J., Nolte S., Ostendorf A. (2002) Femtosecond laser ablation and nanostructuring. Proc. SPIE. 4760: 19-27.

[8] Hamad S., Krishna Podagatlapalli G., Sreedhar S., Tewari S. P., Rao S. V. (2012) Femtosecond and picosecond ablation of aluminum for synthesis of nanoparticles and nanostructures and their optical characterization. Proc. SPIE 8245: 82450L.

[9] Ostendorf A., Kamlage G., Klug U., Korte F., Chichkov B. N. (2005) Femtosecond versus picosecond laser ablation. Proc. SPIE 5713:1- 8.

[10] Kamlage G., Bauer T., Ostendorf A., Chichkov B.N. (2003) Deep drilling of metals by femtosecond laser pulses. Applied Physics A: Materials Science & Processing. 77: 307-310.

[11] Lopez J., Kling R., Torres R., Lidolff A., Delaigue M., Ricaud S., Hönninger C., Mottay E. (2012) Comparison of picosecond and femtosecond laser ablation for surface engraving of metals and semiconductors. Proc. SPIE 8243: 82430O.

[12] Bulgakov A.V., Bulgakova N. M. Burakov I.M. (2009) Nanosized Material Synthesis by Action of High – Power Energy Fluxes on Matter. Novosibirsk: Institute Thermophysics SB RAS. 461 p.

[13] Mazhukin V.I., Lobok M.G., Chichkov B.N. (2009) Modeling of fast phase transitions dynamics in metal target irradiated by pico and femto second pulsed laser. Applied Surface Science. 255: 5112-5115.

[14] Herlach D.M., Galenko P.K., Holland-Moriz D. (2005) Metastable materials from undercooled metallic melts. Amsterdam: Elsevier. 485 p.

[15] Eliezer S., Eliaz N., Grossman E., Fisher D., Gouzman I., Henis Z., Pecke S., Horovitz Y., Fraenkel M., Maman S., Lereah Y. (2004) Synthesis of nanoparticles with femtosecond laser pulses. Phys. Rev. B 69: 144119/1-6.

[16] Mene´ndez-Manjo´n A., Barcikowski S., Shafeev G.A., Mazhukin V.I., Chichkov B.N. (2010) Influence of beam intensity profile on the aerodynamic particle size distributions generated by femtosecond laser ablation. Laser and Particle Beams. 28: 45–52.

[17] Stratakis E., Barberoglou M., Fotakis C., Viau G., Garcia C., Shafeev G. A. (2009) Generation of Al nanoparticles via ablation of bulk Al in liquids with short laser pulses. Opt. Express. 17: 12650-12659.

[18] Mazhukin V.I., Samarskii A.A. (1994) Mathematical Modeling in the Technology of Laser Treatments of Materials. Review. Surveys on Mathematics for Industry. 4: 2/85-149.

[19] Ivanov D.S., Zhigilei L.V. (2003) Combined atomistic-continuum modeling of short-pulse laser melting and disintegration of metal films. Phys. Rev. B. 68: 064114-1 – 064114-22.

[20] . Landau L. D., Lifshic E. M. (1976) Teoreticheskaja fizika, t. V. Statisticheskaja fizika. Chast' 1. M.: Nauka. 616 p.

[21] Christian J. W. (1965) The theory of transformations in metals and alloys: an advanced textbook in physical metallurgy. Oxford: Pergamon Press. 975 p.

[22] Kirkpatrick R. J. (1975) Crystal Growth from the Melt. Review. American Mineralogist A: 60: 798-614.

[23] Asta M., Beckermann C., Karma A., Kurz W., Napolitano R., Plapp M., Purdy G., Rappaz M., Trivedi R. (2009) Solidification microstructures and solid-state parallels: Recent developments, future directions. Acta Materialia: 57: 146/941–971.

[24] Jackson K. A. (2004) Kinetic Processes:Crystal Growth, Diffusion, and Phase Transitions in Materials. Weinheim: Wiley-VCH Verlag GmbH & Co. KGaA. 409 p.

[25] Hoyt J. J., Asta M. (2002) Atomistic computation of liquid diffusivity, solid-liquid interfacial free energy, and kinetic coefficient in Au and Ag. Phys. Rev. B. 65: 21/214106-1 - 214106-11.

[26] Rodway G.H., Hunt J.D. (1991) Thermoelectric investigation of solidification of lead I. Pure lead. Journal of Crystal Growth. 112: 2–3/554–562.

[27] Hoyt J.J., Asta M., Karma A. (2002) Atomistic Simulation Methods for Computing the Kinetic Coefficient in Solid-Liquid Systems. Interface Science. 10: 2-3/181-189.

[28] Buta D., Asta M., Hoyt J. J. (2007) Kinetic coefficient of steps at the Si(111) crystal-melt interface from molecular dynamics simulations. Journal of Chemical Physics. 127: 7/074703 – 074713.

[29] Amini M., Laird B. B. (2006) Kinetic Coefficient for Hard-Sphere Crystal Growth from the Melt. Physical Review Letters. 97: 216102-1 - 216102-4.

[30] Monk J., Yang Y., Mendelev M. I., Asta M., Hoyt J. J., Sun D. Y. (2010) Determination of the crystal-melt interface kinetic coefficient from molecular dynamics simulations. Modelling and Simulation in Materials Science and Engineering. 18: 015004-1 - 015004-18.

[31] Wilson H.A. (1900) On the velocity of solidification and viscosity of supercooled liquids. Philos. Mag. 50: 238-250.

[32] Frenkel Ja. I. (1932) Note on the relation between the speed of crystallization and viscosity. Phys. Z. Sowjet Union. 1: 498 – 499.

[33] Frenkel' Ja.I. (1975) Kineticheskaja teorija zhidkostei. L.: Nauka. 592 p.

[34] Jackson K.A., Chalmers B. (1956) Kinetics of solidification. Can. J. Phys. 34: 473 – 490.

[35] Chalmers B. (1964) Principles of Solidification., New York: John Wiley &Sons. 129 p.

[36] Klude M. D., Ray J.R. (1989) Velocity versus Temperature Relation for Solidification and Melting of Silicon: A Molecular-Dynamics Study. Phys. Rev. B. 39: 3/1738-1746.

[37] Tymczak C.J. Ray J.R. (1990) Interface response function for a model of sodium: A molecular dynamics study, J. Chem. Phys. 92: 12/7520 – 7530.

[38] Tymczak C.J. Ray J.R. (1990), Asymmetric crystallization and melting kinetics in sodium: A molecular-dynamics study. Phys. Rev. Lett. Vol. 64, Issue 11, pp. 1278–1281.

[39] MacDonald C. A., Malvezzi A. M., Spaepen F. (1989) Picosecond time-resolved measurements of crystallization in noble metals. J. Appl. Phys. 65: 1/129 -137.

[40] Hertz H. (1882) Uber die Verdunstung der Flussigkeiten, insbesondere des Quecksilbers, im luftleeren Raume. Ann. Phys. und Chemie. 17: 177-190.

[41] Knudsen M. (1915) Die Maximale Verdampfungs-geschwindigkeit Quecksilbers. Ann. Phys. und Chemie. 47: 697-705.

[42] Siewert C.E., Thomas J.R. (1981) Strong evaporation into a half space. J. of Applied Mathem. and Physics(ZAMP). 32: 421 – 433.

[43] Siewert C.E. (2003) Heat transfer and evaporation/condensation problems based on the linearized Boltzmann equation. European Journal of Mechanics B/Fluids. 22: 391–408.

[44] Frezzotti A. (2007) A numerical investigation of the steady evaporation of a polyatomic gas. European Journal of Mechanics B/Fluids. 26: 93–104.

[45] Titarev V. A., Shakhov E. M. (2001) Teplootdacha i isparenie s ploskoi poverkhnosti v poluprostranstvo pri vnezapnom povyshenii temperatury tela. Mekhanika zhidkosti i gaza. 1/141 - 153.

[46] Abramov A. A., Butkovsky A. V. (2009) Effect of the Mode of Gas-Surface Interaction on Intensive Monatomic Gas Evaporation. Fluid Dynamics. 44: 1/80–87.

[47] Bond M., Struchtrup H. (2004) Mean evaporation and condensation coefficients based on energy dependent condensation probability. Phys. Rev. E. 70: 061605-1 – 20.

[48] Ford I J, Lee T-L. (2001) Entropy production and destruction in models of material evaporation. J. Phys. D: Appl. Phys. 34: 413–417.

[49] Rose J. W. (2000) Accurate approximate equations for intensive sub-sonic evaporation. Int. Journal of Heat and Mass Transfer.. 43: 3869 - 3875.

[50] Bedeaux D., Kjelstrup S., Rubi J. M. (2003) Nonequilibrium translational effects in evaporation and condensation. J. Chemical Physics. 119: 17/9163 – 9170.

[51] Frezzotti A., Gibelli L., Lorenzani S. (2005) Mean field kinetic theory description of evaporation of a fluid into vacuum. Phys. Fluids. 17: 012102-1 - 012102-12.

[52] Rahimi P., Ward Ch. A. (2005) Kinetics of Evaporation: Statistical Rate Theory Approach Int.J. of Thermodynamics. 8: 1/1-14.

[53] Meland R., Frezzotti A., Ytrehus T., Hafskjold B. (2004) Nonequilibrium molecular-dynamics simulation of net evaporation and net condensation, and evaluation of the gas-kinetic boundary condition at the interphase. Phys. Fluids. 16: 2/223 – 243.

[54] Ishiyama T., Yano T., Fujikawa S. (2005) Kinetic boundary condition at a vapor–liquid interface. Phys. Rev. Lett. 95: 0847504.

[55] Frezzotti A. (2011) Boundary conditions at the vapor-liquid interface. Phys. Fluids. 23: 030609-(1 -9).

[56] Kartashov I.N., Samokhin A.A., Smurov I.Yu. (2005) Boundary conditions and evaporation front instabilities. J. Phys. D: Appl. Phys. 38: 3703-3714.

[57] Crout D. (1936) An application of kinetic theory to the problem of evaporation and sublimation of monatomic gases. J. Math. Physics. 15: 1-54.

[58] Labuntsov D.A., Kryukov A.P. (1979) Analysis of Intensive Evaporation and Condensation. Inter. J. Heat and Mass Transfer. 22: 7/989–1002.

[59] Mazhukin V.I., Samokhin A.A. (1987) Matematicheskoe modelirovanie fazovykh perekhodov i obrazovanija plazmy pri deistvii lazernogo izluchenija na pogloshhajushhie sredy. In: Samarskii A.A., Kurdjumov S.P., Mazhukin V.I., editors. Nelineinye differencial'nye uravnenija matematicheskoi fiziki. Moskva: Nauka. pp. 191-244.

[60] Samokhin A.A. (1990) Effect of laser radiation on absorbing condensed matter. In: Prokhorov A.M. Series editor. Fedorov V.B. editor. Proceedings of the Institute of General Physics Academy of Sciences of the USSR. New-York: Nova Science Publishers. 13: 203 p.

[61] Anisimov S.I. (1968) Ob isparenii metalla, pogloshhajushhego lazernoe izluchenie. ZhEhTF. 54: 1/339-342.

[62] Khight C.J. (1979) Theoretical Modeling of Rapid Surfact Vaporation with Back Pressure. AIAA J. 17: 5/81-86.

[63] Mazhukin V.I., Prudkovskii P.A., Samokhin A.A. (1993) O gazodinamicheskikh granichnykh uslovijakh na fronte isparenija. Matematicheskoe modelirovanie. 5: 6/3-9.

[64] Mazhukin V.I., Samokhin A.A. (2012) Boundary conditions for gas-dynamical modeling of evaporation processes. Mathematica Montisnigri. XXIV: 8-17.

[65] Lifshic I.M., Kaganov M.I., Tanatarov L.V. (1956) Relaksacija mezhdu ehlektronami i kristallicheskojj reshetkojj. ZhEhTF. 31: 2(8)/232-237.

[66] Lifshic I.M., Kaganov M.I., Tanatarov L.V. (1959) K teorii radiacionnykh izmenenijj v metallakh. Atomnaja ehnergija. 6/391-402.

[67] Martynenko Ju.V., Javlinskii Ju.N. (1983) Okhlazhdenie ehlektronnogo gaza metalla pri vysokojj temperature. DAN SSSR. 270: 1/88-91.

[68] Beer A.C., Chase M.N., Choquard P.F. (1955) Extension of McDougall-Stoner tables of the Fermi-Dirac functions. Helvetica Physica Acta. XXVII: 5/529-542.

[69] Kittel C. (1978) Introduction to Solid State Physics, 6h Edition. New York: J. Wiley & Sons. 789 p.

[70] Landau L.D., Lifshic E.M. (1963) Teoreticheskaja fizika. Kvantovaja mekhanika. Nereljativistskaja teorija. T.III. Moskva: Fizmatgiz. 702 p.

[71] Frenkel' Ja.I. (1958) Vvedenie v teoriju metallov. Moskva: Fizmatlit. 368 p.

[72] Touloukian Y. S., Powell R. W., Ho C. Y. and Klemens P. G (1970) Thermophysical Properties of Matter Volume 1: Thermal Conductivity: Metallic Elements and Alloys, New York, Washington: IFI/Plenum Data Corp.

[73] Ziman J.M (1974) Principles of the theory of solid. Gambridge: University Press. 472 p.

[74] Cheynet, B., J.D. Dubois, M. Milesi (1996) Données thermodynamiques des éléments chimiques. In : Techniques de L'Ingenieur, Materiaux metalliques. No. M153. Forum M64. pp. 1-22

[75] Aluminum: Properties and physical metallurgy (1984) In: John E. Hatch editor. American society for metals. Ohio: Metals Park. 421 p.

[76] Fizicheskie velichiny. Spravochnik (1991) Grigor'eva I.S., Mejjlikhova E.Z. editors. Moskva: Ehnergoatomizdat. 1232 p.

[77] Iida T., Guthrie R.I.L. (1988) The Physical Properties of Liquid Metals. Oxford: Clarendon Press. 288 p.

[78] Pines D (1965) Ehlementarnye vozbuzhdenija v tverdykh telakh. Moskva: Mir. 383 p.

[79] Qiu T.Q., Tien C.L (1994) Femtosecond Laser Heating of Multi-Layer Metals. I. Analysis. Int. J. Heat Mass Transfer. 37: 17/2789-2797.

[80] Ginzburg V.L., Shabanskii V.L (1955) Kineticheskaja temperatura ehlektronov v metallakh i anomal'naja ehlektronnaja ehmissija. Dokl. AN SSSR. 100: 3/445 - 448.

[81] Martynenko Ju.V., Javlinskii Ju.N (1987) Vozbuzhdenie ehlektronov metalla oskolkom delenija. Atomnaja ehnergija. 62: 2/80-83.

[82] Lin Z., Zhigilei L.V., Celli V (2008) Electron-phonon coupling and electron heat capacity of metals under conditions of strong electron-phonon nonequilibrium. Phys. Rev. B. 77: 7/075133-1 - 075133-17.

[83] Mazhukin V.I., Mazhukin A.V., Demin M.M., Shapranov A.V. (2011) Ehffekty neravnovesnosti pri vozdejjstvii impul'snogo lazernogo izluchenija na metally. Opticheskiy zhurnal. 78: 8/29-37.

[84] Mazhukin V.I., Mazhukin A.V., Demin M.M., Shapranov A.V. (2011) Mathematical modeling of short and ultrashort laser action on metals. In: Sudarshan T.S., Beyer Eckhard, Berger Lutz-Michael editors. Surface Modification Technologies (STM 24). Dresden: Fraunhofer Institute for Material and Beam Technology IWS. 24: 201 -208.

[85] Samarskii A.A., Moiseenko B.D (1965) Ehkonomichnaja skhema skvoznogo scheta dlja mnogomernojj zadachi Stefana. ZhVM i MF. 5: 5/816-827.

[86] Meyer G.H (1973) Multidimensional Stefan Problems. SIAM J. Numer. Anal. 10: 3/522-538.

[87] White B.R.E (1983) A modified finite difference scheme for the Stefan problem. Mathem. of Comput. 41: 164/337-347.

[88] Klimontovich Ju.L (1982) Statisticheskaja fizika. Moskva: Nauka. 608 p.

[89] Mazhukin V.I., Smurov I., Flamant G (1995) Overheated Metastable States in Pulsed Laser Action on Ceramics. J. Applied Physics. 78: 2/1259-1270.

[90] Mazhukin V.I., Smurov I., Dupuy C., Jeandel D (1994) Simulation of Laser Induced Melting and Evaporation Processes in Superconducting. J. Numerical Heat Transfer Part A. 26: 587-600.

[91] V.I.Mazhukin, M.G.Lobok, B.N.Chichkov (2009) Modeling of fast phase transitions dynamics in metal target irradiated by pico and femto second pulsed laser. Applied Surface Science. 255/5112-5115.

[92] Mazhukin V.I., Samarskii A.A., Chujjko M.M (1999) Metod dinamicheskojj adaptacii dlja chislennogo reshenija nestacionarnykh mnogomernykh zadach Stefana. Doklady RAN. 368: 3/307 - 310.

[93] Mazhukin A.V., Mazhukin V.I (2007) Dinamicheskaja adaptacija v parabolicheskikh uravnenijakh. Zhurnal vychislitel'nojj matematiki i matematicheskojj fiziki. 47: 11/1911 – 1934. Mazhukin A.V., Mazhukin V.I (2007) Dynamic Adaptation for Parabolic Equations. Computational Mathematics and Mathematical Physics. 47: 11/1833 – 1855.

[94] Breslavsky P.V., Mazhukin V.I (2008) Metod dinamicheskojj adaptacii v zadachakh gazovojj dinamiki s nelinejjnojj teploprovodnost'ju. Zhurnal vychislitel'nojj matematiki i matematicheskojj fiziki. 48: 11/2067–2080. Breslavskii P. V. and Mazhukin V. I (2008) Dynamic Adaptation Method in Gasdynamic Simulations with Nonlinear Heat Conduction. Computational Mathematics and Mathematical Physics. 48: 11/2102–2115.

[95] Koroleva O.N., Mazhukin V.I (2006) Matematicheskoe modelirovanie lazernogo plavlenija i isparenija mnogoslojjnykh materialov. Zhurnal vychislitel'nojj matematiki i matematicheskojj fiziki. 46: 5/910 – 925. Koroleva O.N., Mazhukin V.I (2006)

Mathematical Simulation of Laser Induced Melting and Evaporation of Multilayer Materials. Computational Mathematics and Mathematical Physics. 46: 5/848 – 862.

[96] Mazhukin V.I., Samarskii A.A., Kastel'janos Oplando, Shapranov A.V (1993) Metod dinamicheskojj adaptacii dlja nestacionarnykh zadach's bol'shimi gradientami. Matematicheskoe modelipovanie. 5: 4/32-56.

[97] Samarskii A.A (1989) Teorija raznostnykh skhem. Moskva: Nauka. 616 p.

[98] Samarskii A.A., Popov Ju.P (1992) Raznostnye metody reshenija zadach gazovojj dinamiki. Moskva: Nauka. 421pp.

[99] Mazhukin V.I., Mazhukin A.V., Koroleva O.N (2009) Optical properties of electron Fermi-gas of metals at arbitrary temperature and frequency. Laser Physics. 19: 5/1179 – 1186.

[100] Mazhukin A.V., Koroleva O.N (2010) Raschjot opticheskikh kharakteristik aljuminija. Matematicheskoe modelirovanie. 22: 5/15-28.

Chapter 9

Direct Writing in Polymers with Femtosecond Laser Pulses: Physics and Applications

Kallepalli Lakshmi Narayana Deepak, Venugopal Rao Soma
and Narayana Rao Desai

Additional information is available at the end of the chapter

1. Introduction

Nonlinear optical phenomena in the optical spectral range followed by the invention of laser in early 1960's directed the generation of optical pulses using Q switching and mode locking techniques. Ultrafast lasers with extremely short pulse duration (<100 fs) opened a new avenue towards fabrication of integrated photonic and signal processing devices in a variety of transparent materials. A new approach for the local modification of transparent materials through nonlinear optical processes has been investigated due to extraordinarily high peak intensities of short pulses. A variety of materials including metals, dielectrics, polymers, and semiconductors have been successfully processed by the use of fs pulses [1-14]. Bulk refractive index change in transparent materials is found to be useful in applications of waveguides. Various applications resulting from fs laser writing of different materials, especially in polymers, have been successfully demonstrated in the fields of micro-fluidics, bio-photonics, and photonics etc. [15-32]. The minimal damage arising from the generation of stress waves, thermal conduction, or melting has proved to be one of the main responsible mechanisms for various applications demonstrated using fs laser micromachining. In the present chapter we discuss the formation of free radicals and defects which are responsible for emission in polymer systems. The impact of fs lasers pulses causing minimal damage can be utilized to fabricate emissive micro-craters, especially in polymers. These emissive micro-craters find prospective applications in memory based devices.

2. Experimental

Herein we present detailed micro-structure fabrication procedures and spectroscopic investigations of those structures in four different polymers (a) Poly Methyl Methacrylate

(PMMA) (b) Poly Di Methyl Siloxane (PDMS) (c) Polystyrene (PS) and (d) Poly Vinyl Alcohol (PVA). The spectroscopic investigations were carried out in both bulk and thin films of polymers. We used a Ti: Sapphire laser delivering 100 fs pulses at 800 nm with an energy of 1 mJ per pulse at 1 kHz repetition rate. The input energy was varied using half wave plate and polarizer combination. Three Newport stages with 15 nm resolution were arranged three dimensionally and the translation was controlled using computer controlled program. We used 40X and 20X microscope objectives with 0.65 and 0.4 numerical apertures, respectively, to focus the laser beam in our experiments. CCD camera was used while Z stage of the micro-fabrication set up was adjusted for fabricating structures either on the surface or inside the bulk. Figure 1 shows schematic diagram of the micro-fabrication setup. As shown in figure 1 the laser pulses were passed through half wave plate (HWP) and polarizer (BP) so that energy could be varied using this combination. Apertures were introduced in middle to align the beam. M1-M3 are the mirrors used to make the beam incident vertically onto the three dimensionally arranged Newport stages. We fabricated several micro-structures, micro-craters, surface and subsurface diffraction gratings, surface grids, microfluidic channels and arbitrary shapes in these polymers. Since these polymers are transparent and have large band gap, we calculated the Keldysh parameter for these polymers to assess the dominant ionization mechanism. Tunneling ionization was found to be the main responsible mechanism for all the investigated polymers. The spectroscopic investigations were carried out in both bulk and thin films of polymers. Bulk PMMA and PS were purchased from Goodfellow, USA and UK. PVA thin films were prepared by dissolving 8.56 grams of PVA beads in 100 ml of water and thin films were prepared using spin coating technique. Thin films of PS are made by preparing the PS solution first. Solution of PS was prepared by mixing 1 gram of polystyrene beads (ACROS) in 8 ml toluene and stirred for 48 hours for complete miscibility. We prepared thin films of PS on a glass plate by spin coating the solution of PS. All these polymers were cut into 1 cm × 1 cm square pieces, polished using different grades of polishing sheets and sonicated in distilled water before micro fabrication experiments were carried out.

The energy of an 800 nm photon corresponds to 1.55 eV while the optical band gap of pure PMMA being 4.58 eV implies that the nonlinear process involving at least three photons is responsible for structural modification at the focal volume [33]. Schaffer et al. [34] have shown that there are three possible mechanisms viz. tunneling, intermediate, and multi-photon ionizations that take place when transparent material interacts with femtosecond pulses. The Keldysh parameter which tells us the mechanism that is dominant is defined as $\gamma = (\omega/e) (m \times c \times n \times \varepsilon_0 \times E_g/I)$ where 'ω' is the laser frequency, 'I' is the laser peak intensity at the focus, 'm' and 'e' are the reduced mass and charge of the electron, respectively, and 'c' the velocity of light, n is the refractive index of the material, E_g is the band-gap of the material and ε_0 is the permittivity of free space. For our studies that Keldysh parameter was <1.5 illustrating tunneling as responsible mechanism for structures written using 40X and 20X microscopic objectives. Figure 2 shows the plot obtained for Keldysh parameter with different energies ranging from 150 µJ to 30 nJ (equivalent to 83 to 0.017 PW/cm²). Results obtained with different polymers using fs laser direct writing are explained in detail in references 35-41.

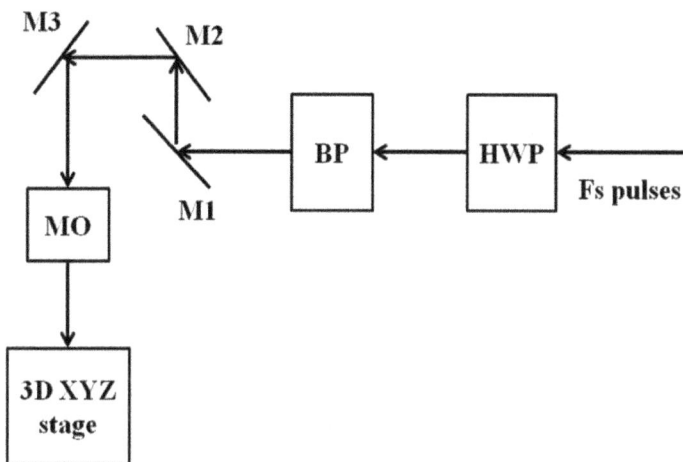

Figure 1. Experimental setup for microfabrication

Our initial studies suggested an increase in structure width with number of scans, energy, and focusing conditions. Figure 3(a) shows some of the Field Emission Scanning Electron Microscope (FESEM) images of the fabricated microstructures on PVA thin film (60-80 μm thickness). We clearly observed the formation of a trough in the central portion of each structure. It is evident as central portion of the incident Gaussian pulse has more intensity (thereby affecting the polymer more) resulting in void formation. This phenomenon was observed for energies ranging from 100 μJ to 10 μJ. Figure 3(b) shows evidently the formation of trough in the central portion of the microstructure fabricated at 100 μJ energy with 1 mm/s speed. For structures fabricated at low energies (typically less than 10 μJ) formation of trough was not observed. Figure 3(c) shows a plot of width of microstructure as well as trough obtained at different energies for different sets. Formation of micro-craters was seen at 1 μJ energy and with 1 mm/s speed. Structures fabricated with 635 and 564 nJ obviously demonstrated formation of micro-craters. Figure 4(a) shows the structure fabricated at 1 μJ energy, while 4(b) and 4(c) show the FESEM images of the microstructures fabricated at 635 and 564 nJ energies with 1 mm/s speed. In our experiments with different polymers we observed that the effect of fs pulses on these polymers, leading to the formation of defects resulting in emission, formation of paramagnetic centers, and broadening of Raman vibrational modes in the fs laser modified regions. Figure 5 shows the part of shape 8 fabricated in PMMA. Since the image was big only a part of 8 shape, which is curved, is shown. We captured the emission from the fs laser modified region using a confocal microscope. We could clearly observe pseudo-green color from the fs laser modified regions which is an indication of emission. Similar results were obtained in case of other polymers also, details of which are summarized in references 35-41. By combining the formation of micro-craters (which took place at high scan speed >3 mm/s and low energies of ~nJ) and the emission coming from the modified regions we establish the possibility of using the fluorescent micro-craters towards memory based applications.

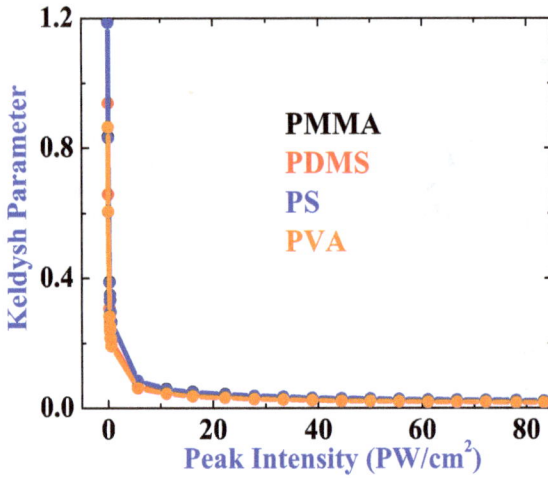

Figure 2. Keldysh parameter of investigated polymers with different peak intensities.

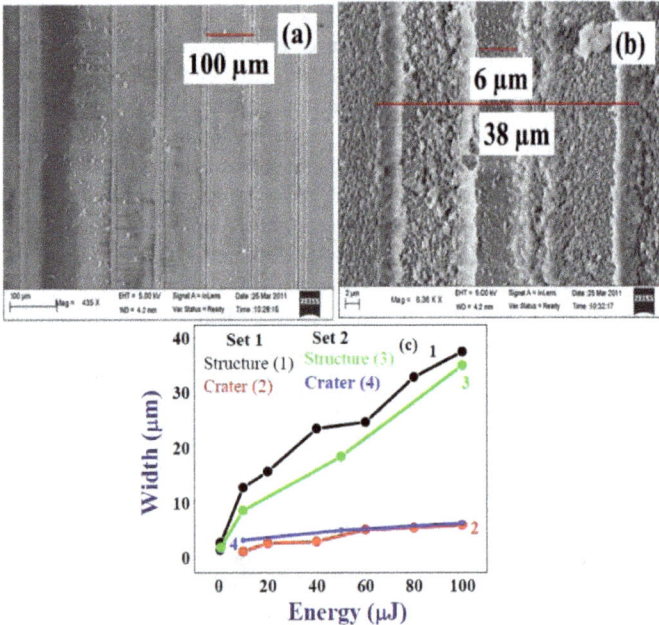

Figure 3. (a) FESEM images of structures fabricated. Structures from right to left were fabricated with energies from 100-10 μJ in steps of 20 μJ. (b) Microstructure fabricated at 100 μJ energy, 1 mm/s speed. (c) Plot of structure and trough widths with energy.

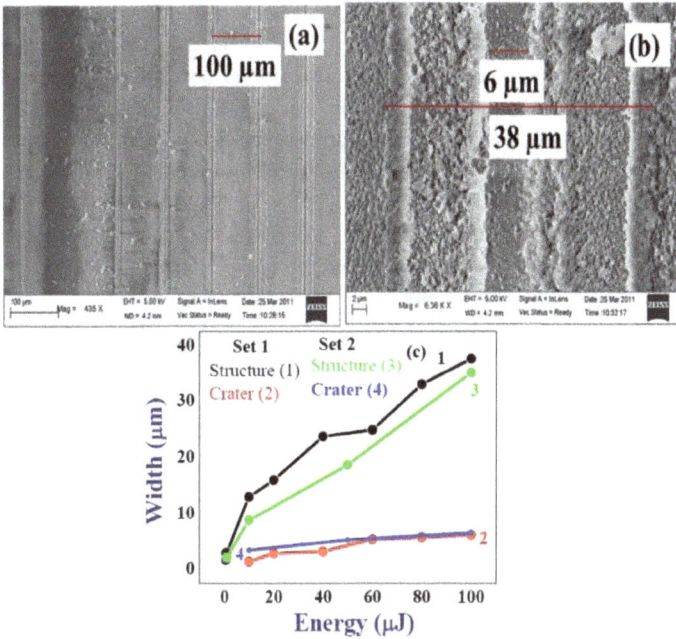

Figure 4. FESEM images microstructure fabricated at 1 mm/s speed, with energy of (a) 1 µJ, (b) 635 nJ, and (c) 564 nJ.

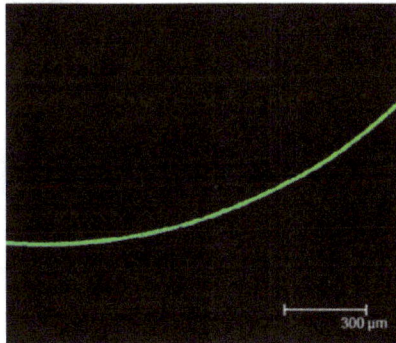

Figure 5. Part of the shape 8 fabricated in PMMA. Pseudo green color shows emission from the modified region excited at 488 nm. Scale bar is 300 µm.

Structures were also fabricated at different energies (1 µJ, 635 nJ, 168 nJ, and 33 nJ) and different scanning speeds to study the formation of micro-craters in PS. Figure 6(a) shows the confocal microscope image of a buried single microstructure in PS fabricated at 1 µJ energy and 1 mm/s speed. For the same energy we observed the appearance of craters when the scanning speed was increased to 2 mm/s. This is depicted clearly in figure 6(b), where the formation of micro-craters was just onset. The process of formation of micro-craters is evident from the edges of figure 6(b). Figures 6(c) and 6(d) illustrate the confocal microscope images of the craters formed at still higher scanning speeds of 3 and 4 mm/s, respectively. From these data we confirmed that by increasing the scanning speed one can reduce the pulse to pulse overlap and the micro-craters can be obtained similar to observations of periodic refractive index modulation observed earlier in glasses. The same trend was observed even at other energies also. However, we found that the formation of craters at higher scanning speeds and lower energies exhibited better profiles. This is obvious because the intensity of heat waves generated at each position where the pulse impinges depends on the pulse energy.

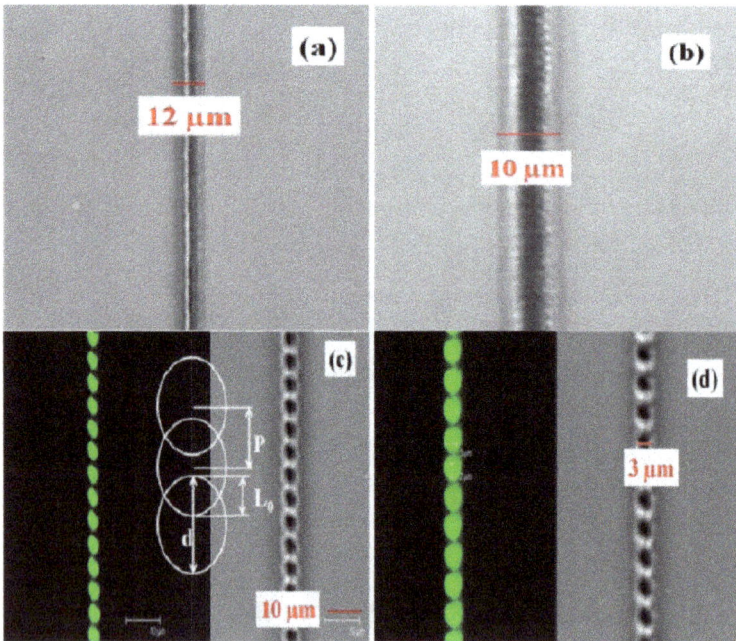

Figure 6. (a) Confocal microscope image of a buried single micro-structure in PS fabricated at 1 µJ energy and 1 mm/s speed. Structure width is ~12 µm. (b) Beginning of formation of micro- craters at the same energy when scanning speed is increased to 2 mm/s. Structure width 10 µm. (c) Scanning speed increased to 3 mm/s and the inset shows model used to fabricate micro- craters (d) Confocal microscope image of the craters formed at still higher scanning speed of 4 mm/s. Pseudo green color indicates emission from the modified micro-craters when excited at 488 nm wavelength. Crater size is 3 µm.

S. No.	Energy (nJ)	Speed (mm/s)	Observation (Line/Crater)	Structure/Crater Size (μm)
1	1000	1	Line	12
		1.5	Line	10
		2	Line	10
		2.5	Crater formation started	6
		3	Crater	6
		3.5	Crater	4
		4	Crater	3
		4.5	Crater	4
		5	Crater	3
2	635	1	Line	14
		1.5	Line	12
		2	Crater formation started	10
		2.5	Crater	3
		3	Crater	3
		3.5	Crater	3
		4	Crater	3
		4.5	Crater	3
		5	Crater	3
3	168	1	Line	4
		1.5	Crater formation started	3
		2	Crater	2
		2.5	Crater	2
		3	Crater	2
		3.5	Crater	2
		4	Crater	2
		4.5	Crater	2
		5	Crater	2
4	33	1	Line	3
		1.5	Crater formation started	3
		2	Crater	2
		2.5	Crater	2
		3	Crater	2
		3.5	Crater	2
		4	Crater	2
		4.5	Crater	2
		5	Crater	2

Table 1. Details of micro crater structures fabricated at different energies and scanning speeds.

Therefore, at higher energies these intense heat waves travel and overlap with the waves generated at other positions. This leads to the formation of a line instead of a crater. This aspect is highlighted in table 1. As energy is reduced, micro-crater formation occurred at lower scan speeds, indicating the fact that the effect of pulse to pulse overlap controlled by scan speed and the phenomena of generation/propagation of heat waves at each position are important factors in the fabrication of micro-craters. Inset of figure 6(c) shows a model that represents the formation of micro-craters. The pitch 'p' which is the distance from center to center can be varied by setting the scanning speed according to the relation **p=s/f** where 's' is scanning speed and 'f' is laser repetition frequency, which is 1 kHz in our case. L_0 is the overlap region between two consecutive focused spots which is (d-p) where d is the diameter of the spot. The spot size is calculated using the relation D=1.22 λ/NA where D is the diameter of the focused spot, λ is the wavelength and NA is the numerical aperture of the microscope objective used. We had used a 40X microscope objective with NA of 0.65 in our experiments. The estimated spot size was ~ 1.55 μm. In our experiments on micro-crater fabrication we observed clear micro-crater formation from 3 mm/s speed onward with different energies as tabulated in table 1. Hence the minimum pitch (**p**) can be taken as 3 μm.

As the radius of the focused spot was ~0.75 μm, the micro craters start appearing for speeds beyond 3 mm/s. However, we could obtain minimum crater size of ~2 μm in our experiments owing to the size of the focused spot and continuous translation. The fabrication of these micro craters is useful in the areas of memory storage devices and photonic crystals.

For better understanding about the emission originating from modified regions, we fabricated microstructures over a large area to record the emission and excitation spectra using conventional fluorescence spectrometer. To accomplish this, an array of closely spaced lines were fabricated in PMMA (energy 10 μJ, speed 1mm/s, period 30 μm), PDMS (50 μJ, 1mm/s, 10 μm), PS (1 μJ, 0.5 mm/s, 30 μm), and PVA (10 μJ, 1mm/s, 20 $\circledast$m) using a 40X objective. All the four polymers are transparent to visible light before fs laser irradiation and hence do not show any emission when excited with visible light. However, due to multi-photon absorption overcoming the large band gaps of these polymers, they get modified and optical centers are produced that exhibit emission when excited at different excitation wavelengths. We observed this kind of emission when the fs laser fabricated channels were irradiated with visible wavelengths such as 458 nm, 488 nm, 514 nm wavelengths. The observation of emission and change in emission peak with the excitation wavelength were observed and could be interpreted in the context of formation of myriad optical centers or defects. The increase in emission intensity with irradiation dose in bulk and thin films of polymers are due to the formation of more number of optical centers with irradiation dose.

Nurmukhametov et al. [42] have exposed PS films and solutions with UV laser light beam (λ = 248 nm) and observed changes in absorption and luminescent properties. They observed formation of optical centers with absorption band from 280 nm to 460 nm with fluorescence band from 330 nm to 520 nm. They established different optical centers being responsible for such a behavior and hypothesized their results more close to the spectrum of diphenylbutadiene (DPBD). They figured out three optical centers responsible for the behavior of emission namely trans-stilbene, DPBD and diphenylhexatriene (DPHT). The absorption and emission spectra due to the fs laser irradiation in the present study match nearly with their reported work indicating the creation of similar optical centers. Figures 7(a) – 7(c) show the emission peaks are different for different excitation wavelengths. This emission was recorded using a confocal microscope. We feel that the modification of PS was due to 800 nm (infrared) laser through multi-photon absorption process. Consequently, fs laser irradiation results in bond scission followed by the formation of large number of defects that act as optical centers for emission. This resulted in shift in emission peak with excitation wavelength. The observed spectral peaks are mentioned in table 2. Observed spectral peaks from fluorescence spectrometer and the confocal microscope are also compared in the table.

S. No.	Excitation Wavelength (nm)	Emission peak (nm) observed in confocal microscope	Emission peak (nm) observed in fluorescence spectrometer
1	458	509, 544	505,536
2	488	535	550
3	514	565	563
4	400	Excitation source was not available	454, 462, 476

Table 2. Comparison of emission data obtained in irradiated PS (PSG1).

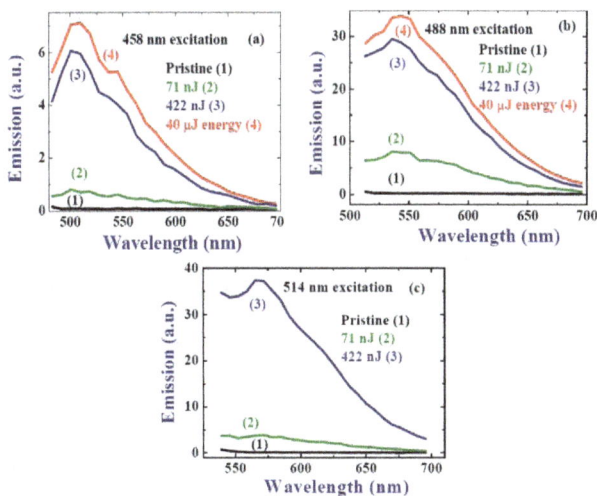

Figure 7. (a) Plots of emission when structures fabricated at 40 µJ, 422 nJ, 63 nJ and pristine region of PS excited at 458 nm; (b) Plots of emission when structures fabricated at 40 µJ, 422 nJ, 63 nJ and pristine region of PS excited at 488 nm; (c) Plots of emission when structures fabricated at 422 nJ, 63 nJ and pristine region of PS excited at 514 nm.

The effect of change in emission peak with excitation wavelength was observed in all polymers and is attributed to the Red Edge Effect (REE). The change in emission peak with excitation wavelength in case of PS is shown in figure 8(a). The excitation spectra collected at different monitoring wavelengths looked alike which is depicted in figure 8(b). Inset of figure 8(b) shows the transmission spectra of un-irradiated and irradiated PS. Varieties of fluoropores in different media with frozen or relatively slow structural dynamics from vitrified and highly viscous solutions to polymer matrices have shown similar effects. These phenomena did not account for concepts of independence of emission energy on excitation energy within the absorption band (Vavilov's law) and the occurrence of emission irrespective of excitation band, always from the lowest electronic and vibrational state of same multiplicity called Kasha's Rule [43-46]. These effects originate not from the violation of fundamental principles but from their operation in specific conditions when the ensemble of excited molecules is distributed in interaction energy with molecules in their surroundings. In a condensed medium, this distribution always exists at the time of excitation, but its display in a variety of spectroscopic phenomena depends on how fast the transitions are between the species forming this ensemble of states. Thus, REE is a site-selective effect which allows probing of the dynamics of redistribution of fluoropores between different environments.

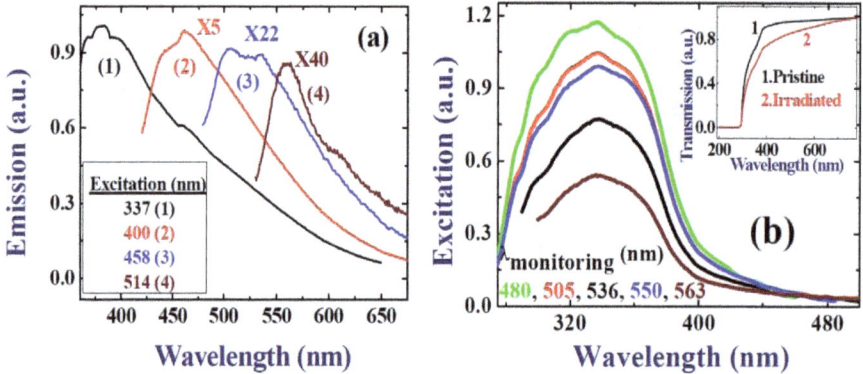

Figure 8. (a) Emission from PSG1 grating excited at different wavelengths. Emissions with longer wavelength excitations are enhanced for clarity (b) Excitation spectra of irradiated PS at different monitoring wavelengths. Inset of (b) shows the transmission spectra of unirradiated and irradiated PS.

The absorption characteristics of these optical centers thus formed showed maximum absorption at same wavelength. As the defect centers are distributed over wide range of energies, the emission spectrum too gets distributed over the entire fluorescence spectrum contribution coming from all types of defects. This results in the spectra being the same in the excitation spectra recorded by monitoring different emission wavelengths. However the emission, when excited at long wavelength side, due to localization of energy we observe different emission spectra. Figure 9(a) shows the plot of excitation wavelength with the observed emission wavelength for all polymers which is an indication of REE effect. Since these polymers contain active functional groups such as aldehyde, ketone etc., the absorption maximum within these functional groups can also lead to the observed emission. In the case of PMMA, the maximum excitation near 370 nm can be ascribed to n $\rightarrow\pi^*$ transition. In case of PS due to the presence of ring, it could be due to $\pi\rightarrow\pi^*$ transition. After these functional groups are excited they can get de-excited to any of the electronic states by emission process. Figure 9(b) shows a schematic diagram for the functional groups within these polymers undergoing different transitions. We also investigated the role of possible vibrational levels in emission spectra we recorded. For this, we excited all irradiated polymers from 250 nm-560 nm in the range of 20 nm and collected emission spectra. Interestingly, some of the emission peaks coincided with Raman peaks of these polymers. Plots 10 (a)-(b) show the emission spectra recorded for fs laser irradiated PMMA at different excitation wavelengths. The energy difference between the excitation and the emission peak is calculated in the wave numbers using the relation $\Delta E/hc$ (cm^{-1}) = $1/\lambda_{ext}$ – $1/\lambda_{em}$, where λ_{ext} and λ_{em} are excitation and emission wavelengths respectively. For PMMA, the emission peaks due to excitation wavelengths at 458, 488, 530, and 560 nm nearly match the Raman mode of characteristic peak at 1736 cm^{-1} of γ (C=O) of (C-COO) mode [47]. Excitations at 380, and 400 nm nearly match with Raman mode of 3454 cm^{-1} which is $2\gamma_2$ overtone of 1730 cm^{-1}. Emission peak at 543 nm matched with another Raman peak of combination band involving γ (C = C) and γ (C-COO) of PMMA. The spectra therefore contain the peaks due to transition from the excited state to the vibrational states. The shift in the emission peaks with excitation wavelength is shown in figure 10 (b). This shift in the peak of the emission with excitation could be due to the excitation of different localized states created during laser irradiation. Table 3 shows excitation, emission peaks and the calculated energy difference between the excitation and the emission peaks.

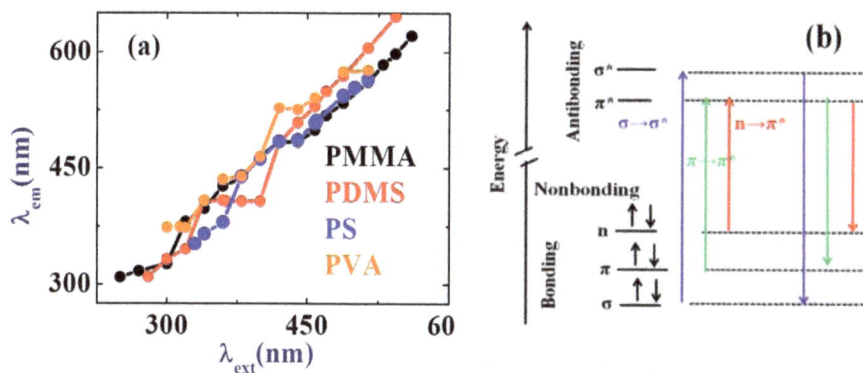

Figure 9. (a) REE in investigated polymers (b) Energy level diagram.

Different research groups have worked on electron spin resonance (ESR) of polymers [48-50]. However, reports on ESR analysis of fs laser irradiated polymers are sparse as the field has gained momentum recently and many theoretical and experimental results need to be explored further. In our endeavor towards understanding ESR of fs irradiated polymers, we fabricated two dimensional grids to increase effective area of irradiation for ESR analysis. We observed peroxide type free radicals when these polymers were treated with fs laser. Pure polymers such as PMMA are not paramagnetic substances and hence do not contain any paramagnetic centers. So, there were no peaks observed in ESR spectrum. When polymers PMMA, PDMS, and PS are treated with fs laser, they showed ESR signal which is an indication of existence of peroxide type free radicals. Figure 11 shows ESR signal in PS. In case of PS, the radicals are alkoxy radicals as reported in literature [51]. Further, confocal micro-Raman studies were carried out for fabricated structures in the channels. Formation of defects such as optical centers and free radicals led to broadening of Raman peaks and reduction in Raman intensity due to high intense shock waves formed at the center of the Gaussian pulse. In order to study the local effect, a microstructure was fabricated on a PVA thin film at two different energies (635 nJ and 10 μJ) with speed of 1 mm/s. Figure 12 shows the Raman spectra collected in the middle and end regions of the structures fabricated. Raman spectra recorded for higher energies showed the broadening of Raman modes along with decrease in Raman intensity compared with structures fabricated at low energies and pristine regions of the polymer from which we could conclude the formation of defects. We observed similar effects even in other polymers also.

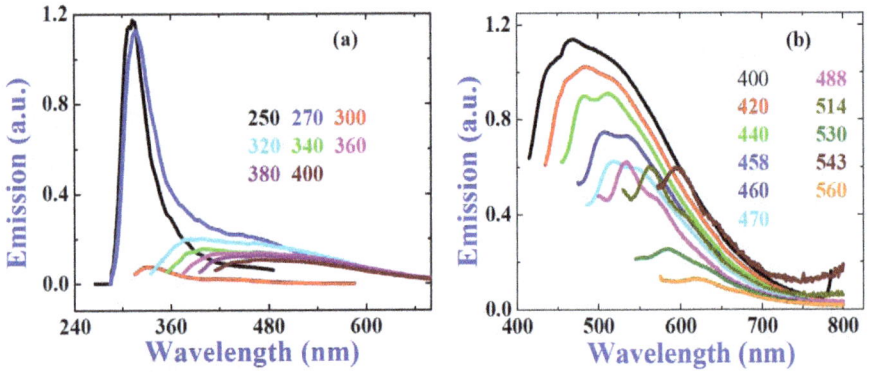

Figure 10. Plot of emission spectra of (a) PMMA at excitation wavelengths from 250-400 nm (b) PMMA at excitation wavelengths from 400-560 nm

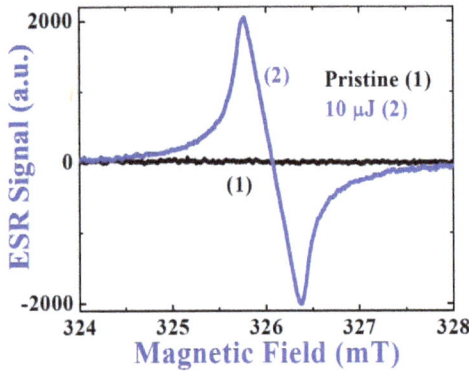

Figure 11. ESR signal observed in irradiated PS.

S. No.	λ_{ext} (nm)	λ_{em} (nm)	$\Delta E/hc$ (cm^{-1}) = λ_{ext}^{-1} - λ_{em}^{-1}
1	250	309	7638
		313	8051
2	270	317	5491
3	300	326	2658
4	320	381	5003
		398	6214
		461	9558
5	340	398	4286
		464	7860
6	360	427	4359
		465	6272
7	380	437	3432
		467	4903
8	400	464	3448
9	420	484	3148
10	440	483	2023
		511	3158
11	458	499	1794
		529	2930
12	460	507	2015
		535	3048
13	470	518	1972
		551	3128
14	488	534	1765
		568	2886
15	514	561	1630
16	530	584	1745
17	543	598	1694
18	560	621	1754

Table 3. Excitation, emission, and Raman shift observed in fs laser irradiated PMMA.

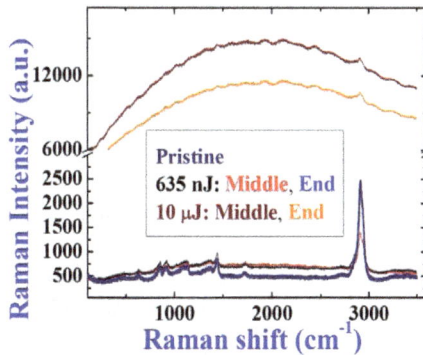

Figure 12. Confocal Raman spectra for structures fabricated on PVA thin film.

3. Conclusions

We presented our results on the physical and spectroscopic investigations of fs laser irradiated polymers. The unusual behavior of emission observed in case of these polymers is attributed to the red edge effect. The transitions involved with in the functional groups such as n→π*, and π→π* are responsible for the emission observed. The role of coincidence of Raman signals with emission observed is illustrated. The presence of paramagnetic centers is confirmed through ESR studies. Both the defects which are responsible for emission and ESR signal are further confirmed through confocal micro Raman studies by observing the broadening of Raman modes.

Author details

Kallepalli Lakshmi Narayana Deepak
School of Physics, University of Hyderabad, Gachibowli, Hyderabad, India
Laboratoire LP3, UMR 6182, CNRS-Universite Aix-Marseille, Pole Scientifique et Technologique de Luminy, Case 917, Marseille, France

Soma Venugopal Rao
Advanced Center for Research in High Energy Materials (ACRHEM), University of Hyderabad, Gachibowli, Hyderabad, India

Desai Narayana Rao*
School of Physics, University of Hyderabad, Gachibowli, Hyderabad, India

Acknowledgement

We are extremely grateful to DRDO, India for all the financial support during the tenure of this project. K. L. N. Deepak acknowledges CSIR for financial support to carry out his doctoral work. DNR and SVR acknowledge Department of Science and Technology (DST), India for financial support through a project **SR/S2/LOP-11/2005**.

4. References

[1] Gattass R R, Mazur E (2008) Femtosecond laser micromachining in transparent materials. Nat. Phot. 2: 219-225.
[2] Juodkazis S, Mizeikis V, Misawa H (2009) Three-dimensional Microfabrication of materials by femtosecond lasers for photonics applications. J. Appl. Phys. 106: 051101.
[3] Nolte S, Will M, Burghoff J, Tuennermann A (2003) Femtosecond waveguide writing: a new avenue to three-dimensional integrated optics. Appl. Phys. A 77: 109-111.

---

* Corresponding Author

[4]   Ams M, Marshall G D, Dekker P, Dubov, Mezentsev V K, Bennion I, Withford M J (2008) Investigation of ultrafast laser–photonic material interactions: challenges for directly written glass photonics. IEEE J. Sel. Top. Quant. Electron., 14(5): 1370-1381.

[5]   Qiu J, Miura K, Hirao K (2008) Femtosecond laser-induced micro features in glasses and their applications. Journal of Non-Crystalline Solids. 354: 1100–1111.

[6]   Della Valle G, Osellame R, Laporta P (2009) Micromachining of photonic devices by femtosecond laser pulses. J. Opt. A: Pure Appl. Opt. 11: 013001.

[7]   Qiu J, Miura K, Hirao K (2008) Femtosecond laser-induced micro features in glasses and their applications," Journal of Non-Crystalline Solids. 354, 1100–1111, 2008.

[8]   Della Valle G, Osellame R, Laporta P (2009) Micromachining of photonic devices by femtosecond laser pulses. J. Opt. A: Pure Appl. Opt. 11: 013001-.

[9]   Taccheo S, Della Valle G, Osellame R, Cerullo G, Chiodo N, Laporta P, Svelto O, Killi A, Morgner U, Lederer M, Kopf D (2004) Er : Yb-doped waveguide laser fabricated by femtosecond laser pulses. Opt. Lett. 29: 2626-2628.

[10] Florea C, Winick K A (2003) Fabrication and characterization of photonic devices directly written in glass using femtosecond laser pulses. Journal of Lightwave Technology 21: 246-253.

[11] Minoshima K, Kowalevicz A M, Ippen E P, Fujimoto J G (2002) Fabrication of coupled mode photonic devices in glass by nonlinear femtosecond laser materials processing. Optics Express 10: 645-652.

[12] Streltsov A M, Borrelli N F (2001) Fabrication and analysis of a directional coupler written in glass by nanojoule femtosecond laser pulses. Optics Letters 26: 42-43.

[13] Davis K M, Miura K, Sugimoto N, Hirao K (1996) Writing Waveguides in Glass with a Femtosecond Laser. Optics Letters 21: 1729-1731.

[14] Glezer E N, Milosavljevic M, Huang L, Finlay R J, Her T H, Callan J P, Mazur E (1996) Three-dimensional optical storage inside transparent materials. Opt. Lett. 21(24): 2023-2025.

[15] Watanabe W, Sowa S, Tamaki T, Itoh K, Nishii J (2006) Three-Dimensional Waveguides Fabricated in Poly(methyl methacrylate) by a Femtosecond Laser. Jap. J. Appl. Phys., Part 2, 45: L675-L767.

[16] Zoubir A, Lopez C, Richardson M, Richardson K (2004) Femtosecond laser fabrication of tubular waveguides in poly(methyl methacrylate). Opt. Lett. 29(16): 1840-1842.

[17] Sowa S, Tamaki T, Itoh K, Nishii J, Watanabe W (2006) Three-dimensional waveguides fabricated in poly(methyl methacrylate) by a femtosecond laser. Jap. J. Appl. Phys., Part 2: Letters 45: 29-32, L765-L767.

[18] Wang K, Klimov D, Kolber Z (2007) Waveguide fabrication in PMMA using a modified cavity femtosecond oscillator. Proc. SPIE 6766: 67660Q.

[19] Watanabe W, Sowa S, Tamaki T, Itoh K, Nishii J(2006) Three-Dimensional Waveguides Fabricated in Poly(methyl methacrylate) by a Femtosecond Laser. Jap. J. Appl. Phys., Part 2, 45: L675-L767.

[20] Zoubir A, Lopez C, Richardson M, Richardson K (2004) Femtosecond laser fabrication of tubular waveguides in poly (methyl methacrylate). Opt. Lett. 29(16): 1840-1842.

[21] Sowa S, Tamaki T, Itoh K, Nishii J, Watanabe W (2006) Three-dimensional waveguides fabricated in poly(methyl methacrylate) by a femtosecond laser. Jap. J. Appl. Phys., Part 2: Letters 45: 29-32, L765-L767.

[22] Cumpston B H, Ananthavel S P, Barlow S, Dyer D L, Ehrlich J E, Erskine L L, Heikal A A, Kuebler S M, Lee I Y S, McCord-Maughon D, Qin J Q, Rockel H, Rumi M, Wu X L, Marder S R, Perry J W (1999) Two-photon polymerization initiators for three dimensional optical data storage and microfabrication. Nature 398: 51-54.

[23] Nie Z, Lee H, Yoo H, Lee Y, Kim Y, Lim K S, Lee M (2009) Multilayered optical bit memory with a high signal-to-noise ratio in fluorescent polymethylmethacrylate. Appl. Phys. Lett. 94: 111912.

[24] Farson D F, Choi H W, Lu C, Lee L J (2006) Femtosecond laser bulk micromachining of microfluidic channels in poly (methyl methacrylate). J. Laser Appl. 18(3): 210-215.

[25] Day D, Gu M (2005) Microchannel fabrication in PMMA based on localized heating by nanojoule high repetition rate femtosecond pulses. Opt. Exp. 13(16): 5939-5946.

[26] Ding L, Blackwell R I, Kunzler J F, Knox W H (2008) Femtosecond laser micromachining of waveguides in silicone-based hydrogel polymers. Appl. Opt. 47(17): 3100-3108.

[27] Si J, Meng Z, Kanehira S, Qiu J, Hua B, Hirao K (2004) Multiphoton-induced periodic microstructures inside bulk azodye-doped polymers by multibeam laser interference. Chem. Phys. Lett. 399: 276-279.

[28] Si J H, Qiu J R, Zhai J F, Shen Y Q, Hirao K (2002) Photoinduced permanent gratings inside bulk azodye-doped polymers by the coherent field of a femtosecond laser. Appl. Phys. Lett. 80(3): 359-361.

[29] Kim T N, Campbell K, Groisman A, Kleinfeld D, Schaffer C B (2005) Femtosecond laser-drilled capillary integrated into a microfluidic device. Appl. Phys. Lett. 86: 201106.

[30] Mendonca C R, Cerami L R, Shih, Tilghman R W, Baldacchini T, Mazur E (2008) Femtosecond laser waveguide micromachining of PMMA films with azoaromatic chromophores. Opt. Exp. 16(1): 200-206.

[31] Zhou G, Ventura M J, Vanner M R, Gu M (2004) Use of ultrafast-laser-driven microexplosion for fabricating three-dimensional void-based diamond-lattice photonic crystals in a solid polymer material. Opt. Lett. 29: 2240-2242.

[32] Nolte S, Will M, Burghoff J, Tuennermann A (2003) Femtosecond waveguide writing: a new avenue to three-dimensional integrated optics. Appl. Phys. A 77: 109-111.

[33] Baum A, Scully P J, Perrie W, Jones D, Issac R, Jaroszynski D A (2008) Pulse-duration dependency of femtosecond laser refractive index modification in poly (methyl methacrylate). Opt. Lett. 33(7): 651-653.

[34] Schaffer C B, Brodeur A, Mazur E (2001) Laser-induced breakdown and damage in bulk transparent materials induced by tightly focused femtosecond laser pulses. Meas. Sci. Technol. 12: 1784–1794.

[35] Deepak K L N, Narayana Rao D, Venugopal Rao S (2009) Fabrication and Optical Characterization of microstructures in PMMA and PDMS using femtosecond pulses for photonic and micro fluidic applications. Appl. Opt. 49 (13): 2475.

[36] Deepak K L N, Venugopal Rao S, Narayana Rao D (2010) Femtosecond laser-fabricated microstructures in bulk poly(methylmethacrylate) and poly(dimethylsiloxane) at 800 nm towards lab-on-a-chip applications. Pramana 75(6): 1221-1232.

[37] Deepak K L N, Kuladeep R, Venugopal Rao S, Narayana Rao D (2011) Luminescent microstructures in bulk and thin films of PMMA, PDMS, PVA, and PS fabricated using femtosecond direct writing technique. Chem. Phys. Lett. 503: 57-60.

[38] Deepak K L N, Kuladeep R, Narayana Rao D (2011) Emission properties of femtosecond (fs) laser fabricated microstructures in Polystyrene (PS). Opt. Commun. 284: 3070-3073.

[39] Deepak K L N, Kuladeep R, Praveen Kumar V, Venugopal Rao S, Narayana Rao D (2011) Spectroscopic investigations of femtosecond laser irradiated Polystyrene and fabrication of microstructures. Opt. Commun. 284: 3074-3078.

[40] Deepak K L N, Kuladeep R, Venugopal Rao S, Narayana Rao D (2012) Studies on Defect Formation in Femtosecond Laser Irradiated PMMA and PDMS," Radiation Effects and Defects in Solids 67: 88-101.

[41] Deepak K L N, Venugopal Rao S, Narayana Rao D (2011) Effect of heat treatment to efficient buried diffraction gratings in Polystyrene. Appl. Surf. Sci. 257: 9299-9305.

[42] Nurmukhametov R N, Volkova L V, Kabanov S P (2006) Fluorescence and absorption of polystyrene exposed to UV laser radiation. J. Appl. Spectr. 73: 55-60.

[43] Galley W C, Purkey R M, Role of Heterogeneity of the Solvation Site in Electronic Spectra in Solution, Proc. Natl Acad. Sci. USA 1970; 67, 1116-1121.

[44] Rubinov A N, Tomin VI (1970) Bathochromic luminescence of organic dyes, Opt. Spektr. 29: 1082-1086.

[45] Terenin A N, *Fotonika molekul krasitelei,* Photonics of Dye Molecules (1967) Nauka:Leningrad, 616.

[46] Birks JB. Photophysics of Aromatic molecules. Wiley-Inter-science: London 1970.

[47] Thomas K J, Sheeba M, Nampoori V P N, Vallabhan C P G, and Radhakrishnan P (2008) Raman spectra of polymethyl methacrylate optical fibres excited by a 532 nm diode pumped solid state laser.J. Opt. A: Pure Appl. Opt. 10: 055303-055307.

[48] Velter-Stefanescu M, Duliua O G, Preda N (2005) On the relaxation mechanisms of some radiation induced free radicals in polymers. J. Optoelectron. Advanced Mat. 7(2): 985–989.

[49] Kaptan H Y, Tatar L (1997) An Electron Spin Resonance study of mechanical fracture of poly (methyl methacrylate). J. Appl. Polym. Sci. 65: 1161-1167.

[50] Abdelaziz M (2008) Electron spin resonance and optical studies of poly (methylmethacrylate) doped with CuCl2. J. Appl. Polym. Sci. 108: 1013-1020.

[51] Ohnishi S, Tanei T, Nitta I (1962) ESR Study of Free Radicals Produced by Irradiation in Benzene and Its Derivatives. J. Chem. Phys. 37: 2402-2407.

Chapter 10

Holographic Fabrication of Periodic Microstructures by Interfered Femtosecond Laser Pulses

Zhongyi Guo, Lingling Ran, Yanhua Han, Shiliang Qu and Shutian Liu

Additional information is available at the end of the chapter

1. Introduction

In recent years, with the developments of the laser technology, femtosecond laser technology is becoming more and more consummate as a novel technology. Femtosecond laser pulse is also becoming into a powerful tool for microfabrication and micro-machining of various multi-functional structures in dielectric materials through multi-photon absorption because of its high-quality and damage-free processing. Up to now, many high-quality material processing techniques have been achieved by using femtosecond laser pulses with the methods of directly writing [1-10] and holographic fabrication [11-22], such as waveguide [1], special diffractive optical elements (DOE) [4-10], micro-gratings [11-15], and photonic crystals [16-20]. Because multiphoton nonlinear effects play a major role in this process, the resulting change in refractive index or cavity formation can be highly localized only in the focal volume where the fluence is above a certain material dependent threshold, which makes it possible to micro-fabricate devices inside the bulk of transparent materials. These structures were usually fabricated with a focused beam and written dot by dot by translation of the sample with respect to the focal point.

Compared to directly writing technology by femtosecond laser pulses, holographic lithography has been considered as a more effective method for fabricating periodic structures because it can be controlled easily by the number of the beam, angles between every two beams, energies of the laser beams, and most importantly, only one pulse needed for holographic fabrication. Especially, holographic lithography is considered to be the most effective method for the fabrication of the photonic crystals. And Cai *et al.* [21-22] have proven that all fourteen Bravais lattices can be formed by interference of four noncoplanar beams (IFNB), and have quantitatively obtained the solutions including the required wavelength and the four wave vectors for each special lattice. The relative simulated results are shown in Fig. 1.

There are also many groups who have attempted to fabricate the periodic structures in photosensitive transparent materials or on the surface of the silica glass and the metal film by interfered multiple femtosecond laser pulses. Especially, it is very easy to fabricate the periodic structures inside of the photosensitive transparent materials with the aid of diffractive beam-splitter (DBS) [18-20]. However it is limited by the angles between two beams and the energy of the pulses. It is therefore difficult to fabricate microstructure with smaller period, especially in the materials with big band gap, such as silica glass. So there are also some researching groups [13, 16-17] focusing on the realization of periodic structures on the surface of the silica glass by a single shot of two or three femtosecond laser pulses originating from one pulse by the beam splitters.

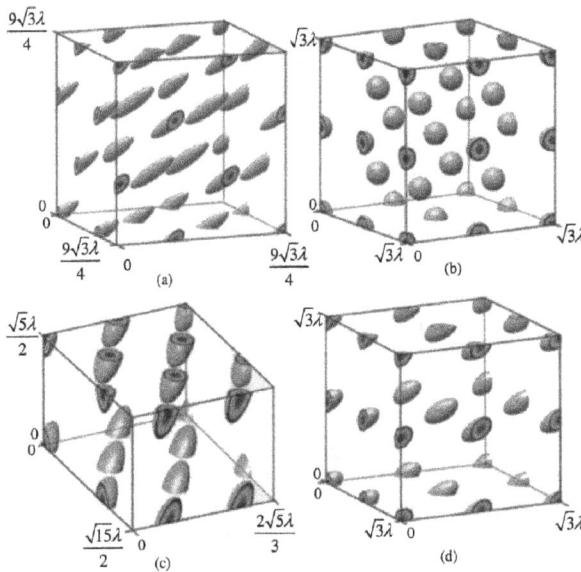

Figure 1. Computer simulations of four Bravais lattices formed by interference of four noncoplanar beams: (a) face-centered cubic lattice, (b) body-centered cubic lattice, (c) hexagonal lattice ($c = 3a/2$), and (d) simple cubic lattice.

In this chapter, we have reviewed the progress of fabrication of the periodic structures on the surface or inside of the transparent silica glass by a single shot of several (two, three, four) femtosecond laser pulses. When a single shot of two pulses interfered with each other, there will be one dimensional grating structures being formed inside of the photosensitive transparent materials or on the surface of the materials. When a single shot of three coplanar pulses interfered with each other, the one-dimensional M-shape grating can be formed on the surface of the silica glass. However, when a single shot of noncoplanar three or four pulses interfered with each other, two-dimensional periodic microstructure can be obtained, which distributed as a hexagonal lattice or tetragonal lattice.

2. Holographic fabrication by a single shot of two femtosecond pulses

2.1. The formation of the common grating and the extraordinary grating on the surface of the silica glass

The experimental setup for the holographic fabrication of microgratings by a single shot of two interfered femtosecond laser pulses is depicted in Fig. 2 (a). A regeneratively amplified Ti:sapphire laser (Coherent. Co.) with a central wavelength of 800 nm, pulse duration of 120 fs, and pulse repetition of 1–1000 Hz was used. A single femtosecond laser pulse with a beam diameter of 6 mm could be selected and split into two pulses that were then redirected at controllable incident angles on the surface of the fused silica glass (K9 glass) which is transparent for the laser with a wavelength of 800 nm. These split pulses generating from the single pulse were focused on the glass surface by two lenses with focal lengths of 20 cm to give a spot size of ~50 μm at the focal plane. Attenuators could be used to obtain a proper energy of the pulses for controlling the results of the holographic fabrication. The two pulses could be adjusted both spatially and temporarily by the optical delay device perfectly. The angle between two pulse and the pulse energies could be controlled easily.

When we set the angle between the split beams as 40° and the pulse energy as 45 μJ, the fabricated grating is shown in the Fig. 2 (b-e) by optical microscopy and atomic force microscopy (AFM) respectively. From the Fig. 2 (d), we can obtain the period of the fabricated grating is about 1.06 μm which agree well with the calculated result according to the formula $d=\lambda/[2\sin(\theta/2)]$ in which θ is colliding angle ($\theta=40°$ in this experiment) between two incident beams and λ is the incident laser wavelength of 800 nm . The periods of the micrograting encoded by holographic technology should be controlled easily by varying the angle. Analogously, the depth of the fabricated grating could also be tuned by controlling the pulse energy.

However, when we set the angle θ as 20° and the pulse energy as 65 μJ, the fabricated grating is shown in the Fig. 3. Not only did we get the ordinary grating whose periods accorded with the theoretic equation $d=\lambda/[2\sin(\theta/2)]$, but also obtained the extraordinary grating [15, 23] whose period is a half of the ordinary grating. The extraordinary grating formed at the middle of each bulge of the ordinary grating as shown in Fig. 3 (a-b), so the period of the extraordinary grating is a half of the ordinary grating's observed easily from Fig. 3(c) which also shows that the depth of the extraordinary grating is a half of the ordinary grating's nearly.

In the experiments, with the decrease of the incident energy in the same angle between two beams, the modulation depths of the extraordinary grating are decreasing gradually. With the increase of the angle between two beams with the same energy, the modulation depths of the extraordinary grating are also decreasing gradually. At last the modulation will be vanishing from the central part to the edge of the ordinary grating gradually.

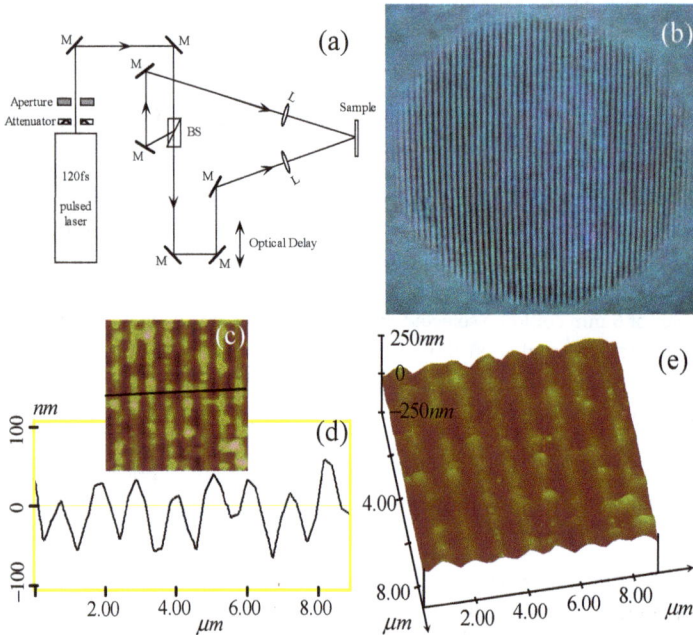

Figure 2. The experimental setup for the holographic fabrication of micrograting, (b) the fabricated grating observed by optical microscopy, (c) top-viewing AFM image of the fabricated grating, (d) cross-section view in the direction of black line shown in (c), (e) three dimentional view of the fabricated structure.

The formation of the modulation grating could be attributed to the higher-order modulation arising from second-harmonic generation (SHG) when the femtosecond laser pulse was incident to the surface of silica glass. As a rule, because of the inversion symmetry of the silica glass, there should be no second order nonlinearity in the silicon glass. However, a layer of plasma could be formed on the surface of the glass when the pulse incident to the sample because of the ultra high electrical field of the femtosecond laser in a time given by the duration of the laser pulse[24,25]. Therefore, when the femtosecond laser pulse with high intensity is incident on this thin plasma layer, there will be the higher harmonic generation because of the electrons quivering nonlinearly, and the SHG can reach to 2% of the fundamental laser [26]. The second-harmonic radiation is emitted in the direction of the reflected fundamental laser direction as depicted in Fig. 2 (d) [24]. So the period of the modulation grating is a half of the common grating because of the same of the angle between the two beams but a half of the wavelength. Although the intensity of the SHG is much smaller than the fundamental laser, the modulated depth of the silica glass can reach to a correspondingly large depth because it is much easier for the silica glass to be ionized by a mechanism of single or two photons absorption than multiphoton absorption.

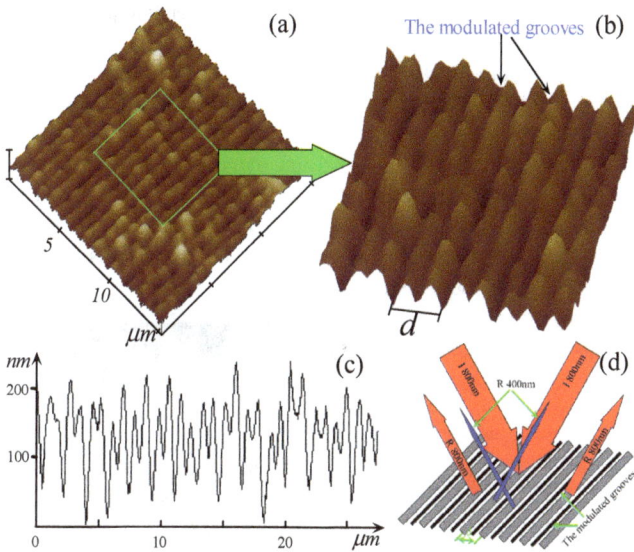

Figure 3. AFM photos of the resulted micrograting encoded with an energy of ~65 μJ for each interfered beam and a colliding angle of ~20°. (a) Image of the central portion of the grating, (b) an enlarged version for the chosen part in (a), and *d* is the period of the common grating. (c) cross-section view of the fabricated grating. (d) schematic illustration for the formation mechanism of the modulating grating.

2.2. Fabrication of multiple layers of grating

Li *et al.* [11] have reported that the multiple layers of grating can be fabricated inside soda–lime glass by a single shot of two interfered femtosecond laser pulses at different depths of the sample. A top view of experimental setup is shown in Fig. 4 (a), which is similar to the Fig. 2 (a). The two beams, with diameters of 5 mm, were focused by two lenses with focal length of 150 mm to give a spot size of 30 30μm at the focal plane.

The multiple gratings were written inside of the soda–lime glass one after another by focusing the beam 1 and beam 2 into special position inside of the glass sample with the recording plane. Without loss of generality, three layers of micrograting could be recorded at depths of 200, 400, and 600 μm, respectively. The grating at depth of 600μm should be encoded firstly. Then the sample was translated along the z direction to a depth of 400μm and the second grating could be recorded. After that, the third grating was fabricated after moving the sample to a depth of 200μm lastly.

Because the micrograting was formed inside of the samples around the focal point without damage to the surface or other parts of the sample, multiple layers of grating can be recorded successfully. And the images of the fabricated multiple layers of microgratings could be read out by beam 2 and taken over three different recording planes as depicted in

Fig. 4 (b) and (c). The schematics on the right illustrate the relative positions of the sample, three layers of grating, the recording plane, and the objective. The experimental results demonstrated that the readout image of the first layer (not shown here) and the second layer (as depicted in Fig. 4 (b)) of the recorded grating consisted of well-defined straight bright lines alternating with black lines. However, for the readout of the third layer of the micrograting, some of the straight lines became curved as shown in Fig. 4 (c). This aberration may be caused by the wave-front distortion of the incident beams due to the accumulation of nonlinear effects when the focused high peak-power pulse propagated through the sample. The longer the optical path inside the sample was, the more severe the wave-front distortion and the more obvious the resulting grating aberration.

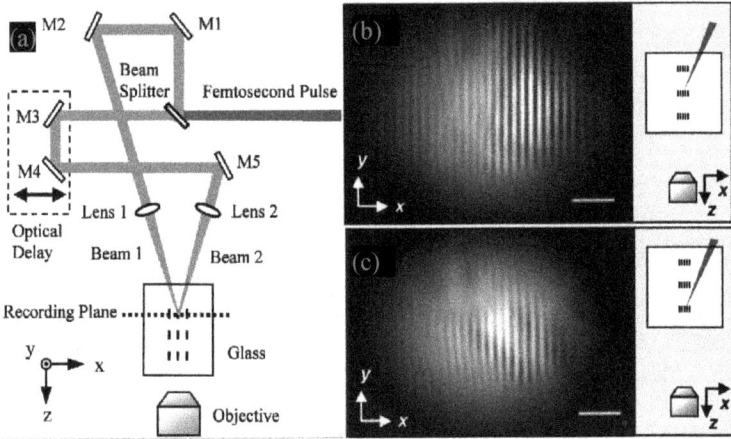

Figure 4. (a) Top view of the experimental setup for the formation of multiple gratings inside glass. The recording plane is in the $x - y$ plane and through the overlapping focal points of the two incident beams: beam 1 and beam 2. (b)(c). Readout images of the second and the third layers of grating inside a slide glass plate by beam 2. The images were taken through a 100X optical objective that was focused onto the recording planes at depths of 400μm (b) and 600μm (c) respectively. The schematics on the right indicate the relative positions of the sample, grating layers, the readout beam, and the 100X optical objective. Scale bar: 5μm.

2.3. Fabrication of micrograting inside the glass doped Au nanoparticles

Noble metal nanoparticle-contained glasses exhibit large third-order nonlinear susceptibility and ultrafast nonlinear response due to the local field effect near surface Plasmon resonance and quantum size effect [27, 28]. In recent years, many studies have been carried out on the fabrication of nanoparticle-doped glasses [29-31]. Shiliang Qu et al. realized the formation of micrograting constituted with Au nanoparticles in Au_2O-doped glasses induced by two interfered femtosecond laser pulses followed by successive heat treatment [12].

A typical silicate glass is composed of $70SiO_2.20Na_2O.10CaO$ doped with $0.1Au_2O$ (mol%). Reagent grade SiO_2, $CaCO_3$, Na_2CO_3, and $AuCl_3.HCl.4H_2O$ were used as starting materials.

An approximately 40g batch was mixed and placed into a platinum crucible. Melting was carried out in an electric furnace at 1550 °C for 1 h. The glass sample was obtained by quenching the melt to room temperature. The sample thus obtained was transparent and colorless. The annealed sample was cut and polished, and then subjected to successive experiments for femtosecond laser.

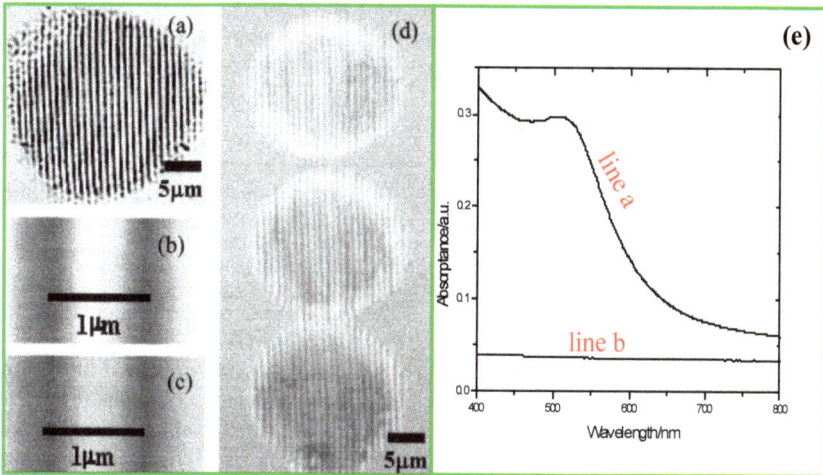

Figure 5. Optical microscopic photos of Au nanoparticles precipitation in periodic arrays in silicate glass (microgratings), taken by a 100X transilluminated optical microscope. (a) Energy is 30 µJ per pulse. (b) Magnified view of (a). (c) Energy is 38µJ per pulse. (d) Part of a group of formed microgratings array inside of the sample. (e) Absorption spectra of the glass samples after holographic irradiation by femtosecond laser pulses with (line a) and without (line b) heat treatment.

The used laser system and the experimental setup are the same as the used setup as shown in Fig. 2 (a). The two incident beams were first focused onto the front surface of Au_2O-doped glass to optimize the incident pulse energy. In the case of sufficiently high energy, the two coherent beams can induce periodic ablation, forming a grating in the glass. Herein, we reduced the incident pulses' energy to a certain lower level at which the two coherent beams cannot directly induce periodic ablation on the surface of the glass. Then the sample was moved 50µm to the lens in the direction of angular bisector of two incident light paths and made the laser pulses be focused inside the glass. After irradiation by a single shot of two interfered pulses, the micrograting can be recorded inside the glass, which can not be observed immediately because there is no obvious changes in the focusing place, however, after heat treating the sample at 550 °C for 1 h, the formed micrograting can be observed because of the Au nanoparticle precipitation. Such grating was constituted by the laser-heating induced Au nanoparticle precipitation in the Au_2O-doped glass.

In Qu's experiments, this lower pulse energy was selected to be 30 and 38 µJ for comparison, and the colliding angle θ between the two incident beams was fixed at 45°. The fabricated micrograting is shown in Fig.5 (a-d). The period d of the obtained gratings was about ~1μm which was agreeing well with the value calculated from the colliding angle θ and the laser wavelength λ according to the formula d=λ/[2sin(θ/2)].

The absorption spectrum of the grating was measured by a spectrophotometer (JASCOV-570) as shown in Fig. 5 (e) (line a). Apparently, a weak peak occurs around 508 nm, which is induced by the surface plasmon resonance of Au nanoparticles in the glass. The Au nanoparticles with 3 nm average size in the glass were observed in the grid of the fabricated micrograting from a transmitted electronic microscopy (TEM). However, if the glass sample was irradiated only by the interfered pulses but not heat treated, neither could absorption peak in the range of 500–600 nm be observed in the absorption spectrum as shown in Fig. 5 (e) (line b), nor could Au nanoparticles be observed by the TEM. This indicates that the Au nanoparticles can be precipitated in the periodic one-dimensional arrays in the glass through the irradiation of two coherent beams with the aid of heat treatment.

2.4. Fabrication of multiple gratings by a single shot of two interfered femtosecond laser pulses

The interference of two laser beams can create microstructures with one-dimensional periodic patterns in certain materials due to the periodic modulation of the laser intensity with a period scale of the order of the laser wavelength. As stated as above, there has been many groups focusing the fabrication of microgratings in glasses [11, 12], thin films [32], polymers[33], and inorganic–organic hybrid materials [34] by use of two interfered femtosecond laser pulses in a single shot. However, there is a disadvantage for micrograting fabrication with this method: Only a single micrograting can be formed for one pulse. Shiliang Qu *et al.* have reported holographic fabrication of multi-microgratings on silicate glass by a single shot of two interfered femtosecond laser beams with the aid of a mask [14].

In the experimental setup as shown in Fig. 6 (a) which is similar with the Fig. 2 (a) and Fig. 4 (a), a regeneratively amplified Ti:sapphire laser (Spectra-Physics) with a wavelength of 800 nm, pulse duration of 120 fs, and pulse repetition of 1–1000 Hz was used. A single laser pulse with a beam diameter of 8 mm was selected and split into two beams that were then redirected at approximately equal incident angles on a silicate glass surface. The two beams were focused on the glass surface by two lenses with focal lengths of 10 cm (L1) and 20 cm (L2) to yield spot sizes of 40μm and 80μm, respectively.

The colliding angle θ between the two beams was fixed at 40°. After the optical paths were adjusted to realize perfect overlap of the two beams both spatially and temporally, the surface of the glass was adjusted to be approximately normal to the perpendicular bisector of the two incident beams, so that the glass surface became the laser interfering plane. A mask used for laser beam modulation was placed in the optical path in which lens L1 is located. The mask consists of three equilaterally and triangularly arrayed apertures, whose diameters are 2.5 mm and the spaces between them are 3.5 mm as shown in Fig. 6(b).

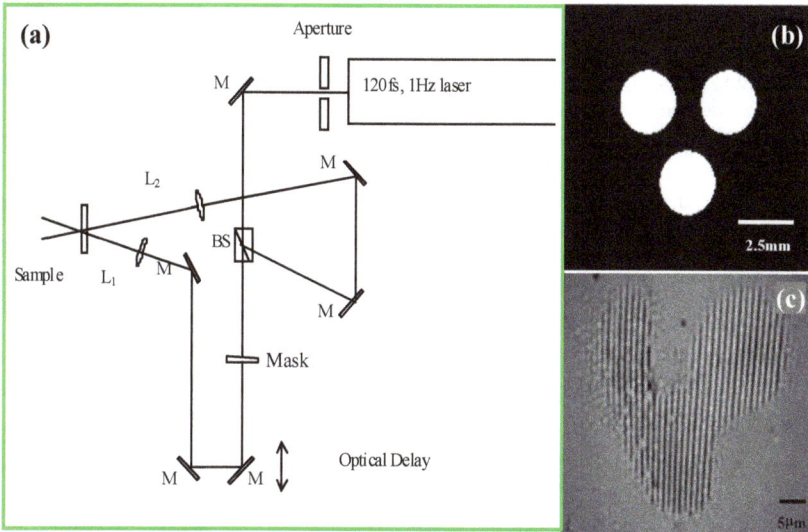

Figure 6. Experimental setup for the one-off writing of multi-micrograatings by a single shot of two femtosecond laser pulses. M's, mirrors; BS, beam splitter. (b) Data mask for multimicrograting formation. (c) Optical microscopic observation of a multi-micrograting formed on silicate glass with a period of 1.1μm ($E_{1,2} \sim 70$ μJ, $\theta \sim 40°$).

After irradiation by one single shot of two interfered pulses, the multi-micrograting comprises three micrograatings as shown in Fig. 6(c), which has a high fidelity to the configuration of the mask used. The multi-micrograting was formed through periodic ablation resulting from the interference of the reference laser beam with the three beams caused by the mask. And the period of the formed micrograatings is also agree well with the theoretical expectation of the common grating $d=\lambda/[2\sin(\theta/2)]$. This means that a multi-micrograting comprising even more micrograatings and configurations can also be one-off written by changing only the mask structure. However, multi-micrograting formation can be realized only when the two interfered beams are overlapped at an appropriate position out of their focus on the front surface of the glass.

3. Holographic fabrication by a single shot of multiple femtosecond pulses

3.1. The formation of the M-shaped micrograting by three coplanar interfered beams

As stated above, two interfered femtosecond pulses' interference can induce one and two-dimensional periodic structures by single and double-exposure techniques, respectively [11–15, 40]. However, usually the second pulse could not overlap completely with the microstructure formed by the first pulse in the double-exposure technique due to the rather small size of the focal spot. Here, we will show that the fabrication of M-shape gratings with controllable modulation depth could be realized by adding the third beam into a two-beam interference system. The experimental results show that the depth ratio between neighbor grooves can be conveniently controlled by changing the pulse energy of the third beam. Morphology characterizations of as-fabricated periodic M-shape gratings with a period of 2.6 µm are presented by optical microscopy and atomic force microscopy (AFM). A theoretical simulation has also been done for explaining the concrete experimental results, which shows a higher fidelity to the experiment.

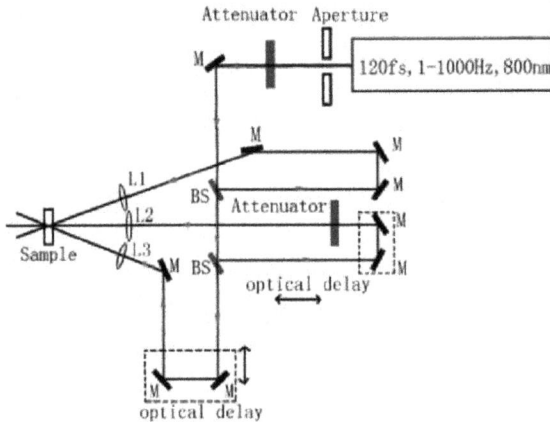

Figure 7. Experimental setup of three-beam interference optical system (AT and AP stand for attenuator and aperture, respectively).

The schematic experimental setup for fabricating M-shape grating is shown in Fig.7. The used ultrafast pulses (with a pulse width of 120 fs, central wavelength of 800 nm, and repetition rate of 1–1000 Hz) are produced by a Ti: sapphire regenerative amplified laser system (Coherent Inc). A laser pulse with a diameter of ~5 mm is split into three beams by two beam-splitters. Then three beams are focused by three lenses (L1; L2, and L3), each with a focal length of 150 mm, to give a spot size of ~50 µm at the focal plane. The collision angles between L1–L2, L2–L3, and L1–L3 are θ, θ, and 2θ ($\theta = 18°$ and $30°$), respectively. Two time-delay devices are employed for time adjustment of the beams collision. A BBO frequency-double crystal is located at the focal spot to check time superposition by SHG. When three collinear blue points appear on the white screen located on the back of frequency-double crystal, it indicates that three beams have been superposed on each other both spatially and temporarily. Two attenuators are used to adjust the energy of femtosecond laser pulse. The sample silica glass is located in the direction perpendicular to beam L2.

The optical microscope images of the gratings formed on the silica glass at different pulse energies of L2 ($\theta = 18°$, the pulse energies of L1 and L3 are both 50 µJ) are shown in Fig. 8 (a) and Fig. 8 (b). The neighbor grooves with different widths alternatively can be observed in the grating as shown in Fig. 8 (a). However, in Fig. 8 (b), the neighbor grooves of the as-formed grating have the same widths. Fig. 8 (c) shows the AFM characteristic results of as-formed grating (in Fig. 8 (a)). Fig. 8 (c) is a top-view AFM image of the formed micrograting, where the neighbouring grooves with different widths alternately appear more clearly. Fig. 8 (d) displays the three dimensional morphology, which clearly shows that the M-shape grating is formed by the alternate appearing of two kinds of grooves: the deeper grooves and the shallower grooves. Fig. 8 (e) indicates a cross-section scanning picture along the horizontal line 'L' shown in Fig. 8 (c).

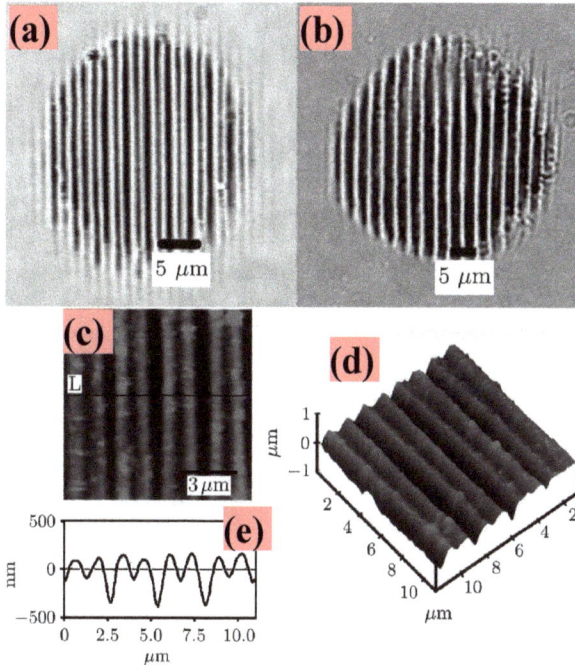

Figure 8. Optical microscope images of as-formed gratings by coplanar three-beam interference with different pulse energies of L2, i.e. (a) 50 μJ and (b) 100 μJ. AFM images of the M-shape grating formed on silica glass by three coplanar interfering beams, each with equal energy of 50 μJ, the collision angles of three beams are 18°, 18°, and 36°, respectively. Panel (c) is for planar image, panel (d) for three-dimensional image, and panel (e) for cross-section pattern along line L.

It is obviously shown that the structure is formed by periodically arranging the M-shape units with a size of about 2.6 μm. The modulation depths of the deeper grooves and shallower grooves of the M-shape grating are ~500 nm and ~240 nm, respectively. The deeper grooves and the shallower grooves of the M-shape grating each have a period of ~2.6 μm revealed in Fig.8 (c). This result accords well with the value calculated from the universally known period formula $d = \lambda/\sin \theta$ (in our case λ = 800 nm, θ = 18°, and d = 2.59 μm).

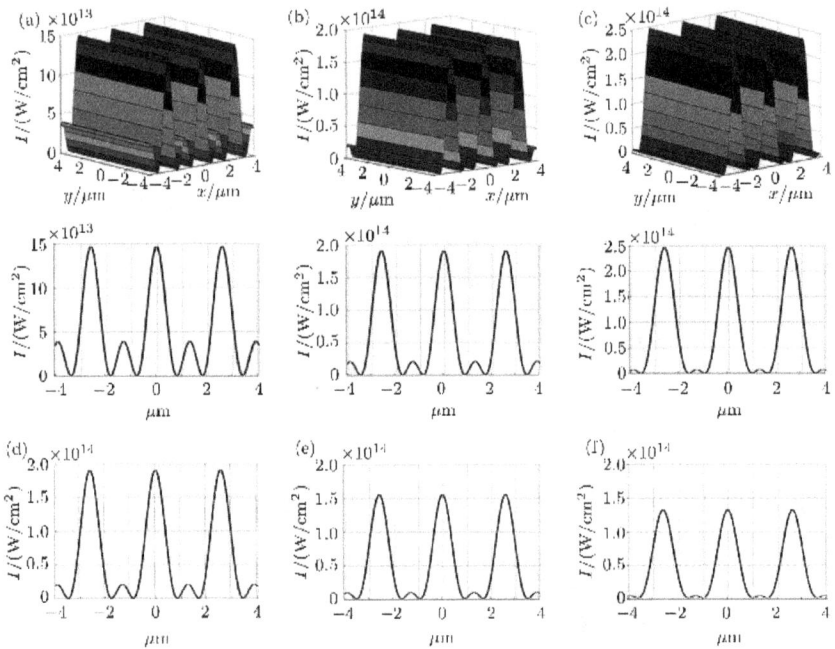

Figure 9. Intensity distributions simulated by using three interfering coplanar beams with different pulse energies of L_2 ((a), (b), and (c)) and L_3 ((d), (e), and (f)), where $k = 2\pi/\lambda$, $\lambda = 800$ nm, $\theta = 18°$ and different values of I_{01}, I_{02} and I_{03}, i.e. (a) $I_{01} = 2.12$, $I_{02} = 0.85$, and $I_{03} = 2.12$, (b) $I_{01} = 2.12$, $I_{02} = 2.12$, and $I_{03} = 2.12$, (c) $I_{01} = 2.12$, $I_{02} = 4.24$, and $I_{03} = 2.12$, (d) $I_{01} = 2.12$, $I_{02} = 2.12$, and $I_{03} = 2.12$, (e) $I_{01} = 2.12$, $I_{02} = 2.12$, and $I_{03} = 1.06$, and (f) $I_{01} = 2.12$, $I_{02} = 2.12$, and $I_{03} = 0.53$, which are all in units of 10^{13} W/cm².

The simulated intensity distributions in the interfering region formed by three coplanar interfering beams are shown in Figs. 9 (a), (b) and (c), which correspond to the middle beam (L2) pulse energies of 20 μJ, 50 μJ and 100 μJ, respectively. In Figs. 9 (a), (b) and (c), the upper images show the three-dimensional patterns and the nether curves display the two-dimensional cross-section patterns. From the cross-section images of Figs. 9 (a-c), we can see clearly that the periodic intensity patterns are formed by the periodic arrangement of inversely M-shape structures, which is attributed to stronger intensity peaks and weaker ones arrayed with the same period of about 2.6 μm alternately and periodically. These results are consistent with the AFM results as shown in Fig. 8. We notice that with the energy of the middle pulses increasing, the stronger intensity peaks in the interfering region increase, while the weaker intensity peaks decrease. It indicates that the intensity ratio of the stronger peak and the weaker one increases. If laser pulses with such a periodic inversely M-shape intensity distribution irradiate the materials, the stronger intensity peaks will lead to the formation of deeper grooves, and the weaker intensity peaks will induce the formation of shallower ones. As a result, the M-shape surface structures will be resultantly formed. This fabricating technology for the formation of M-shape gratings provides a fabricating method for special gratings with special use in industrial applications, such as the calibration for 3D reconstruction in computer vision application. And the fabricated M-shape grating can also be used in microfluidic chip devices as transport channels with different flowing speeds.

3.2. The formation of the triangular lattice by three noncoplanar interfered beams

When a single shot of three coplanar pulses interfered with each other, the M-shaped gratings could be fabricated as above. However, when a single shot of three non-coplanar pulses interfered with each other as depicted in Fig. 10 (a) by adjusting two time-delay for obtaining perfect overlap of the three pulses both spatially and temporarily, the two-dimensional periodic microstructure have been obtained, which distributed as a hexagonal lattice as shown in Fig. 10 (b) and agreed well with the simulated results [see Fig. 10 (c)] [16, 17]. In experiments, the geometric angles were kept as $\theta=30^0$ and $\varphi=35^0$ respectively as shown in Fig. 10 (a). When we set the pulse energy as 75 µJ, we can obtain two-dimensional periodic hexagonal lattice of microholes [see Fig. 10 (d-f)], however, when we set the pulse energy as 30 µJ, the two dimensional periodic microstructures present doughnut orbicular platform [see Fig. 10 (g-i)]. The period in the direction of line "L" is nearly 2.358 µm according to Fig. 10 (f) and (i) which agrees well to the calculated result [16].

The different microstructures in our experiments were attributed to the formation of plasma and molten liquid at different pulse energy levels. Generally it is hard to form plasma and molten liquid on the surface of silica glass by a laser with a wavelength of 800 nm because of the bigger band gap. However, the intensity at the focal point of the femtosecond laser beam where three beams interfere together could reach to 100 TW/cm^2 nearly. Such a high-energy influence within the focal volume ionized the silica glass quickly through the combined action of the avalanche and multiphoton processes [17, 35]. A layer of plasma formed on the surface of the silica glass at the time of the laser pulse duration. While the intensity decreased to be lower than a certain value, which can be called the ionized threshold as shown in Fig. 11 (a), the plasma vanished and the molten liquid of the material appeared. With the decrease of the intensity to be lower than a certain value, which could be called the molten threshold [see Fig. 11 (a)], the molten liquid of the material disappeared, so there would be a doughnut molten liquid formed in every enhanced spot of the interfered field [see Fig. 11 (b)]

When the pulse energy was set as 75 µJ, a layer of modulated plasma was formed by the interfered field with hexagonal lattice depicted in Fig. 10 (c) after the anterior part of the pulse was incident to the surface of the silica glass. Higher intensity induced a relatively larger plasma area in every enhanced spot of the interfered field. The subsequent posterior part of the pulse removed the plasma very swiftly because of the high light pressure originating from the reflection. Therefore, the microholes formed on the surface of the silica glass, as depicted in Fig. 10 (d-f). When the pulse energy decreased to 30 µJ, there was just a layer of the plasma with smaller areas than that of 75 µJ in the center of every enhanced spot in the interfered field after the anterior part of the pulse was incident to the surface. The

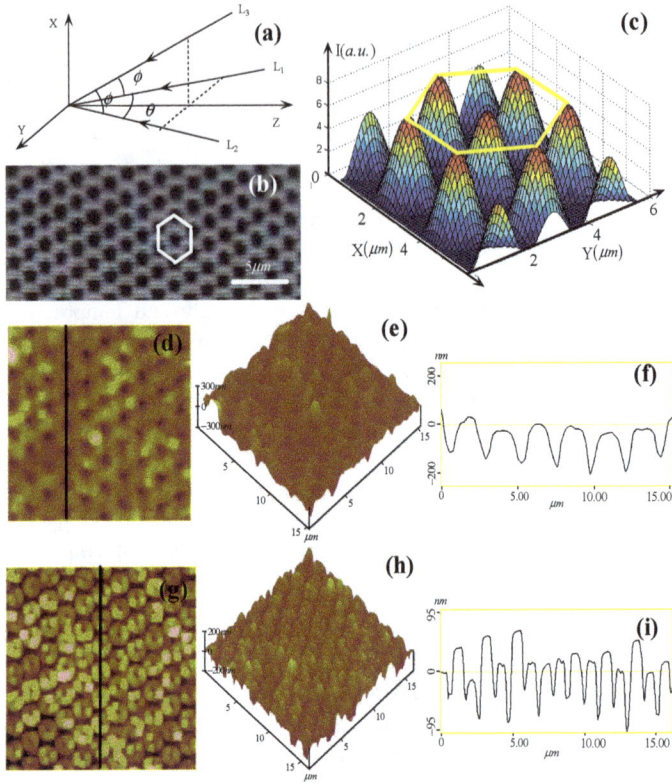

Figure 10. (a) Geometric sketch of the three non-coplanar interfered beams, and the sample is laid in the x–y plane, (b) Photo of the periodic microstructure fabricated in silica glass, taken by an optical microscopy, (c) Calculated intensity distribution of the interference by three non-coplanar beams. The induced periodic microstructure depicted by an AFM: (d) Energy is 75 μJ per pulse, (g) Energy is 30 μJ per pulse, (f) (i) Three-dimensional image of the microstructure of (d) and (g) respectively, (e) (h) The cross section of the microstructure in the direction of black line showed in (d) and (g) respectively

subsequent posterior part of the pulse also removed the central plasma very swiftly by means of light pressure, so there is a tiny hole formed in the center of the enhanced spot. However, in the region of the molten liquid, there were two distinct interaction components because of Marangoni effect, thermocapillary and chemicapillary, which resulted from the thermal potential of a temperature gradient and the chemical potential of a compositional gradient, respectively [17, 36-37]. The thermocapillary force moved the molten material outward from the center, while the chemicapillary force moved the molten material toward the center [38-39] as depicted in Fig. 10 (b). When the chemicapillary force dominated, a platform formed in the center of the spot. The combined action of the light pressure to the plasma and the chemicapillary force to the molten liquid induced a periodic orbicular platform on the surface of the silica glass as depicted in Fig. 10 (g-i).

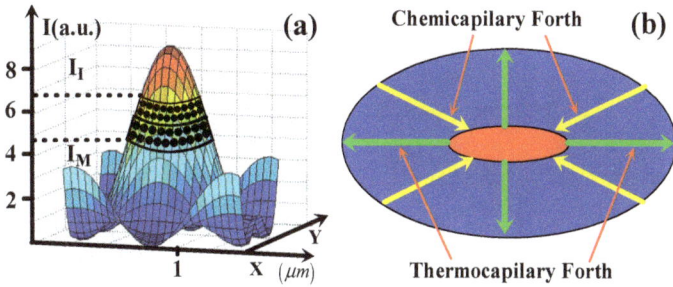

Figure 11. (a) The intensity distribution of the enhanced spot in the interfered field by three non-coplanar beams, I_I and I_M represent the densities of the ionized and molten thresholds respectively, (b) schematic diagram explanation for the mechanism of formed different structures owing to the formation of the plasma and Marangoni effect.

Figure 12. (a) AFM photo of the periodic structure induced by a single shot of three pulses with pulse energies of 50 μJ. Three-dimensional analyzed photos of the three selected parts I, II, and III in (a) are shown in (b)–(d) respectively.

In order to verify our explanation on the different induced microstructures, we set the pulses energy as 50 μJ, and the resulted microstructure is shown in Fig. 12 (a). Because of the Gaussian type intensity distribution of the pulse, three regions are selected from the central part (I), outer part (II) and the edge part (III) of the microstructure as depicted in Fig. 12 (a). The corresponding three dimensional images of the selected areas are shown in Fig. 12 [(b)-(d)], respectively. For the region I, the intensity is high enough to make the light pressure dominate, therefore the periodic microholes formed [see Fig. 12 (b)]. In contrast, in the region II, where the intensity is comparatively low, the chemicapillary force and the light pressure dominate, so that the microstructure present a orbicular platform [see Fig. 12 (c)].

However, in the selected area III, the intensity is very low and there is no plasma layer but just a layer of liquid formed on the surface of the silica glass. In this case, if the chemicapillary force dominates, the microcones can be observed. As shown in Fig. 12 (d), just several microcones (indicated by arrows) formed in the edge of the interfered districts.

3.3. The formation of the tetragonal lattice by four noncoplanar interfered beams

The interference of two beams creates a one-dimensional (1D) periodic pattern. By increasing the number of beams [16-20] or the double-exposure techniques [40], in principle, two-dimensional (2D) and three-dimensional (3D) periodic patterns can be designed. Although, as stated above, the interference of the three beams [13, 16-17] can create a 1D [13] or 2D [16, 17] periodic pattern on the surface of the materials of inside the transparent materials, the complicated optical setup is required for the interference of multiple laser beams, and its precise adjustment is difficult. Kondo *et al.* [18, 19] and Si *et al.* [20] have reported the fabrication of the 2D tetragonal lattice by four noncoplanar interfered pulses originating from the single pulse with the aid of the special DBS (diffractive beam splitter).

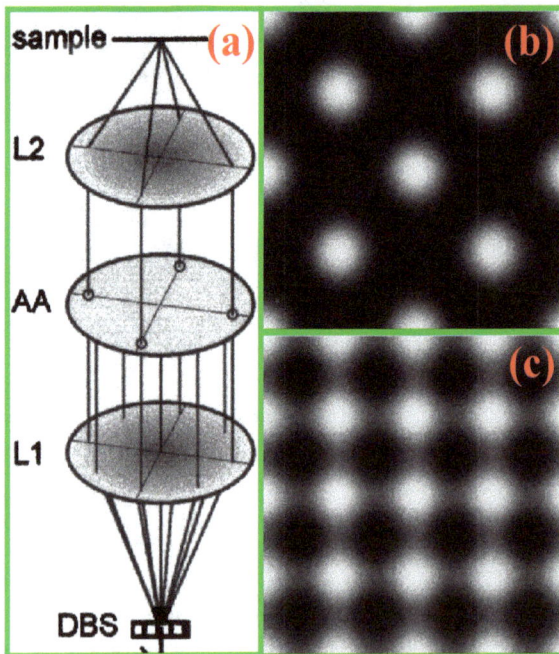

Figure 13. (a) Optical setup. DBS: diffractive beam splitter, L1 and L2: lenses,AA: aperture array. The inset shows the absorption spectrum of 4-mm-thick SU-8 film spin-coated on a coverglass. (b) Calculated intensity distribution by the interference of four beams which have same phase, (c) phase of one beam is shifted by π.

The optical setup used for the present experiments is shown in Fig. 13(a). Briefly, a DBS (G1023A or G1025A; MEMS Optical Inc.) divides the input laser beam into several beams, and the beams are collected on the sample by two lenses. Temporal overlap of the divided pulses is achieved without adjusting the optical path lengths. Each beam was made to be parallel or slightly focused by the adjustment of the distance between the two lenses. Slight focusing increased laser power density and helped to make the MPA efficient. The beams meant to form interference were selected by an aperture array, which is placed between the two lenses. Negative photoresist SU-8 (Microlithography Chemical Corp.) was used as an initial material for the fabrication. The absorption spectrum of SU-8 indicates that one-photon absorption is negligible at an 800-nm wavelength. Consequently, it is expected that photopolymerization, if occurring, is due to a multiphoton reaction. The layer of SU-8 was spin-coated on a coverglass plate having a thickness of about 4 µm, and prebaked before exposure to fs pulses. The interference angles θ_{air} (the angle between the main optical axis and the other beams in air) applied in the experiments were measured to be 33.6°, 21.9°, and 10.8°.

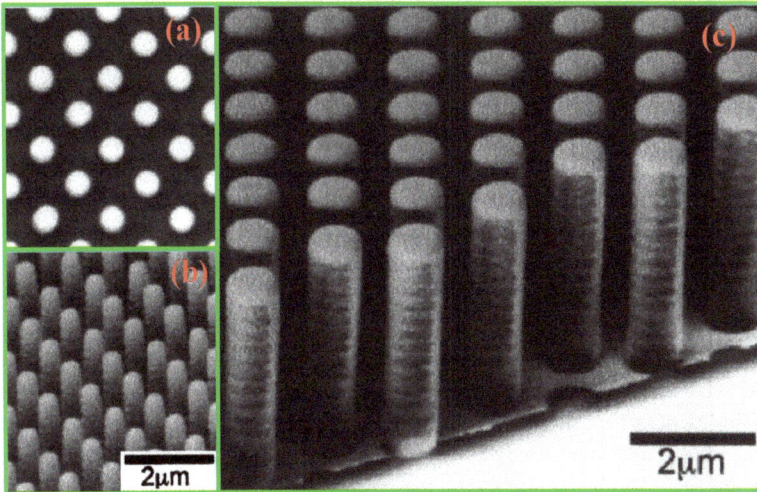

Figure 14. (a)(b) Top view and oblique view SEM images of the structure fabricated by the four-beam interference of fs pulses, (c) Close-up view of rods in the structure fabricated by four-beam interference of fs pulses with an interference angle of 21.9°.

Femtosecond pulses from a Ti:sapphire regenerative amplifier (wavelength of 800 nm, pulse duration of 150 fs, repetition rate of 1 kHz) were used for experiments. By using this method, periodic structures were fabricated. Periodic structures fabricated with an interference angle of 33.6° are presented in Fig. 14. In this figure, (a) and (b) show SEM images of the same sample from different perspectives. The oblique view shown in (b) clearly demonstrates that the periodic structure consists of high-aspect-ratio rods. The height and diameter of the rods are about 4 µm and 0.6 µm, respectively; thus, an aspect ratio of about 7 was achieved. It should be noted that the height of the obtained structure is not limited by the coherent length of the pulse. A higher structure could be obtained if the stiffness of the material allowed the rods to withstand capillary forces during the development procedure.

Figure 14 (c) provides a close-up view of the rods fabricated with an interference angle at 21.9°. To obtain this image, the coverglass containing the fabricated structures was deliberately broken and a small fragment was observed. As seen, the rods are slightly bellows-shaped. Distinct ring-like features repeat periodically along each rod with a period about 0.3μm. This could be attributed as the result of interference between the incident and reflected beams at the resist–coverglass interface.

4. Conclusion

In conclusion, we have reviewed the fabrications of the one-dimensional and two-dimensional periodic microstructures on the surface or inside of materials by multiphoton absorption using a single shot of two or multiple interfered femtosecond laser pulses.

Firstly, we have introduced the fabrication of the microgratings by a single shot of two interfered femtosecond laser pulses. When two interfered pulses overlapped on the surface of the silica glass, not only did we get the ordinary grating whose periods accorded with the theoretic equation $d=\lambda/[2\sin(\theta/2)]$, but also obtained the extraordinary grating whose period is a half of the ordinary grating. The formation of the modulation grating could be attributed to the higher-order modulation arising from second-harmonic generation (SHG) when the femtosecond laser pulse was incident to the surface of silica glass. The multiple layers of the microgratings have been fabricated in the different depths of the silica glass sample by a single shot of two interfered femtosecond laser pulses, and the results show that the fabricated gratings can be read out by one of the recorded beams very easily. The noble metal nanoparticles consisted microgratings have also been realized in silicate glasses by two interfered femtosecond laser pulses with the aid of heat treatment because the noble metal nanoparticles in silica glass can be precipitated after the irradiation of the femtosecond laser and the successive heat treatment. At the same time, the multiple gratings can also be realized on the surface of the silica glass samples by a single shot of two interfered femtosecond laser pulses with the aid of a mask, which is very significative for enhancing the processing efficiency of the fabricated microgratings.

Secondly, we have also introduced the fabrication of the 1-D or 2-D periodic microstructures by a single shot of multiple interfered femtosecond laser pulses. When a single shot of three coplanar interfered femtosecond laser pulses interfered with each other on the surface of the silica glass, M-shaped gratings can be formed and the morphologies of the M-shaped gratings can also be modulated by tuning the incident pulse energy. However, when a single shot of noncoplanar three or four pulses interfered with each other, two-dimensional periodic microstructure can be obtained, which distributed as a hexagonal lattice. Different morphologies of the induced structures such as microvoid, orbicular platform and nanotip, could be formed with the changes of the incident pulse energy. The fabrication of the 2D tetragonal lattice have also been fabricated in Negative photoresist SU-8 by four noncoplanar interfered pulses originating from the single pulse with the aid of the special DBS (diffractive beam splitter). Although the special DBS provided an easy way for realizing multiple beams interference, it also had some shortages, such as difficulty for fabricating microstructure with a smaller period, especially in the materials with big bandgap, such as silica glass.

Author details

Zhongyi Guo*
School of Computer and Information, Hefei University of Technology, Hefei, China
Department of Optoelectronic Science, Harbin Institute of Technology at Weihai, Weihai, China
Department of Physics, Harbin Institute of Technology, Harbin, China

Lingling Ran
Department of Optoelectronic Science, Harbin Institute of Technology at Weihai, Weihai, China
College of Electronic Engineering, Heilongjiang University, Harbin, China

Yanhua Han and Shiliang Qu
Department of Optoelectronic Science, Harbin Institute of Technology at Weihai, Weihai, China

Shutian Liu
Department of Physics, Harbin Institute of Technology, Harbin, China

Acknowledgement

This work was supported by the National Science Foundation of China (NSFC: 10904027; 61108018), the China Postdoctoral Science Foundation (AUGA41001348) and the Heilongjiang Province Postdoctoral Science Foundation (AUGA1100074), and development program for outstanding young teachers in Harbin Institute of Technology, HITQNJS. 2009. 033.

5. References

[1] C. Schaffer, A. Brodeur, J. García, and E. Mazur, "Micromachining bulk glass by use of femtosecond laser pulses with nanojoule energy," Opt. Lett. 26, 93-95 (2001)

[2] J. Koch, F. Korte, T. Bauer, C. Fallnich, A. Ostendorf, B. Chichkov. Nanotexturing of Gold Films by Femtosecond Laser-induced Melt Dynamics. Appl. Phys. A. 2005, 81:325–328.

[3] S. Juodkazis, H. Misawa, T. Hashimoto, E. Gamaly, and B. Luther-Davies, Laser-induced microexplosion confined in a bulk of silica: Formation of nanovoids, *Appl. Phys. Lett.* 88 (2006), 201909

---

* Corresponding Author

[4] Y. Li, Y. Dou, R. An, H. Yang, and Q. Gong,, "Permanent computer-generated holograms embedded in glass by femtosecond laser pulses," *Opt. Express*, 13 (2005) 2433-2438.

[5] Q. Zhao, J. Qiu, X. Jiang, E. Dai, C. Zhou, and C. Zhu, "Direct writing computer-generated holograms on metal film by an infrared femtosecond laser," *Opt. Express*, 13 (2005) 2089-2092.

[6] Z. Guo, S. Qu, Z. Sun, and S. Liu, "Superposition of orbit angular momentum of photons by combined computer-generated hologram fabricated in silica glass with femtosecond laser pulses", *Chin. Phys. B.* 17 (2008) 4199.

[7] Z. Guo, S. Qu, S. Liu, Generating optical vortex with computer-generated hologram fabricated inside glass by femtosecond laser pulses, Optics Communications, 273, 2007, 286-289.

[8] Z. Guo, H. Wang, Z. Liu, S. Qu, J. Dai, and S. Liu, Realization of holographic storage on metal film by femtosecond laser pulses micromachining, Journal of Nonlinear Optical Physics and Materials, 18(4), 2009, 617-623.

[9] Z. Guo, W. Ding, S. Qu, J. Dai, and S. Liu, Self-assembled volume grating in silica glass induced by a tightly focused femtosecond laser pulses, Journal of Nonlinear Optical Physics and Materials, 18(4), 2009, 625-632.

[10] K. Zhou, Z. Guo, W. Ding, and S. Liu, Analysis on volume grating induced by femotosecond laser pulses, Opt. Express 18(13), 2010, 13640-13646.

[11] Y. Li · W. Watanabe, K. Yamada, T. Shinagawa, K. Itoh, J. Nishii, and Y. Jiang "Holographic fabrication of multiple layers of grating inside soda–lime glass with femtosecond laser pulses". *Appl.Phys.Lett*, 80 (2002) 1508.

[12] S. Qu, J. Qiu, C. Zhao, X. Jiang, H. Zeng, C. Zhu, and K. Hirao, "Metal nanoparticle precipitation in periodic arrays in Au2O-doped glass by two interfered femtosecond laser pulses," *Appl. Phys. Lett.* 84, 2046-2048, (2005).

[13] Y. Han, S. Qu, Q. Wang, Z. Guo, and X. Chen, Controllable grating fabrication by three interfering replicas of single femtosecond laser pulse, Chinese physics B, 18, 5331, (2009).

[14] S. Qu, C. Zhao, Q. Zhao, J. Qiu, C. Zhu, and K. Hirao, One-off writing of multimicrogratings on glass by two interfered femtosecond laser pulses, Opt. Lett. 29, 2058 (2004).

[15] Z. Guo, S. Qu, L. Ran, and S. Liu, Modulation grating achieved by two interfered femtosecond laser pulses on the surface of the silica glass, Appl. Surf. Sci., 253, 8581, (2007).

[16] Z. Guo, S. Qu, Y. Han, and S. Liu, Multi-photon fabrication of two-dimensional periodic structure by three interfered femtosecond laser pulses on the surface of the silica glass, Optics Communications, 280, 23-26, (2007)

[17] Z. Guo, L. Ran, Y. Han, S. Qu, and S. Liu, "The formation of novel two-dimensional periodic microstructures by a single shot of three interfered femtosecond laser pulses on the surface of the silica glass", *Opt. Lett.* 33 (2008) 2383.

[18] T. Kondo, S. Matsuo, S. Juodkazis, and H. Misawa, "Femtosecond laser interference technique with diffractive beam splitter for fabrication of three-dimensional photonic crystals" *Appl.Phys.Lett.* 79 (2001) 725-727.

[19] T. Kondo, S. Matsuo, S. Juodkazis, V. Mizeikis, and H. Misawa, "Multiphoton fabrication of periodic structures by multibeam interference of femtosecond pulses," *Appl.Phys.Lett.* 82 (2003) 2758-2760.

[20] J. Si, Z. Meng, S. Kanehira, J. Qiu, B. Hua and K. Hirao, "Multiphoton-induced periodic microstructures inside bulk azodye-doped polymers by multibeam laser interference", *Chem. Phys. Lett.* 399 (2004) 276-279

[21] L. Z. Cai, X. L. Yang, and Y. R. Wang, "All fourteen Bravais lattices can be formed by interference of four noncoplanar beams," Opt. Lett. 27, 900-902 (2002)

[22] L. Z. Cai, X. L. Yang, and Y. R. Wang, "Formation of three-dimensional periodic microstructures by interference of four noncoplanar beams," J. Opt. Soc. Am. A 19, 2238-2244 (2002)

[23] Zhongyi Guo, Shiliang Qu, Shutian Liu, and Jung-Ho Lee, Periodic microstructures induced by interfered femtosecond laser pulses, Proc. of SPIE Vol. 7657 76570K-1, 2010.

[24] D. Von der Linde, H. Schulz, T. Engers and H. Schuler, Second harmonic generation in plasmas produced by intense femtosecond laser pulses, *IEEE J. Quantum Electron.* 28, 2388, (1992).

[25] Y. Shen, The Principles of Nonlinear Optics, Wiley, New York (1984) 553.

[26] U. Österberg and W. Margulis, "Dye laser pumped by Nd:YAG laser pulses frequency doubled in a glass optical fiber," Opt. Lett. 11, 516-518 (1986).

[27] S. Qu, C. Zhao, X. Jiang, G. Fang, Y. Gao, H. Zeng, Y. Song, J. Qiu, C. Zhu, and K. Hirao, Optical nonlinearities of space selectively precipitated Au nanoparticles inside glasses, Chem. Phys. Lett. 368, 352 (2003).

[28] H. B. Liao, R. F. Xiao, J. S. Fu, H. Wang, K. S. Wong, and G. K. L. Wong, "Origin of third-order optical nonlinearity in Au:SiO$_2$ composite films on femtosecond and picosecond time scales," Opt. Lett. 23, 388-390 (1998)

[29] J. Qiu, M. Shirai, T. Nakaya, J. Si, X. Jiang, C. Zhu, Space-selective precipitation of metal nanoparticles inside glasses, Appl. Phys. Lett. 81, 3040 (2002).

[30] James R. Adleman, Helge A. Eggert, Karsten Buse, and Demetri Psaltis, "Holographic grating formation in a colloidal suspension of silver nanoparticles," Opt. Lett. 31, 447-449 (2006).

[31] E. Valentin, H. Bernas, C. Ricolleau, and F. Creuzet, Ion Beam "Photography": Decoupling Nucleation and Growth of Metal Clusters in Glass, Phys. Rev. Lett. 86, 99 (2001).

[32] Y. Nakata, T. Okada, and M. Maeda, Fabrication of dot matrix, comb, and nanowire structures using laser ablation by interfered femtosecond laser beams, Appl. Phys. Lett. 81, 4239 (2002).

[33] Yan Li, Kazuhiro Yamada, Tomohiko Ishizuka, Wataru Watanabe, Kazuyoshi Itoh, and Zhongxiang Zhou, "Single femtosecond pulse holography using polymethyl methacrylate," Opt. Express 10, 1173-1178 (2002)

[34] G. Qian, J. Guo, M. Wang, J. Si, J. Qiu, K. Hirao, Holographic volume gratings in bulk perylene-orange-doped hybrid inorganic-organic materials by the coherent field of a femtosecond laser, Appl. Phys. Lett., 83 (2003), p. 2327

[35] S. Juodkazis, H. Misawa, T. Hashimoto, E. Gamaly, and B. Davies, Laser-induced microexplosion confined in a bulk of silica: Formation of nanovoids, Appl. Phys. Lett. 88, 201909 (2006).

[36] Y. Lu, S. Theppakuttai, and S. Chen, Marangoni effect in nanosphere-enhanced laser nanopatterning of silicon, Appl. Phys. Lett. 82, 4143 (2003).

[37] J. Eizenkop, I. Avrutsky, G. Auner, D. Georgiev, and V. Chaudhary, Single pulse excimer laser nanostructuring of thin silicon films: Nanosharp cones formation and a heat transfer problem, J. Appl. Phys. 101, 094301 (2007).

[38] S. Huang, Z. Sun, B. Luk'yanchuk, M. Hong, and L. Shi, Nanobump arrays fabricated by laser irradiation of polystyrene particle layers on silicon, Appl. Phys. Lett. 86, 161911 (2005).

[39] R. Piparia, E. Rothe, and R. Baird, Nanobumps on silicon created with polystyrene spheres and 248 or 308 nm laser pulses, Appl. Phys. Lett. 89, 223113 (2006).

[40] K. Kawamura, N. Sarukura, M. Hirano, N. Ito, and H. Hosono, Periodic nanostructure array in crossed holographic gratings on silica glass by two interfered infrared-femtosecond laser pulses, Appl. Phys. Lett. 79, 1228 (2001).

Permissions

The contributors of this book come from diverse backgrounds, making this book a truly international effort. This book will bring forth new frontiers with its revolutionizing research information and detailed analysis of the nascent developments around the world.

We would like to thank Igor Peshko, for lending his expertise to make the book truly unique. He has played a crucial role in the development of this book. Without his invaluable contribution this book wouldn't have been possible. He has made vital efforts to compile up to date information on the varied aspects of this subject to make this book a valuable addition to the collection of many professionals and students.

This book was conceptualized with the vision of imparting up-to-date information and advanced data in this field. To ensure the same, a matchless editorial board was set up. Every individual on the board went through rigorous rounds of assessment to prove their worth. After which they invested a large part of their time researching and compiling the most relevant data for our readers. Conferences and sessions were held from time to time between the editorial board and the contributing authors to present the data in the most comprehensible form. The editorial team has worked tirelessly to provide valuable and valid information to help people across the globe.

Every chapter published in this book has been scrutinized by our experts. Their significance has been extensively debated. The topics covered herein carry significant findings which will fuel the growth of the discipline. They may even be implemented as practical applications or may be referred to as a beginning point for another development. Chapters in this book were first published by InTech; hereby published with permission under the Creative Commons Attribution License or equivalent.

The editorial board has been involved in producing this book since its inception. They have spent rigorous hours researching and exploring the diverse topics which have resulted in the successful publishing of this book. They have passed on their knowledge of decades through this book. To expedite this challenging task, the publisher supported the team at every step. A small team of assistant editors was also appointed to further simplify the editing procedure and attain best results for the readers.

Our editorial team has been hand-picked from every corner of the world. Their multi-ethnicity adds dynamic inputs to the discussions which result in innovative outcomes. These outcomes are then further discussed with the researchers and contributors who give their valuable feedback and opinion regarding the same. The feedback is then collaborated with the researches and they are edited in a comprehensive manner to aid the understanding of the subject.

Apart from the editorial board, the designing team has also invested a significant amount of their time in understanding the subject and creating the most relevant covers. They scrutinized every image to scout for the most suitable representation of the subject and create an appropriate cover for the book.

The publishing team has been involved in this book since its early stages. They were actively engaged in every process, be it collecting the data, connecting with the contributors or procuring relevant information. The team has been an ardent support to the editorial, designing and production team. Their endless efforts to recruit the best for this project, has resulted in the accomplishment of this book. They are a veteran in the field of academics and their pool of knowledge is as vast as their experience in printing. Their expertise and guidance has proved useful at every step. Their uncompromising quality standards have made this book an exceptional effort. Their encouragement from time to time has been an inspiration for everyone.

The publisher and the editorial board hope that this book will prove to be a valuable piece of knowledge for researchers, students, practitioners and scholars across the globe.

List of Contributors

Igor Peshko
Department of Physics and Computer Science, Wilfrid Laurier University, Canada

Zhiyi Wei, Binbin Zhou, Yongdong Zhang, Yuwan Zou, Xin Zhong, Changwen Xu and Zhiguo Zhang
Beijing National Laboratory for Condensed Matter Physics and Institute of Physics, Chinese Academy of Sciences, Beijing, China

E. Nava-Palomares, F. Acosta-Barbosa, S. Camacho-López and M. Fernández-Guasti.
Lab. de Óptica Cuántica, Depto. de Física, Universidad A. Metropolitana - Iztapalapa, 09340 México D.F., Ap. postal. 55-534, Mexico

Kazuyuki Uno
University of Yamanashi, Japan

Ricardo Elgul Samad, Leandro Matiolli Machado, Nilson Dias Vieira Junior and Wagner de Rossi
Instituto de Pesquisas Energéticas e Nucleares – IPEN-CNEN/SP, Brazil

Roman V. Dyukin, George A. Martsinovskiy, Olga N. Sergaeva, Galina D. Shandybina,
Vera V. Svirina and Eugeny B. Yakovlev
Department of Laser-Assisted Technologies and Applied Ecology,
National Research University of Information Technologies, Mechanics and Optics, Saint-Petersburg, Russia

A. Yu. Ivanov and S. V. Vasiliev
Grodno State University, Grodno, Belarus

V.I. Mazhukin
Keldysh Institute of Applied Mathematics of RAS, Moscow, Russia

Kallepalli Lakshmi Narayana Deepak
School of Physics, University of Hyderabad, Gachibowli, Hyderabad, India
Laboratoire LP3, UMR 6182, CNRS-Universite Aix-Marseille, Pole Scientifique et Technologique de
Luminy, Case 917, Marseille, France

Zhongyi Guo
School of Computer and Information, Hefei University of Technology, Hefei, China
Department of Optoelectronic Science, Harbin Institute of Technology at Weihai, Weihai, China
Department of Physics, Harbin Institute of Technology, Harbin, China

Lingling Ran
Department of Optoelectronic Science, Harbin Institute of Technology at Weihai, Weihai, China
College of Electronic Engineering, Heilongjiang University, Harbin, China

Yanhua Han and Shiliang Qu
Department of Optoelectronic Science, Harbin Institute of Technology at Weihai, Weihai, China

Shutian Liu
Department of Physics, Harbin Institute of Technology, Harbin, China

www.ingramcontent.com/pod-product-compliance
Lightning Source LLC
Chambersburg PA
CBHW070730190326
41458CB00004B/1110